Optoelectronic Devices and Principles

Optoelectronic Devices and Principles

William J. Mooney
Monroe Community College

Prentice Hall, Englewood Cliffs, New Jersey 07632

Library of Congress Cataloging-in-Publication Data

Mooney, William J., *(date)*
Optoelectronic devices and principles / William J. Mooney.
p. cm.
Includes index.
ISBN 0-13-634486-0
1. Optoelectronic devices. I. Title.
TA1750.M66 1991
621.381'045—dc20 90-25757
CIP

Editorial/production supervision and
interior design: **Kathryn Pavelec**
Cover design: **Wanda Lubelska**
Manufacturing buyer: **Mary McCartney** and **Ed O'Dougherty**
Acquisitions Editor: **Sharon Jacobus**

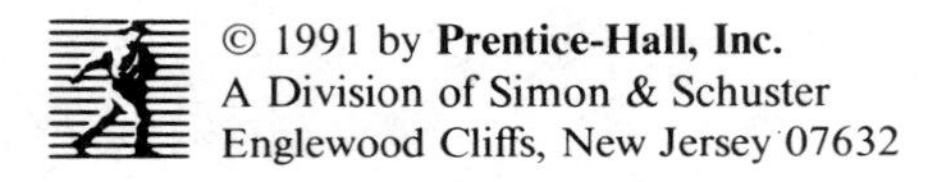

A Division of Simon & Schuster
Englewood Cliffs, New Jersey 07632

Printed in the United States of America

10 9 8 7 6 5 4 3 2 1

ISBN 0-13-634486-0

Prentice-Hall International (UK) Limited, *London*
Prentice-Hall of Australia Pty. Limited, *Sydney*
Prentice-Hall Canada Inc., *Toronto*
Prentice-Hall Hispanoamericana, S.A., *Mexico*
Prentice-Hall of India Private Limited, *New Delhi*
Prentice-Hall of Japan, Inc., *Tokyo*
Simon & Schuster Asia Pte. Ltd., *Singapore*
Editora Prentice-Hall do Brasil, Ltda., *Rio de Janeiro*

In memory of Richard "Rich" Westerburg,
friend, colleague, fellow audio enthusiast,
and consummate technician.

Contents

Preface

Electro-optical devices have been used in various optical and electronic systems for years. Until recently, however, these devices have been treated as peripheral and not very important topics in both optical technology and electrical technology programs. With the advent of practical fiber optic transmission lines and laser diodes, a heightened interest in electro-optical devices has arisen in the electronics technology community. But communications is only a part of the picture. Digital mass storage, process control transducers, medical applications, and imaging systems are all areas where the optical and electronics world interface. These developments have reached the point where a new engineering term, *photonics,* has been coined. This term refers to the technology of dealing with radiant energy systems that manipulate the photon in a manner analogous to the manipulation of electrons in an electronic system.

This book began as a large stack of handouts and laboratory exercises that were originally used in training courses for technicians working in the Rochester, New York optics industry. An outgrowth of this training activity is a four-course sequence covering topics in electronics and photonics for our Associate Degree Optical Technology students. This sequence includes a one-semester course on photonic devices which uses this text. With some modifications, a variant of this devices course has been created which is directed to general electronics technicians and students. The completed book is a distillation of the materials used in all these courses.

The topics covered deal with many of the questions raised by technician trainees. As a result of this input, the coverage is broader than that of many of the optoelectronics communications texts that have appeared in recent years. This is a devices book, and it emphasizes the principles of operation and specifications of the major classes of photonic devices. The text material is directed to a technical, as opposed to a scientific, audience and stresses concepts important to someone involved with device testing and applications. The mathematics used is at a level that is approachable by technicians, although some technician trainees do admit to having to knock the rust off underutilized brain cells. There are a few equations complex enough to make solving them with a calculator a real challenge. Instructors who are fortunate enough to have computers readily available will find that these equations lend themselves to interesting programming assignments.

An important goal of the book is to make the technician and technical student familiar with the sometimes vague and apparently contradictory terminology used to describe and specify photonic devices. Perhaps even more important, technicians need a sense of what is reasonable in terms of device performance and application. Good technicians will recognize that a problem solution, which calls for a 40,000-W resistor or an index of refraction of 27, is in error. They need some concrete idea of practical magnitudes for the parameters of an optical detector or source.

In each chapter dealing with a specific device, the descriptive material utilizes the terminology that seems to be most commonly associated with that device. In addition, and as a method of dealing with the concrete as opposed to the strictly theoretical, one or more representative data sheets are examined. These data sheets provide the reader with an opportunity to see how each of the device characteristics is specified. The data sheets selected often contain graphs and tables that provide empirical illustrations of how device parameters vary with changing conditions. Last but not least, data sheets, even those of obsolete device models, help the reader to develop a sense of what is reasonable in terms of device operating limits.

This book has about four more chapters than I have ever been able to squeeze into one semester. This organization reflects the needs of the different populations to which the book is directed. The introductory chapters, 1 through 5, should be familiar ground to optics professionals, so they can be used for review and reference with these technicians. Electronics technicians, on the other hand, find in these chapters a number of new concepts—flux density, intensity, and color temperature, to name a few—that need to be mastered before proceeding. Some of the devices covered, such as vacuum photodetectors (Chapter 7), thermal detectors (Chapter 8), and photoresistors (Chapter 9), although of significant interest to the optical and general electronics technician, may be of less interest to the electronics communications specialist.

The text user can, and should, select those sections of the book best suited to meet the needs of the target audience. We have found this material to be a flexible instructional tool that can be used to meet a variety of needs in a number of different settings.

ACKNOWLEDGMENTS

I am grateful to my students who have always been candid and helpful in their assessment of these materials as they have evolved over the last several years. Fortunately, a few of them took great delight in finding mistakes in the text and problems; hopefully they found them all, but any that remain are of course my own errors. Not to be overlooked are the many hours that our division secretary Kathy VanAlstyne and her student aides devoted to typing the many early versions of the laboratory exercises and handouts that became the data base for this book. Special thanks go to my wife and daughters, who were always ready with a smile and a hug when the project got bogged down. Finally, great gobs of gratitude to my daughter Colleen, who helped with the sketches and drawings.

WILLIAM J. MOONEY

Optoelectronic Devices and Principles

1

Basic Optical Devices

1-1 INTRODUCTION

In this chapter the basic principles of geometrical ray optics are reviewed. Concepts are emphasized. The operation of typical optical components is described.

1-2 WHAT IS A RAY?

In the past you have probably studied the behavior of light and light rays in a high school or college physics course. In the study of electro-optics we are concerned not only with the visible electromagnetic radiation, called *light,* but also with invisible ultraviolet and infrared radiation. Fortunately for us, all the forms of electromagnetic radiation behave in the same manner, so those things that you have learned and observed about the behavior of light also apply to the invisible forms of radiation.

Rays of light are a mental convenience used to describe the path that a beam of radiation will take. The ray itself is imaginary and is usually represented on a drawing by an arrow or straight line. The straight line of the ray represents the path through space that a beam of radiation will follow. If you hold a flashlight above a flat surface so that the body of the flashlight is at right angles to the surface, a circular image of the light beam will be formed. A ray describing the flashlight beam would be drawn at right angles to the circular image with an arrowhead facing away from the flashlight.

In the case of the flashlight beam, the beam diameter expands as the distance from the flashlight increases. In the case of an expanding beam, more than one ray may be used to describe the beam. In the case of the flashlight beam one ray might be drawn perpendicular to the center of the beam diameter and two others drawn tangent to the sides of the expanding beam. In this way, both the beam direction and its divergence are described by a ray diagram (Figure 1-1).

In summary: A ray is a symbolic way of representing the direction or directions that the front surfaces of a beam of radiation will take.

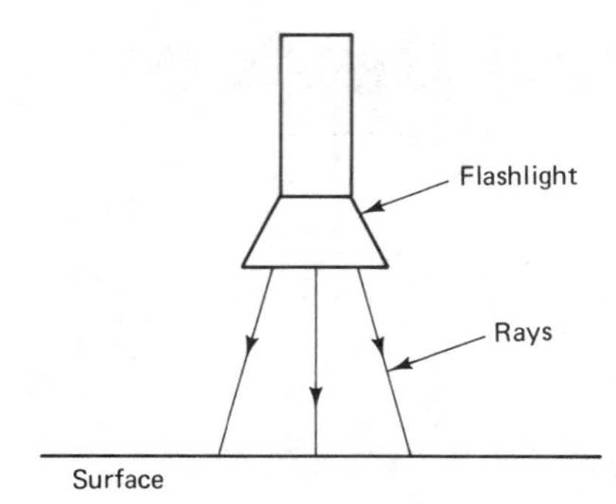

Figure 1-1 Beams and rays.

1-3 REFRACTION AND REFLECTION

In this section we review what happens when a beam of radiation traveling in air is incident on a flat glass surface. Assume for the purposes of discussion that the beam of radiation is a column or tube of light that has a constant diameter regardless of the distance from the source. Such a beam, called a *collimated beam,* is commonly used for optical testing.

The behavior of a collimated beam is described by a single ray. The glass surface is assumed to have a flat, smooth surface. To predict ray behavior completely, a parameter of the glass called the *index of refraction* needs to be known. The index of refraction allows us to predict the path that the ray will take through the glass. Physically, the index of refraction is a dimensionless number that compares the velocity of radiation in the glass to the velocity of radiation in a vacuum.

$$n = \frac{c}{v} \tag{1-1}$$

where n is the index of refraction
c is the velocity of propagation in a vacuum, 3×10^8 m/s
v is the velocity of propagation in the glass

Since the free-space velocity c represents the maximum possible velocity of propagation, the index n will always have a magnitude greater than 1. For simplicity we assume that the index of air is also equal to 1. The actual index of air is about 1.0003. Almost any liquid or glass that you would test in a laboratory will have an index of refraction between 1.3 an 1.8. Water and alcohol have an index of about 1.3 Crown glass has a typical index of 1.51, and flint glass would be expected to have a higher index of about 1.62. Optical glass is assumed to have a nominal index of 1.53. Crystals and semiconductors will have indices of refraction greater than 1.8; diamonds, for example, have an index of 2.42. In general, if you determine that the index of a piece of glass is less than 1.47 or greater than 1.70, you have probably made a mistake.

When a collimated beam of radiation encounters a flat glass surface it is broken into two beams. One beam is reflected off the surface and the other beam enters the material (Figure 1-2). The original beam from the source, called the *incident beam,* is represented by an *incident ray.* The *reflected beam* is represented by the *reflected ray.* The beam that enters the materials is generally called the *refracted beam* or *transmitted beam.* This beam is represented by a ray called the *refracted ray* or *transmitted ray.* The physical locations of these three beams are described by angles measured with respect to a line drawn normal (that is, at right angles) to the glass surface. Each of these rays has a unique fixed angular relationship to the other two rays and to the normal to the surface. The angle of the incident ray and the reflected ray with respect to the normal will be equal. The angle of the transmitted ray is determined by the angle of the incidence and the index of refraction. All these rays lie in the same plane. Equation (1-2) describes the relationship that exists between the incident angle, the index of refraction, and the angle of refraction.

$$\sin \phi = \frac{n}{n'} \sin \theta \tag{1-2}$$

where $\sin \phi$ is the sine of the angle of refraction
n is the index of the material in which the ray is initially being propagated
n' is the index of the material in which the refracted ray will be propagagted
$\sin \theta$ is the sine of the incident angle

These angles are illustrated in Figure 1-2.

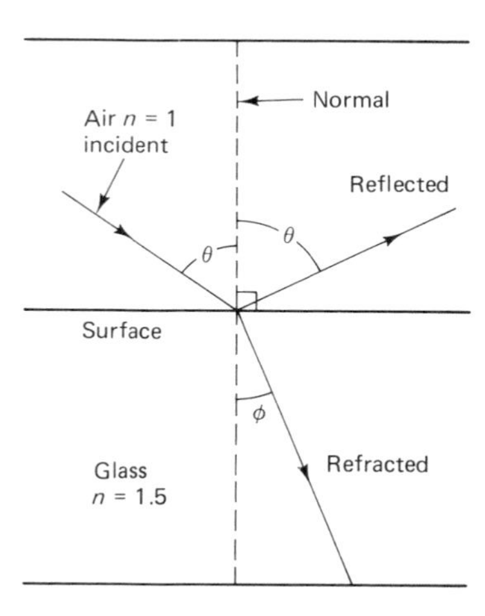

Figure 1-2 Reflection and refraction.

EXAMPLE 1-1

Refracted Angle and Velocity of Propagation

Given: **(1)** $\theta = 30°$
(2) $n = 1$ (air)
(3) $n' = 1.53$ (glass)

Find: **(a)** ϕ the angle of refraction
(b) μ the angle between the reflected beam and the refracted beam
(c) v the velocity of propagation in the glass

Solution: **(a)**

$$\sin \phi = \frac{n}{n'} \sin \theta$$
$$= \frac{1}{1.53} \sin 30°$$
$$= 0.327$$
$$\phi = \sin^{-1} 0.327$$
$$= 19°$$

(b)

$$\mu = 180° - (\theta + \phi)$$
$$= 180° - 49°$$
$$= 131°$$

(c)

$$n = \frac{c}{v}$$
$$v = \frac{c}{n}$$
$$= \frac{3 \times 10^8 \text{ m/s}}{1.53}$$
$$= 1.96 \times 10^8 \text{ m/s}$$

The preceding example assumed that the piece of glass in the system was infinitely long, so that the transmitted ray would just continue through the material without any further reflections or refractions. In practice, however, pieces of glass have a finite thickness, so the ray will be refracted as it passes from the air to the glass, and also as it passes from the glass to the air. In addition there will be reflections at the incident surface and at the existing surface. Figure 1-3 illustrates what happens at the two surfaces. The incident ray I_0 acting on the two surfaces S_1 and S_2 gives rise to a complex system of parallel reflected and transmitted rays caused by *multiple internal reflections.* These parallel rays are separated by a distance d that is a function of the incident angle, the angle of refraction, and the glass thickness. For a fixed glass thickness the separation d will increase as the incident angle is increased. For a fixed incident angle the separation d will increase as the glass

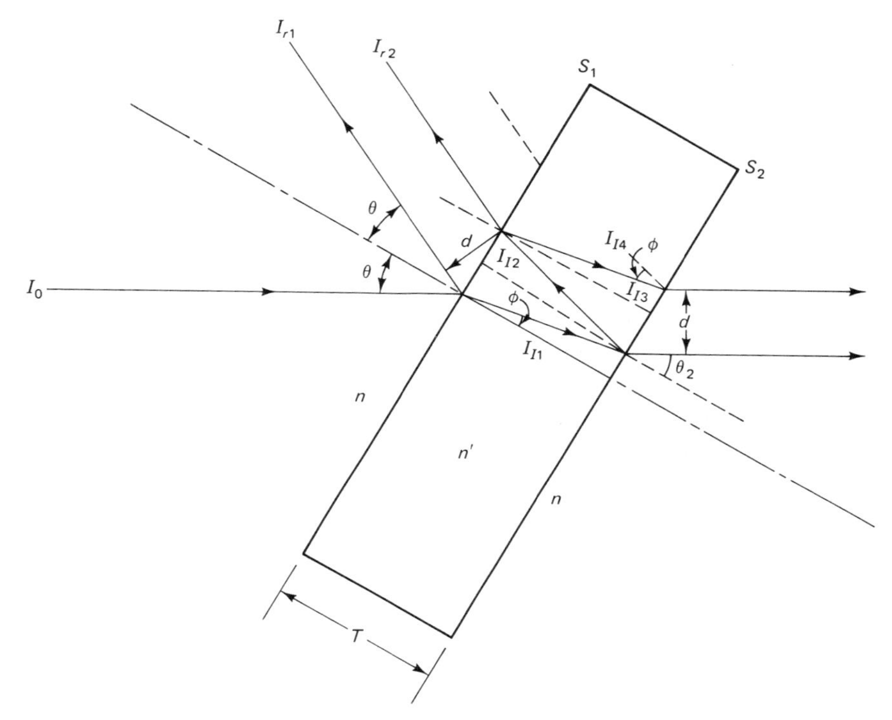

Figure 1-3 Reflection and transmission with a parallel plate.

thickness is increased. A 1-cm-thick piece of optical glass with an index of about 1.5 will have a ray separation of about 0.73 cm for an incident angle of 40° and the distance will decrease as the incident angle decreases.

The intensities of the transmitted and reflected rays are also dependent on the glass index and the angle of incidence. For angles of incidence between 0 and 40° the fraction of the incident beam intensity reflected from one surface can be estimated by

$$r = \frac{(n' - 1)^2}{(n' + 1)^2} \tag{1-3}$$

where r is the ratio of the reflected beam intensity to the incident beam intensity; referring to Figure 1-3, we have

$$r = \frac{I_{r1}}{I_0} \quad \text{for angles} \leq 40°$$

n' is the index of the glass
1 is the index of air

Equation (1-3) applies to the reflection that occurs going from air to glass and from glass to air. If we assume that almost all the reflected energy is contained in reflected beams I_{r1} and I_{r2} of Figure 1-3, an estimate of the total reflected energy is given by

$$\bar{r} \simeq 2r - 2r^2 + r^3 \tag{1-4}$$

where $\bar{r}$ is the net effective reflected intensity ratio viewed from the front surface
r is the reflection ratio for a single surface reflection given by equation (1-3)

$$\bar{r} = \frac{I_{r1} + I_{r2}}{I_0}$$

Essentially all of the transmitted intensity is contained in ray I_{T1} because the transmitted ray I_{T2} represents the intensity that is left after three successive reflections I_{r1}, I_{I2}, and I_{I4} and two successive transmissions I_{T1} and I_{r2}. To a good approximation the transmission ratio is given by

$$T = \frac{I_{T1}}{I_0} \simeq (1 - \bar{r} \simeq (1 - r)^2 \tag{1-5}$$

where T is an estimate of the transmission ratio for incident angles $< 40°$.

Ideally, the reflection ratio $\bar{r}$ and the transmission ratio would add up to 1. These approximate equations will not add to exactly 1, but for indexes near 1.5 and angles less than 40°, the error will be on the order of 0.2%. All of this analysis assumes that all the intensity changes observed are the result of reflection and that no attenuation, absorption, and scattering take place within the glass itself. For good-quality optical glass this is a valid assumption.

EXAMPLE 1-2

Calculation of Transmission and Reflection Ratios

Given: A plane-parallel plate placed in a collimated beam so that its surface is at right angles to the beam path (angle of incidence zero). The index of the glass is 1.520.

Find: The reflection and transmission ratios.

Solution:

1. Compute the first surface reflection factor.

$$r = \frac{(n' - 1)^2}{(n' + 1)^2}$$

$$= \frac{(1.520 - 1)^2}{(1.520 + 1)^2}$$

$$= 0.0426$$

2. Estimate the total reflection factor.

$$\bar{r} = (2r - 2r^2 + r^3)$$
$$= 2(0.0426) - 2(0.0426)^2 + (0.0426)^2$$
$$= 0.0816$$

3. Estimate the total transmission factor.

$$T = (1 - r)^2$$
$$= (1 - 0.0426)^2$$
$$= 0.9166$$

4. Calculate the error from the ideal.

$$\text{Ideal } 1 = T + \bar{r}$$
$$T + \bar{r} = 0.9166 + 0.0816$$
$$= 0.9982$$
$$\text{Change with respect to ideal} = 0.0018$$

As illustrated in Example 1-2, equations (1-3), (1-4), and (1-5) allow an estimate of the reflection and transmission ratios to be made. These calculations assume that both sides of the piece of glass are in air, and that the reflection at the air-to-glass interface is the same as the reflection at the glass-to-air interface. The numerical calculations show that for typical optical glass with an index of 1.52, about 8% of the incident intensity will be reflected and about 92% will be transmitted. These conditions and equations hold for incident angles less than 40°. For angles greater than 40° the reflected intensity will be observed to increase markedly, while the transmitted intensity would show a corresponding decrease in magnitude.

1-4 MIRRORS AND RETRO-REFLECTORS

In this section we review some of the basic principles of mirrors. A *mirror* is an optical device that reflects almost all of the incident radiation. It has a highly polished surface coated with a thin layer of material that will efficiently reflect the wavelengths of interest. A mirror for reflecting visible light will generally have a silvered or aluminized surface. Infrared mirrors will generally have a gold finish. Special mirrors will have dielectric coatings.

Mirrors used in optical setups are called *first-surface mirrors* when the reflecting coating is on the surface toward the source. The reflecting coating is on the first surface the radiation encounters. Bathroom and automotive mirrors are *second-surface mirrors* because the reflecting surface is on the side away from the incident radiation. Second-surface mirrors cause two reflections, one from the glass surface and one from the coated surface. A beam in a test setup might also be adversely affected by traveling through the glass. These are the reasons for using first surface mirrors in optical setups.

In this section only first-surface mirrors are discussed. A ray incident on a mirror surface will reflect at an angle equal and opposite to the angle of incidence. For typical coatings about 90% reflection can be obtained. Since the purpose of a mirror is to reflect radiation, the 10% transmitted/absorbed can be considered lost.

Figure 1-4 illustrates the paths of two rays I_1 and I_2 emitting from point source P and reflecting off a plane mirror. The reflected rays R_1 and R_2 would appear to an observer as if they were being emitted from point P'. The distance s from the real source to the reflecting surface is equal to the distance s' from the mirror surface to the imaginary apparent source at P'. An apparent or virtual image of point source P would appear to exist at P'.

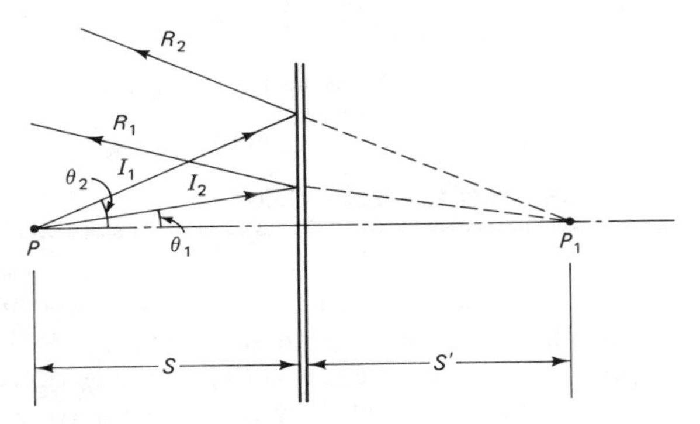

Figure 1-4 First-surface mirror. (*Note:* s and s' are measured from the first surface.)

Figure 1-5 illustrates the way an image of a three-dimensional object located at point P would be reflected. The virtual image of the object would appear to be located at point P' and the size of the object, and its image will be the same as the original object. Mirrors with a spherical radius of curvature are often encountered in detecting and radiating setups. Figure 1-6(a) illustrates a spherical curved mirror with a mechanical radius R. Figure 1-6(b) illustrates that if this mirror is illuminated by a collimated source or a plane wave, all the rays will be focused to point P.

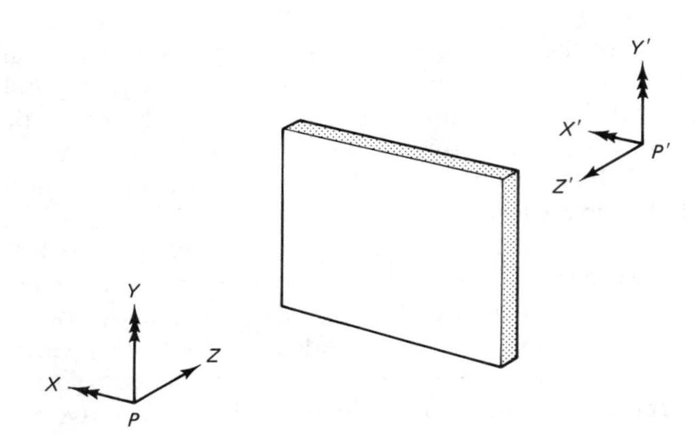

Figure 1-5 First surface forms a virtual three-dimensional image at p'.

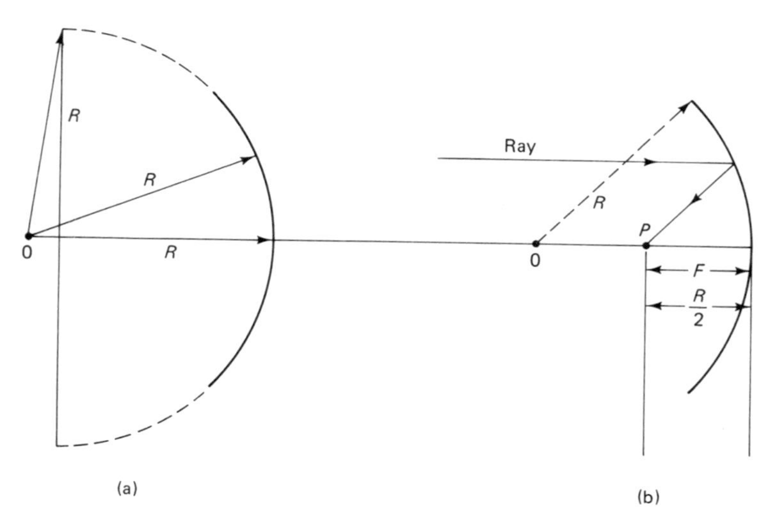

Figure 1-6 Spherically curved mirror with a mechanical radius of R.

This focal point will be located at a distance from the mirror surface equal to one-half the radius of curvature. If a point source is positioned at point P, a collimated beam will be formed by the mirror reflections. Spherical mirrors can be used to concentrate received radiation on a detector placed at the focal point of the mirror or to create and aim a collimated beam of radiation from a point source positioned at the focal point.

If a source is placed at the center of the radius of curvature (point O) all the incident radiation from the source will be reflected back through the center of the source. This reflected radiation adds to the direct radiation from the source and therefore increases the intensity of the beam from the source in the direction away from the mirror. A projection system often uses this technique. All of the foregoing discussion regarding spherical mirrors assumes that the mirror surface represents a small percentage of the spherical surface. Specifically, it assumes that the mechanical side-to-side diameter of the mirror is small compared to the mirror's radius of curvature. A mechanical diameter value larger than $0.1 \times R$ would be too large; diameter values of $0.002 \times R$ are common. The relationship of the object distance s, image distance s', and focal length F can be determined by

$$\frac{1}{s} + \frac{1}{s'} = \frac{1}{F} = \frac{2}{R} \tag{1-6}$$

where s is the distance from the mirror surface to the source or the object
- s' is the distance to the focused image of the source or the object
- F is the focal length of the mirror, the distance to the point at which a collimated beam will be focused
- R is the radius of curvature

If the mechanical side-to-side diameter of a mirror is large with respect to the radius of curvature, all the incident rays from a collimated source will not be focused at the focal point. Where such large mirrors are needed, a compensating lens or different-shaped mirrors are used.

A *retro-reflector* is an optical device that will always reflect radiation back on its own path. This can be very useful in certain laser systems. It is also used in noncontact measurement systems where the source and detector are at a location some distance from the object being monitored. Volcano monitoring systems use retro-reflectors on the mountainside, while the monitoring lasers and detectors are located at what is hopefully a safe distance away.

The retro-reflector is shaped like a pyramid except that it has three triangular sides rather than the four sides of a true pyramid. The base of the retro-reflector is circular and the incident and reflected radiation enter and leave through the base. Incident radiation enters the base and is internally reflected off the triangular sides and finally exits from the base parallel to the incident ray. Figure 1-7 is a drawing of a typical retro-reflector. The triangular sides of the retro-reflector might be coated with a reflecting material, but designers often depend on total internal reflection taking place.

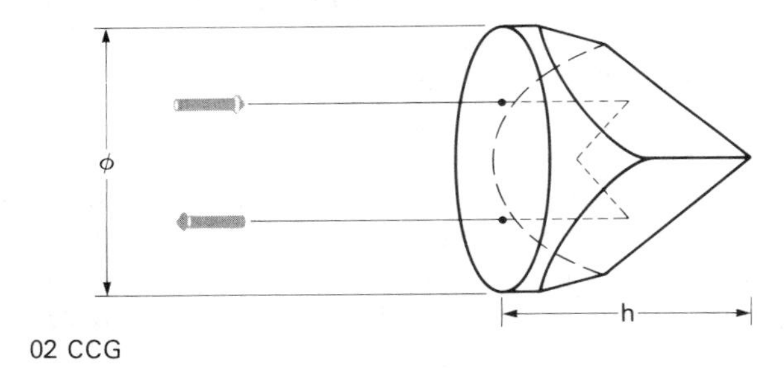

Figure 1-7 Solid glass retroflectors. (*Melles Griot Optics Guide 4,* 1988; courtesy Melles Griot, Irvine, CA.)

Total internal reflection occurs when the ray that is propagating inside the glass object is totally reflected from the second surface of the glass object. The internal angle of incidence that causes a total internal reflection to occur is called the *critical angle.* If the angle of incidence established by the velocity vector of the ray and the second surface of the glass is equal to or greater than the critical angle, total internal reflection will take place. The critical angle can be solved for by setting the sine of the angle of refraction in equation (1-2) equal to 1. For typical optical glass with an index of 1.5, the critical angle will be 42°. Internal angles of incidence equal to or greater than 42° will cause total internal reflection.

Because of their shape, retro-reflectors are often called *corner cubes.* Plane mirrors, spherical mirrors, and corner cubes are all devices used in optical setups that employ reflection properties as their primary mode of operation. Plane mirrors and spherical mirrors have reflection properties that are established primarily by their surface coatings. Corner cubes sometimes have their reflecting surfaces coated, but they generally rely on the process of total internal reflection.

1-5 LENSES

Lenses come in an infinite variety of shapes and sizes. For simplicity, consider the four basic types illustrated in Figure 1-8. The curved surfaces of these four lens types are assumed to be spherical. A lens has a spherical surface when the lens surface could be fitted to the surface of a perfect ball or sphere. The amount of curvature is specified in terms of the radius of the imaginary sphere that the lens would fit. This radius specification is called the *radius of curvature.* If the radius of curvature is very large compared to the diameter of the lens itself, the lens will appear to be almost flat [Figure 1-9(a)]. As the radius of curvature approaches the same order of magnitude as the lens diameter, the lens will have a very noticeable curvature [Figure 1-9(b)]. A limit is reached at the point where the radius of curvature is equal to half the lens diameter [Figure 1-9(c)]. At that limiting value the lens will be a hemisphere. If the physical thickness of the lens divided by the radius of curvature is a small number, the lens is called a *thin lens.* Like plane-parallel plates, the lens will both reflect and transmit light, and a simple glass lens will have a transmission of about 92%, just like the plane-parallel plate. In discussing lens characteristics and drawing lens ray diagrams, the reflected rays are usually ignored and only the incident and transmitted rays are drawn.

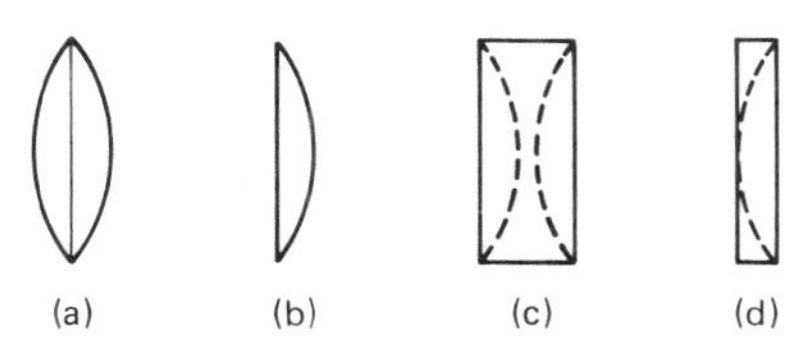

Figure 1-8 Four spherical lens types: (a) convex convex; (b) planoconvex; (c) concave concave; (d) planoconcave.

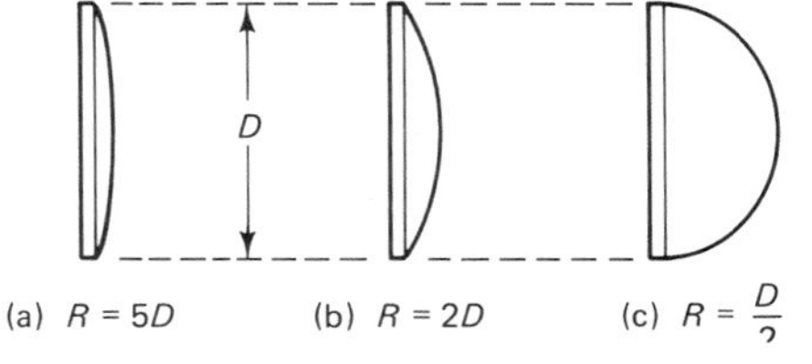

Figure 1-9 Three Planoconvex lenses with different R/D ratios.

Figure 1-10 illustrates the ray and image diagrams for a thin *converging lens.* The ray and image diagrams for a planoconvex lens would be similar. As illustrated in Figure 1-10(a), when the incident radiation is a collimated beam, the beam will be focused to a bright spot at the focal point F. This process can easily be demonstrated in sunlight with a simple magnifying glass. For locations beyond the focal point F, the beam of light will expand in a cone shape. If a point source of radiation is located at the focal point F, the transmitted beam will be a collimated beam, like the collimated beam obtained with a spherical mirror. This process is commonly used on optical benches to form a collimated beam for optical testing and is illustrated in Figure 1-10(b). Figure 1-10(c) illustrates how the image of an object is formed by a lens. The object is represented by the arrow located at point P. A sharp, clear image of the object is formed at the image plane P'. The location of the image plane and the size of the image are a function of the lens focal length, F, and the object distance from the lens, s. Observe on the drawing that the image arrow at

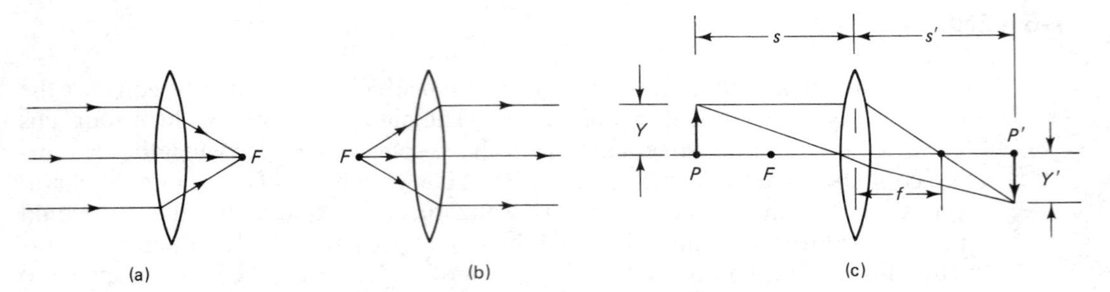

Figure 1-10 Convex-convex lens: (a) collimated source focused to a point; (b) point source collimated; (c) object at P, image at P'.

P' is inverted. The height of the image compared to the original object is called the *lens magnification, m.*

$$m = \frac{Y'}{Y} \tag{1-7}$$

where Y and Y' are heights shown in Figure 1-10(c). The magnification is a function of focal length F and object distance s.

$$m = \frac{-F}{s - F} = \frac{-s'}{s} \tag{1-8}$$

where m is the magnification and is negative if the image is inverted
F is the lens focal length
s is the object distance of Figure 1-10(c)

For the practical problem of obtaining a focused real image on an image plane the object distance s must be greater than the focal length F. If the object is located at a distance from the lens equal to the lens focal length, the denominator of equation (1-8) goes to zero. In this case mathematicians would say that the equation is *undefined.* Physically what happens is that rays from various points on the image become collimated, so an image is not formed. If the object distance is less than the focal length F, the rays drawn from various points on the image expand outward from the lens and a virtual image is formed on the same side as the object. In summary, a focused image can be formed on the image plane only when the object distance is greater than the focal length F. By doing a little algebraic substitution for values of s in equation (1-8), you can show that the absolute value of the magnification m will be greater than 1 when the object distance s is greater than F or less than $2F$. When s is equal to $2F$, the magnification is equal to -1. For values of s greater than $2F$, the magnification will be less than 1.

The location of the image plane P' is also dependent on the image distance and the focal length. Equations (1-9a) and (1-9b) illustrate the relationships that exist between magnification, object distance, and focal length.

$$s' = -ms \tag{1-9a}$$

$$s' = \frac{-F(s)}{s - F} \tag{1-9b}$$

where s' is the distance to the image plane
m is the magnification ratio
s is the distance to the object
F is the focal length

EXAMPLE 1-3

Magnification and Image Distance

Given: A lens with a focal length of 25 mm and a diameter of 22 mm to be used to image an object 2 mm tall.

Find: The image and object distances necessary to obtain a 10-mm real image.

Solution:

1. $m = \dfrac{Y'}{Y}$

$= \dfrac{-10 \text{ mm}}{2 \text{ mm}}$

$= -5$ magnification

2. $m = \dfrac{-F}{s - F}$

Therefore,

$$s = \frac{-F}{m} + F$$

$$= \frac{25 \text{ mm}}{-5} + 25 \text{ mm}$$

$$= 30 \text{ mm} \quad \text{object distance}$$

3. $s' = -ms$

$= -(-5)(30 \text{ mm})$

$= 150 \text{ mm}$ image distance

Summary of Example 1-3: For this lens to yield a magnification of 5, the object must be 30 mm away from the lens on one side and the focused image will be on an image plane 150 mm away from the lens on the opposite side.

Concave lenses are lenses whose center thickness is less than the edge thickness. A collimated beam of radiation incident on a concave lens will be changed into an expanding beam. This lens property can be used to expand laser beams or to increase the focal length of an existing lens system. A converging beam of radiation can be converted into a collimated beam with a concave lens.

The diverging rays from a concave lens caused by an incident collimated beam appear to diverge from a point source located on the incident side. This point represents the negative focal length of the lens. The smaller this number, the faster the beam will diverge.

Figure 1-11 and equation (1-10) illustrate how focal length and beam radius are related for a planoconcave lens and an incident collimated beam.

$$\frac{Y'}{Y} = \frac{d-F}{-F} = \frac{d+|F|}{|F|} \tag{1-10}$$

where Y' is the diverging beam radius (height) at distance d
Y is the incident collimated beam radius (height)
d is the distance from the lens to the plane of measurement of Y'
F is the focal length and will be a negative number

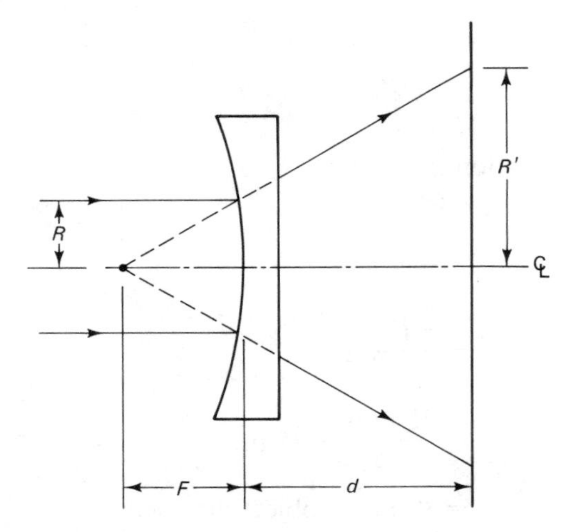

Figure 1-11 Concave lens expanding a collimated beam.

EXAMPLE 1-4

Beam Expander

Given: A laser beam of diameter 0.5 mm. It is desired to expand the beam to a diameter of 10 mm in a distance of 4 cm.

Find: The required planoconcave lens focal length.

Solution:

$$R' = 5 \text{ mm}$$

$$R = 0.25 \text{ mm}$$

$$d = 40 \text{ mm}$$

$$\frac{Y'}{Y} = \frac{d + |F|}{|F|}$$

$$|F| = \frac{d}{(Y'/Y)^{-1}}$$

$$= \frac{40 \text{ mm}}{(5 \text{ mm}/0.25 \text{ mm})^{-1}}$$

$$= 2.1 \text{ mm}$$

In this section it has been shown that convex lenses converge beams, whereas concave lenses can be used as beam expanders. In both cases the lens parameter called the focal length, F, is critical to lens performance.

The focal length of a lens is determined by its radius of curvature and the glass index. This discussion and the equations presented have assumed that the lenses in question are very "thin" with respect to the focal lengths. Thick lenses exhibit similar properties.

1-6 RADIUS OF CURVATURE AND FOCAL LENGTH

In Section 1-5 the lens parameter focal length was used to calculate image distances and magnification. In this section the focal length is shown to be related to the lens radius of curvature and the index of the lens material. The exact formula for a lens is given by

$$\frac{1}{F} = (n-1)\left(\frac{1}{R_1} - \frac{1}{R_2}\right) + \left[\frac{(n-1)^2}{n}\,\frac{t}{R_1 R_2}\right] \tag{1-11}$$

where F is the focal length
n is the glass index
R_1 is the radius on the incident scale
R_2 is the radius on the transmit side
t is the lens thickness on axis

R_1 will have a positive value if convex and a negative value if concave. R_2 will have a negative value if convex and a positive value if concave.

For a thin lens the thickness factor is usually dropped and the equation is simplified as follows:

$$\frac{1}{F} = (n-1)\left(\frac{1}{R_1} - \frac{1}{R_2}\right) \tag{1-12}$$

This equation is commonly called the *lensmaker's formula.* A point of confusion that arises in using equation (1-12) is the algebraic sign ascribed to radius R_2. For a convex convex lens the numerical value of R_2 will be negative, so the R_1 and R_2 ratios will be summed. Example 1-5 illustrates how to use equation (1-12) to estimate focal length for a convex-convex lens.

EXAMPLE 1-5

Lensmaker's Formula

Given: A convex-convex lens with:

(1) R_1 measured as 50 cm
(2) R_2 measured as 30 cm
(3) n specified as 1.52

Find: F from both sides. Use equation (1-12).

(a) $$\frac{1}{F} = (n-1)\left(\frac{1}{R_1} - \frac{1}{R_2}\right)$$

$$= (1.52 - 1)\left(\frac{1}{50\text{ cm}} - \frac{1}{-30\text{ cm}}\right)$$

$$= (0.52)\left(\frac{1}{50\text{ cm}} + \frac{1}{30\text{ cm}}\right)$$

$F = 36.06$ cm

(b) R_B is the incident side; therefore, $R_B = R_1$.

$$\frac{1}{F} = (0.52)\left(\frac{1}{30\text{ cm}} - \frac{1}{-50\text{ cm}}\right)$$

$$= (0.52)\left(\frac{1}{30\text{ cm}} + \frac{1}{50\text{ cm}}\right)$$

$$= 36.06$$

This is the same result as in part (a)!

We can conclude that the focal length is the same regardless of which side is illuminated. If one side of the lens is plano, equation (1-12) reduces to equation (1-13) below because the plano side has a radius of infinity.

$$\frac{1}{F} = (n - 1)\frac{1}{R} \tag{1-13a}$$

or

$$F = \frac{R}{n - 1} \tag{1-13b}$$

Equation (1-13b) is useful in illustrating how focal length changes with changes in radius and index. Focal length can be shortened by decreasing R, which implies a noticeably curved surface. Increasing the glass index will also decrease focal length.

In summary, the focal length of a lens is established by the lens's radius of curvature and index of refraction. Once the focal length is established it can be used to calculate image magnification and image distance.

1-7 PRISMS AND BEAMSPLITTERS

Prisms are used to steer beams along a path in an optical system or to split a beam into two separate beams. In beam-steering applications the prism manufacturer usually strives to achieve a large internal reflection. Sometimes prisms are put together for the purpose of splitting a beam into two beams; in this case the prism combination is called a *beamsplitter.* Figure 1-12 illustrates some common beam-steering prism types. As these prism types illustrate, a prism can be used to change the direction of propagation and the orientation of the image. Both the right-angle prism and the penta-prism change the direction of propagation by 90°, but the orientations of the two projected images are different. The dove prism does not change the direction of propagation, but it does invert the image.

A cube beamsplitter, illustrated in Figure 1-13 on page 19, is fabricated out of two right-angle prisms cemented together along the hypotenuse. A partially reflecting coating is deposited on one of the cemented surfaces. An inexpensive cube beamsplitter would yield about 40% transmission on both the straight path and the reflected right-angle path. About 20% of the incident energy is lost.

Another type of beamsplitter is the plate beamsplitter. A thin (1 mm) plane-parallel plate has its incident surface coated with a partially reflecting coating. The incident beam hits the first surface (reflecting) at a specified angle, often 45°, and causes a reflected beam. A portion of the beam passes through the first surface and is transmitted out the rear of the plate. The ratio of the transmitted and reflected energy to the incident energy is a function of the coatings used and usually varies with wavelength. Beamsplitter ratios are normally specified as R/T, where R is the reflected percent and T is the transmitted percent. Values of 50/50, 30/70, and 70/30 are common. Plate beamsplitters can have very low losses, on the order of 1% or less.

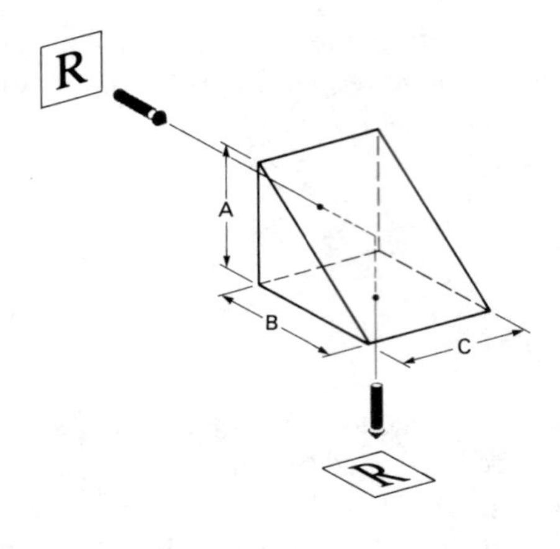

01 PRA/PRS

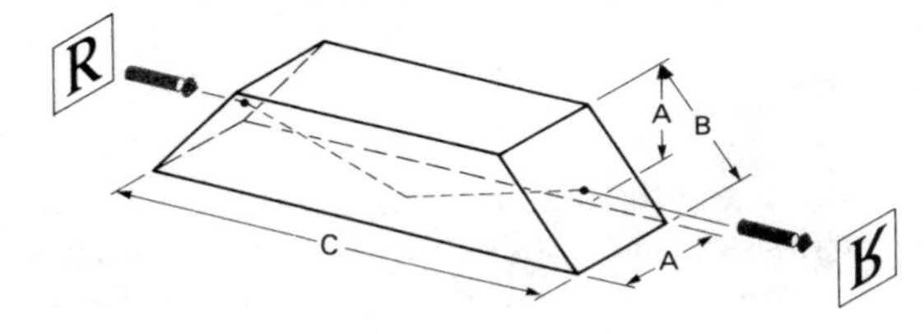

01 PDE

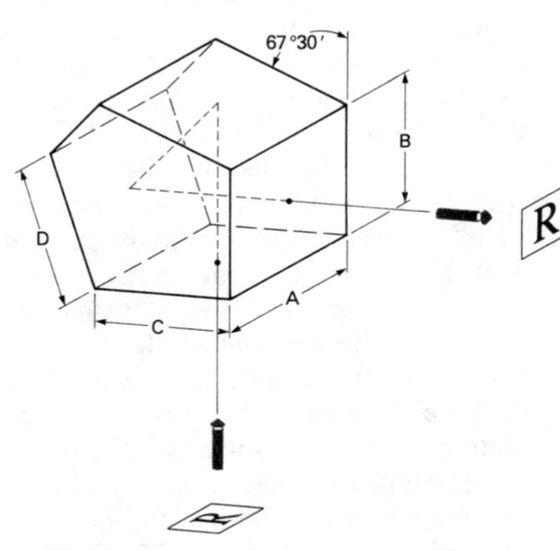

01 PPA/PPS

(c)

Figure 1-12 Three prism types: (a) right angle; (b) dove; (c) penta. (*Melles Griot Optics Guide 4,* 1988; courtesy Melles Griot, Irvine, CA.)

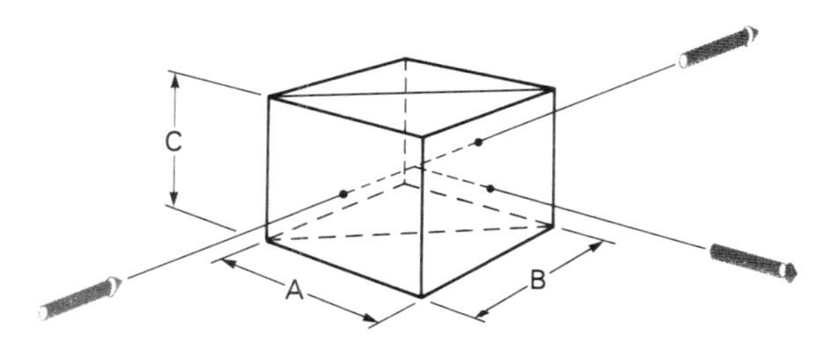

Figure 1-13 Cube beamsplitter. (*Melles Griot Optics Guide 4,* 1988; courtesy Melles Griot, Irvine, CA.)

1-8 SUMMARY

In this chapter the interaction of light with a variety of optical components has been discussed. Where necessary, equations have been used either to illustrate how device characteristics vary with the physical dimensions or to develop a sense of order of magnitude. Three physical phenomena have been discussed: (1) reflection, (2) transmission, and (3) refraction.

Six types of optical components have been discussed: (1) plane-parallel plates, (2) mirrors, (3) retro-reflectors, (4) lenses, (5) prisms, and (6) beamsplitters. Plane-parallel plates were used to develop the ideas of reflection, refraction, and transmission. It was shown that a plane-parallel plate will typically have a transmission of about 92% and a reflection of about 8%. Mirrors are components designed to maximize reflection and minimize transmission of energy. Curved mirrors can be used to focus the beam. A retro-reflector is a special mirror system that always folds the incident beam back toward the incident beam. Lenses are devices that emphasize transmission characteristics. Typical uncoated lenses will exhibit about 92% transmission. Lenses are used to focus or expand beams and to project, focus, and magnify images.

Lens characteristics are established by the index of refraction and the radius of curvature. Lens focal length can be determined from these physical parameters. Prisms are used to steer beams and orient images. Prisms depend on internal reflection to operate. Beamsplitters utilize reflection and transmission properties to subdivide a beam into two separate beams.

PROBLEMS

1. What is the velocity of propagation in glass with an index of 1.62?
2. A pulse of light enters a hollow tube and a glass tube at the same time. Both tubes are 1 m long. The glass tube has an index of 1.55.

(a) From which tube will a pulse of light first emerge?
(b) What will the arrival time difference be between the two pulses?

3. Compute the angle of refraction for incident angles of 10°, 20°, 30°, and 40°. Assume an index of 1.53.
4. Equation (1-2) is true if and only if the incident angle is less than 90°. For a ray of light in air and entering optical glass with an index of 1.53, determine the maximum angle of refraction for which equation (1-2) is valid.
5. Determine the reflection factor for crystals with indexes of 1.53, 1.60, 1.80, and 2.00. What general trend is observed as index increases?
6. Three identical parallel plates are mounted in series on a lens bench. A radiometric instrument measures the flux density of the output beam to be 680 $\mu W/cm^2$. One plate is removed and a reading of 773 $\mu W/cm^2$ is obtained.
   (a) What is the net transmission of one plate?
   (b) Estimate the reflection of one plate. Assume that $\bar{r} + T = 1$.
   (c) Estimate the single-surface reflection factor.
   (d) Estimate the index of the material.
   (e) How much flux is incident on the first surface of the first plate?
7. A 5-cm-diameter spherical mirror has a radius of curvature of 500 cm. What is the focal length?
8. A spherical mirror has a focal length of 10 cm. What is its radius of curvature?
9. A planoconvex lens has a focal length of 5 cm. Calculate the magnification for object distances of 6 cm, 7 cm, and 10 cm.
10. A planoconvex lens has a focal length of 10 cm. Calculate the image distances for object distances of 12 cm, 14 cm, and 20 cm.
11. Calculate the focal length for a planoconvex lens with an index of 1.53 and a radius of curvature of 250 cm.
12. (a) Calculate the focal length for a convex-convex lens with an index of 1.53. Each surface has a radius of curvature of 250 cm.
    (b) How does this result compare with the planoconvex result of Problem 11?
13. Three planoconvex lenses each have a radius of curvature of 250 cm. The index of the three lenses are 1.53, 1.62, and 1.76, respectively.
    (a) Calculate the focal length of each lens.
    (b) Calculate the magnification of each lens for an object placed 560 cm from the lens.
14. Three prototype planoconvex lenses are fabricated from optical glass (index 1.53). The radius of curvature of the three lenses is 130 cm, 177 cm, and 250 cm, respectively.
    (a) Calculate the focal length of each lens.
    (b) Calculate the magnification of each lens for an object located 480 cm from the lens.
15. A plate beamsplitter is placed in the path between a laser and a retro-reflector. The beamsplitter is specified as a 30/70 beamsplitter at the wavelength of interest. The retro-reflector exhibits 95% reflection at the wavelength of interest. What is the magnitude in percent of the retro-reflected beam that is returned to the laser location? (*Hint:* The beam will make two passes through the beamsplitter.)

2

Interference and Diffraction Devices

2-1 INTRODUCTION

In Chapter 1 the operating principles of simple optical components were reviewed using the principles of ray optics. An alternative way of viewing radiation is as a train of sine waves. The wave model of radiation allows us to explain observations that could not be explained by ray concepts. As is the case in any branch of physics, the model selected to solve any given problem depends on the nature of the problem. In some cases both ray and wave models are needed to understand an optical device or system completely. In this chapter, wave concepts and ray concepts are applied to the analysis of a variety of optical components and setups.

2-2 WAVE CHARACTERISTICS

Radiated waves are called *electromagnetic waves* because they are the result of the interaction of electrical and magnetic fields. As these waves are propagated in space, the amplitudes of these electrical and magnetic fields vary sinusoidally with respect to time. The rate at which the wave goes through one complete sinusoidal alternation is fixed when the wave is generated, and it does not change. Frequency and period are the mathematical ways of specifying the rate at which the electromagnetic sine wave varies. *Period* is the length of time required for the sine wave to go through one complete cycle. *Frequency* is the reciprocal of period and has units of cycles per second, called hertz (Hz).

$$f = \frac{1}{T} \tag{2-1}$$

where f is the frequency, in hertz
T is the period, in seconds

Visible light has frequencies of about 10^{15} Hz. The frequency and period of a sine wave are very important because they are the only characteristics of the wave that remain constant as the wave propagates through space.

When an electromagnetic wave propagates, the electrical field and the magnetic field are at right angles to each other and to the direction of propagation. Figure 2-1 illustrates a linearly polarized electromagnetic wave. This wave is said to be *linearly polarized* because the electric field occurs repeatedly along the same axis, in this case the y axis. Other polarizations are circular, elliptical, and randomly polarized. Circular, elliptical, and linearly polarized waves get their names from the path traced by the electric field around the axis of propagation. While the electric field must remain at right angles to the axis of propagation, which is the x axis in Figure 2-1, the electric field may assume any angle with respect to the y and z axes. In a linearly polarized wave the electric field assumes a single fixed angle with respect to the y and z axes, so it traces out a straight line in space—hence the name *linear* [Figure 2-2(a)]. A circularly polarized wave would assume different angles at each instant of time, thus tracing out a circle [Figure 2-2(c)]. Randomly polarized radiation such as that radiated from an incandescent lamp can be thought of as a random mix of all possible polarizations. Beams of randomly polarized radiation are often represented as four equal-length crossed vector lines [Figure 2-2(b)].

Optical phenomena are dependent on the interaction of the electric field and the various optical components. For this reason the magnetic field is usually ignored. In the visible region the frequency of the light causes our eye to perceive it as a particular color, and the amplitude of the electric field controls its apparent brightness. The frequency of the wave does not change, but its amplitude and polarization can be affected by a whole variety of transmission and reflection effects.

Wavelength is another important wave characteristic. One *wavelength* is equal to the distance traveled by the wave during the period of one complete cycle. The wavelength of an electromagnetic wave is dependent on its velocity of propagation and its frequency. As noted in Chapter 1, the velocity of propagation depends on the index of refraction of the material in which the propagation takes place. A wave of fixed frequency (and all waves have a fixed frequency) will have different

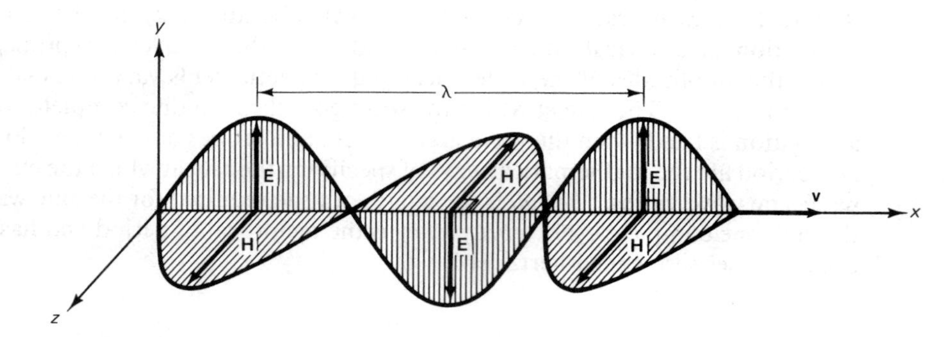

Figure 2-1 Linearly polarized electromagnetic wave illustrating (a) electrical phasor **E**, (b) magnetic phasor **H**, (c) velocity vector **v**, and (d) wavelength λ.

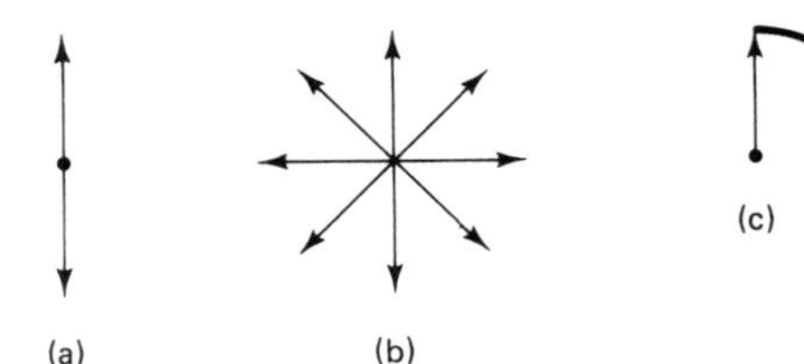

Figure 2-2 Polarization symbols representing electrical phasor: (a) linearly polarized; (b) randomly polarized; (c) circularly polarized.

wavelengths in different materials. Those nice tables of wavelengths that you find listed for various colors in textbooks are the wavelengths in free space. The following equations relate wavelengths to frequency and index and to free-space wavelength.

$$\lambda = \frac{v}{f} \tag{2-2a}$$

$$\lambda_0 = \frac{c}{f} \tag{2-2b}$$

$$\lambda = \frac{c/f}{n} \tag{2-3}$$

$$\lambda = \frac{\lambda_0}{n} \tag{2-4}$$

where λ is the wavelength in any material
v is the velocity in any material
f is the frequency
λ_0 is the wavelength in free space
c is the velocity in free space, 3×10^8 m/s
n is an index

Equations (2-2), (2-3), and (2-4) tell us that the wavelength of a wave is variable and that in an optical material it will be shorter than it is in free space.

2-3 CHROMATIC DISPERSING

A *dispersing prism* is an equilateral prism that allows us to illustrate two interesting facts:

1. White light is composed of a variety of different frequencies.
2. Refractive index is not constant but changes with frequency.

It is usual to describe these phenomena in terms of wavelengths, where the wavelengths being referred to are the free-space wavelengths.

Figure 2-3 illustrates how an incident white light beam breaks into red, orange, yellow, and so on, beams. The red beam, which has the longest wavelength (lowest frequency), deviates the least. The violet beam, which has the shortest wavelength (highest frequency), deviates the most. The obvious conclusion is that the index of refraction increases as the frequency increases. The magnitude of the angular deviation between the colored beams will vary with incident angle. At a fixed angle of incidence the angular deviation is often specified with respect to the red and blue lines.

This effect is not confined to prisms but can occur with all optical devices. Figure 2-4 shows how white light passing through the test lens is separated into red

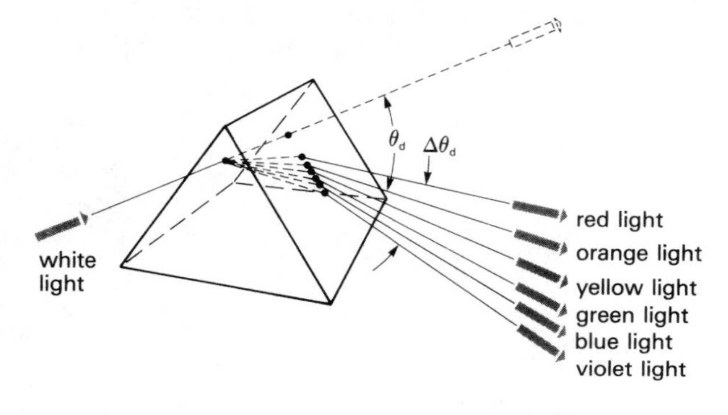

Figure 2-3 Equilateral dispersing prisms. (*Melles Griot Optics Guide 4,* 1988; courtesy Melles Griot, Irvine, CA.)

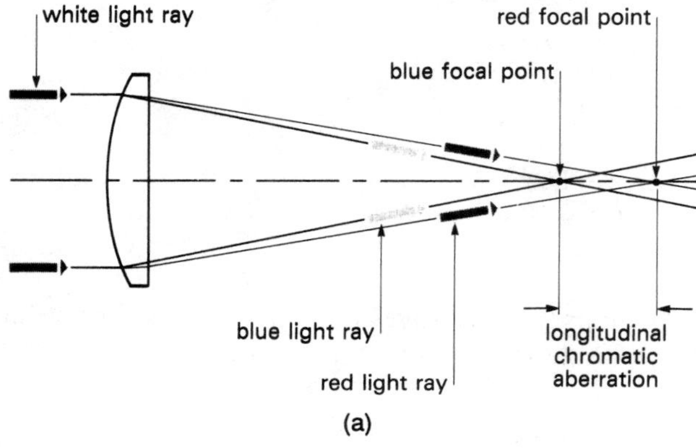

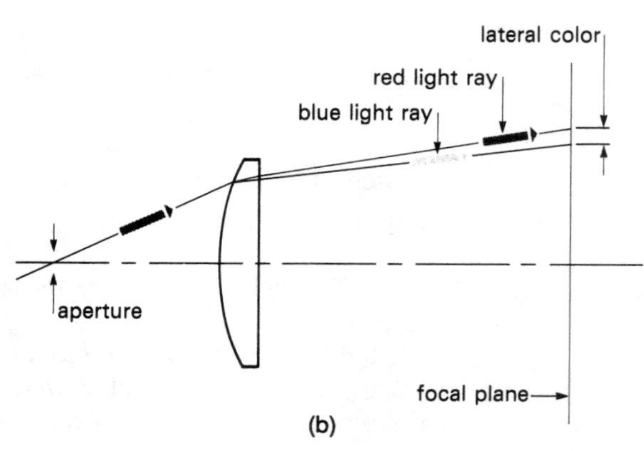

Figure 2-4 Chromatic aberration caused by index variation as a function of wavelength: (a) longitudinal; (b) lateral. *Note:* typical optical crown glass values of index: blue (436 nm), 1.534; green (546 nm), 1.525; red (656 nm), 1.520. (*Melles Griot Optics Guide 4,* 1988; courtesy Melles Griot, Irvine, CA.)

and blue beams, and all those in between, and that each of these separate wavelengths follows its own unique path determined by its frequency-dependent index. The color rendition of the image will be distorted by this process and is called *chromatic aberration.* In systems where it is important to minimize chromatic aberration, two lenses are cemented together to form a lens combination called a *doublet* or *achromat.* These lens combinations decrease the chromatic offsets from millimeter sizes at the image plane to micrometer sizes. A typical achromat would consist of a convex convex lens cemented to a meniscus lens. The index of the two lenses would differ slightly and the combined effects of their shapes and index differences provide the required corrections.

2-4 DIFFRACTION

Figure 2-5(b) illustrates the actual shadow pattern that will be caused by a rectangular object placed between a monochromatic source and an image plane. This shadow pattern varies significantly from the one shown in Figure 2-5(a), which is based on ray-optic assumptions only. The actual shadow is the result of the wave nature of radiation. The actual image exhibits a smooth transition from maximum dark at the center to light regions at the side, with fringes of dark-to-light bands occurring outside the region where shading would be expected.

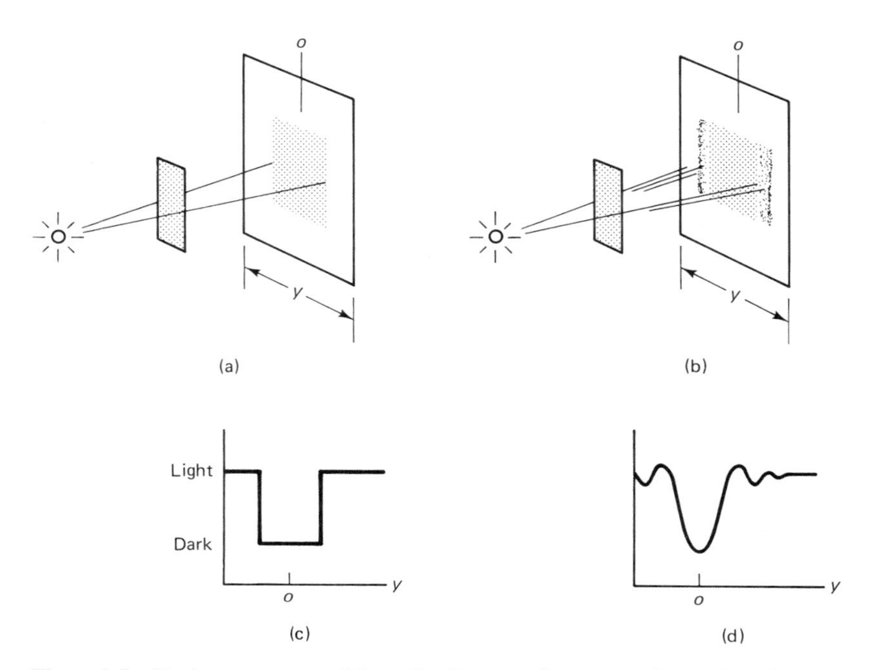

Figure 2-5 Shadow patterns and intensity diagrams for a monochromatic point source and a rectangular obstacle: (a, c) ray concepts only; (b, d) actual patterns due to diffraction.

If a slit is substituted for the rectangular object, a pattern of illuminated and dark regions of the same form will be observed with maximum light at the center (Figure 2-6). The bright maximum will shade off into a region of light and dark fringes that is much wider than the simple ray projection of the slit. These observations illustrate a general principle: *Regularly shaped objects or regularly shaped apertures when illuminated from a monochromatic source will create an image with a regularly shaped intensity distribution pattern that extends beyond the boundary of the image that would be predicted by ray-optics concepts.*

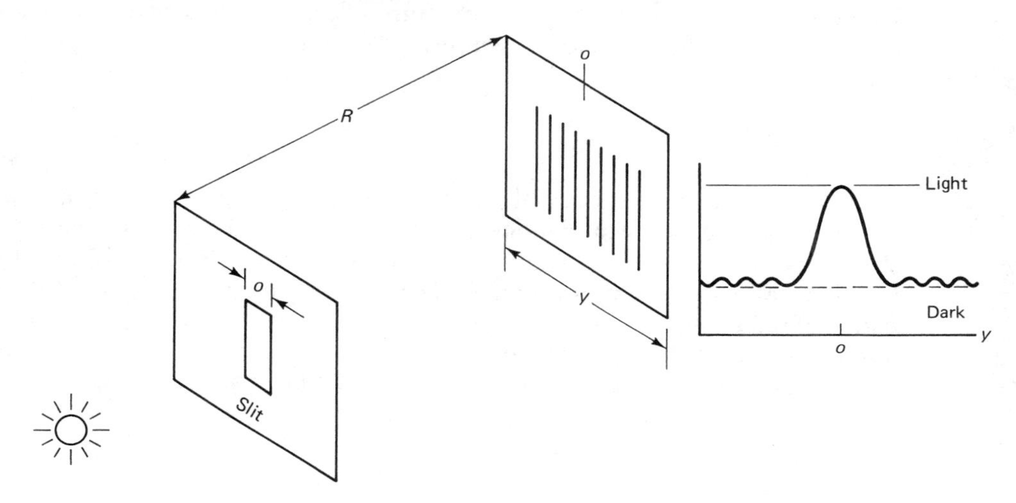

Figure 2-6 Diffraction caused by a single slit.

In practice, to observe this effect readily you would have to work in a room with a low light level so that the fringes could be observed. If the source is not monochromatic, each wavelength would cause its own pattern, which would overlap the others and make observation of the effect difficult. A typical helium–neon laser and a guitar string or slit 1 mm or so in width can be used to illustrate the phenomenon.

The important characteristics of the slit image are the location and spacing of the fringes and the width of the central lobe of intensity. These image characteristics are dependent on slit width D, wavelength λ, and the distance from the slit to the image plane. A typical test setup is illustrated in Figure 2-7. An image plane is mounted at a distance R from a slit of width D. The distance from the central maximum of the image to any minimum or the fringe pattern is represented by the distance y. Line H represents the hypotenuse of a triangle formed by R, y, and H, and α represents the angle between R and H. Intensity minima will be located at positions defined by

$$D \sin \alpha = m\lambda \tag{2-5}$$

where D is the slit width
α is the angle illustrated in Figure 2-7

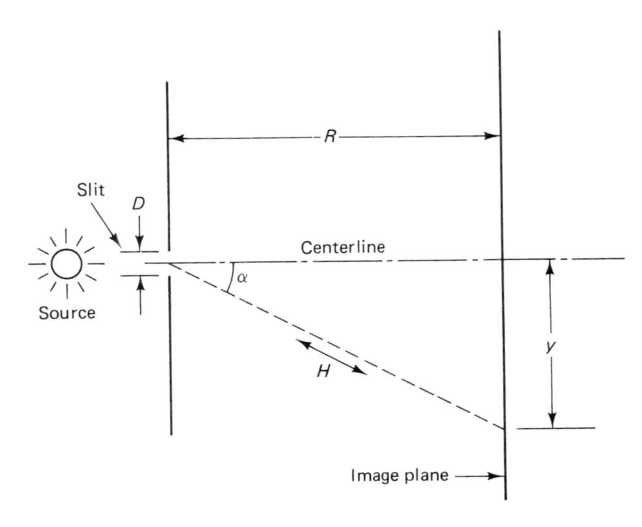

Figure 2-7 Single-slit test setup (top view).

m is any integer
λ is the wavelength of the source

As Figure 2-7 illustrates, sin α could be obtained by measuring y and H and using the trigonometric definition of the sine function.

$$\sin \alpha = \frac{y}{H} \tag{2-6a}$$

If $H \simeq R$, then

$$\sin \alpha \simeq \frac{y}{R} \tag{2-6b}$$

In practice, sin α is approximated using the ratio of y to R. This approximation results in errors of less than 2% for angles of less than 20°. Using the approximation of equation (2-6), equation (2-5) can be rewritten to the forms in equation (2-7).

$$\sin \alpha \simeq \frac{y}{R} \qquad \text{(2-6b)}$$

$$D\frac{y}{R} \simeq m\lambda \tag{2-7a}$$

$$y \simeq \frac{m\lambda R}{D} \tag{2-7b}$$

Equation (2-7b) can be used to determine the fringe spacing, Δy. Let

y_m be the distance from the central maximum of intensity to a minimum intensity fringe of order m

y_{m+1} be the distance from the central maximum of intensity to a minimum intensity fringe of order $m + 1$

Then

$$y_m = \frac{m\lambda R}{D}$$

$$y_{m+1} = \frac{(m+1)\lambda R}{D}$$

$$y_{m+1} - y_m = \Delta y$$

$$y_{m+1} - y_m = \frac{\lambda R}{D}$$

$$\Delta y = \frac{\lambda R}{D} \tag{2-8}$$

where Δy is the distance between two adjacent fringe minima for off-axis angles less than 20°.

As equations (2-7b) and (2-8) illustrate, small slits will cause large patterns and large fringe spacings. Conversely, large slits will cause small patterns and fringe spacings. As was pointed out at the beginning of this discussion, these equations will yield 2% or better results for angles less than 20°. The tangent of 20° is about 0.36, so in a practical setup, if y divided by R is less than 0.36, these equations will exhibit less than 2% error.

EXAMPLE 2-1

Determining Slit Width from a Diffraction Pattern

Given: **(1)** Source helium–neon tunable laser wavelength 611.8 nm
(2) $R = 1.5$ m
(3) $\Delta y = 4.6$ mm average of five spacings
(4) $y \simeq 55$ mm

Find: The slit width.

Solution:

1. Are y and R in an acceptable range?

$$\frac{y}{R} = \frac{55 \text{ mm}}{1.5 \text{ m}} = 0.037$$

$$\frac{y}{R} < 0.36 \qquad \text{therefore, the approximations are valid}$$

2. $\Delta y = \dfrac{\lambda R}{D}$

$$D = \frac{\lambda R}{\Delta y}$$
$$= \frac{(611.8 \text{ nm})(1.5 \text{ m})}{4.6 \text{ mm}}$$
$$= 199.5 \ \mu\text{m}$$

Standard-precision air slits are available in sizes from 5 to 200 μm, so this is a reasonable result.

In some applications you will be interested only in the area of maximum intensity of the fringe pattern, not with the side fringes. The width of the entire central maximum of intensity can be determined using equation (2-7b) for m equal to 1 and width W equal to twice the y value; this is illustrated below.

$$y = \frac{m\lambda R}{D} \tag{2-7b}$$

$$2y = \frac{2m\lambda R}{D}$$

$$W = 2y = \frac{2m\lambda R}{D}$$

$$= \frac{2\lambda R}{D} \qquad m = 1 \tag{2-9}$$

Occasionally, you will be interested in the width of the central maximum at the half-intensity points. This width is given by

$$W_{1/2} = (0.89)\frac{R\lambda}{D} \tag{2-10}$$

EXAMPLE 2-2

Determining the Width of the Central Maximum and the Half-Power Maximum

Given: **(1)** Source = 611.8 nm
(2) $R = 1.5$ m
(3) $D = 200 \ \mu$m

Find: **(a)** The width of the central maximum.
(b) The half-intensity width.

Solution:

(a) $W = \dfrac{2\lambda R}{D}$

$= \dfrac{2(611.8 \text{ nm})(1.5 \text{ m})}{200 \text{ μm}}$

$= 9.18 \text{ mm}$

$\dfrac{W}{D} = 45.9$

(b) $W_{1/2} = (0.89)\dfrac{R\lambda}{D}$

$= 4.09 \text{ mm}$

$\dfrac{W_{1/2}}{D} = 20.5$

As Example 2-2 illustrates, the central maximum of intensity projected on the image plane is much wider than the slit width D. Even the more restrictive case of the one-half intensity width is more than 20 times larger than the slit width.

A second common test aperture is the pinhole. Manufacturers make precision pinholes in sizes ranging from 1 to 1000 μm. Precision pinholes are made by laser-drilling holes in a thin (0.003 to 0.014 mm) foil. Less expensive versions are made by deposition on glass. Like the slit, the pinhole will cause a regularly shaped pattern on the image plane. This pattern causes a round central region of maximum intensity surrounded by alternating circular light and dark rings (Figure 2-8). The central circle, called *Airy's disk,* is named after Sir George Airy, an English astronomer.

The formula for the location of the minimum of the diffraction pattern caused by a pinhole has the same form as the formula for the minimums of the diffraction pattern caused by a slit. The value of m for a slit is the set of integers, but the value of m for a pinhole pattern is a set of numbers that will yield solutions to a first-order

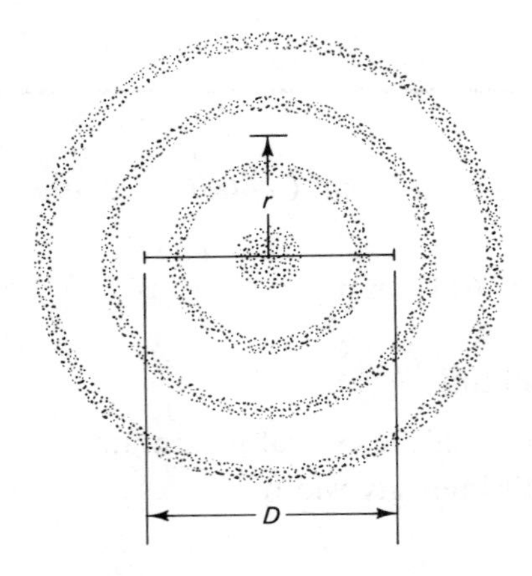

Figure 2-8 Diffraction caused by a pinhole showing Airy disk and concentric rings. Maxima dark, minima light. D and r are shown measured at the second minima where $m = 2.233$.

Bessel function. Also, on the image caused by a slit, a linear distance y is measured from the center line of the image pattern, but for a pinhole pattern the measured dimension is a pattern radius r. The numerical values of m for a pinhole were calculated in 1886 by E. V. Lonmel and they are listed in Table 2-1.

TABLE 2-1 VALUES OF *M* USED TO CALCULATE THE LOCATION OF INTENSITY MINIMUM CAUSED BY SLITS AND PINHOLES

Order	Slit	Pinhole
1	1	1.220
2	2	2.233
3	3	3.238
4	4	4.241
5	5	5.234

When making measurements the diameters of the dark fringes are generally easier to measure than the radii. Equations for the radius distance from the center of the image to the center of the dark fringe and for the diameter measurement of any given dark fringe follow:

$$r = \frac{m\lambda R}{D} \tag{2-11a}$$

$$d = \frac{2m\lambda R}{D} \tag{2-11b}$$

where r is the radius measured from the center of the pattern
m is the order of the ring
λ is the wavelength of the source
R is the distance from the image to the pinhole
D is the diameter of the pinhole
d is the diameter of the dark fringe

EXAMPLE 2-3

Determination of Pinhole Diameter

Given: **(1)** $\lambda = 632.8$ nm
(2) $R = 1.5$ m

Ring Number Order	Ring Diameter d (mm)
2	10.6
3	15.4
4	20.1

$$D = \frac{2m\lambda R}{d}$$

Solutions		Pinhole Diameter
Order	m	D (μm)
2	2.233	400
3	3.238	399
4	4.241	401
$\bar{x}$	—	400

Solution: 400-μm-diameter pinhole.

In summary, we can conclude that regularly shaped apertures illuminated by monochromatic sources will cause regularly shaped patterns on an image plane. For a fixed distance from the image plane and a fixed wavelength, the size of the image will increase as the size of the aperture decreases. The images will exhibit a central area of maximum intensity surrounded by alternating fringes of dark and light bands. The distance of the center of the dark bands from the center of the image or from each other can be used to determine various parameters of the test setup. Slits cause patterns with equally spaced fringes, and pinholes cause unequally spaced fringes. The general equation for diffraction patterns is the same for slits and pinholes.

2-5 MULTISLITS AND GRATINGS

In Section 2-4 image plane patterns caused by a single slit were examined. These patterns were the result of the process of diffraction. In this section patterns are described that are the result of diffraction and interference occurring simultaneously. The effects caused by two slits are discussed and then these effects will be used to describe the effect of a diffraction grating that has many closely spaced slits.

Figure 2-9 illustrates two slits and the resulting diffraction and interference patterns. At the outset it is important to realize that in the practical case both diffraction and interference occur simultaneously in a multislit system. Any observed image will be the result of simultaneous interference and diffraction patterns. The effects are discussed individually only as an intellectual convenience.

Figure 2-9 illustrates the composite diffraction pattern caused by two slits separated by a distance. This composite diffraction pattern is located on the centerline between the two slits and has peaks and minima located with respect to that centerline which have the same spacings that a diffraction pattern from either slit individually would have. Equation (2-12) defines the location for diffraction minimum at the angle θ measured with respect to the centerline. This equation is identical to

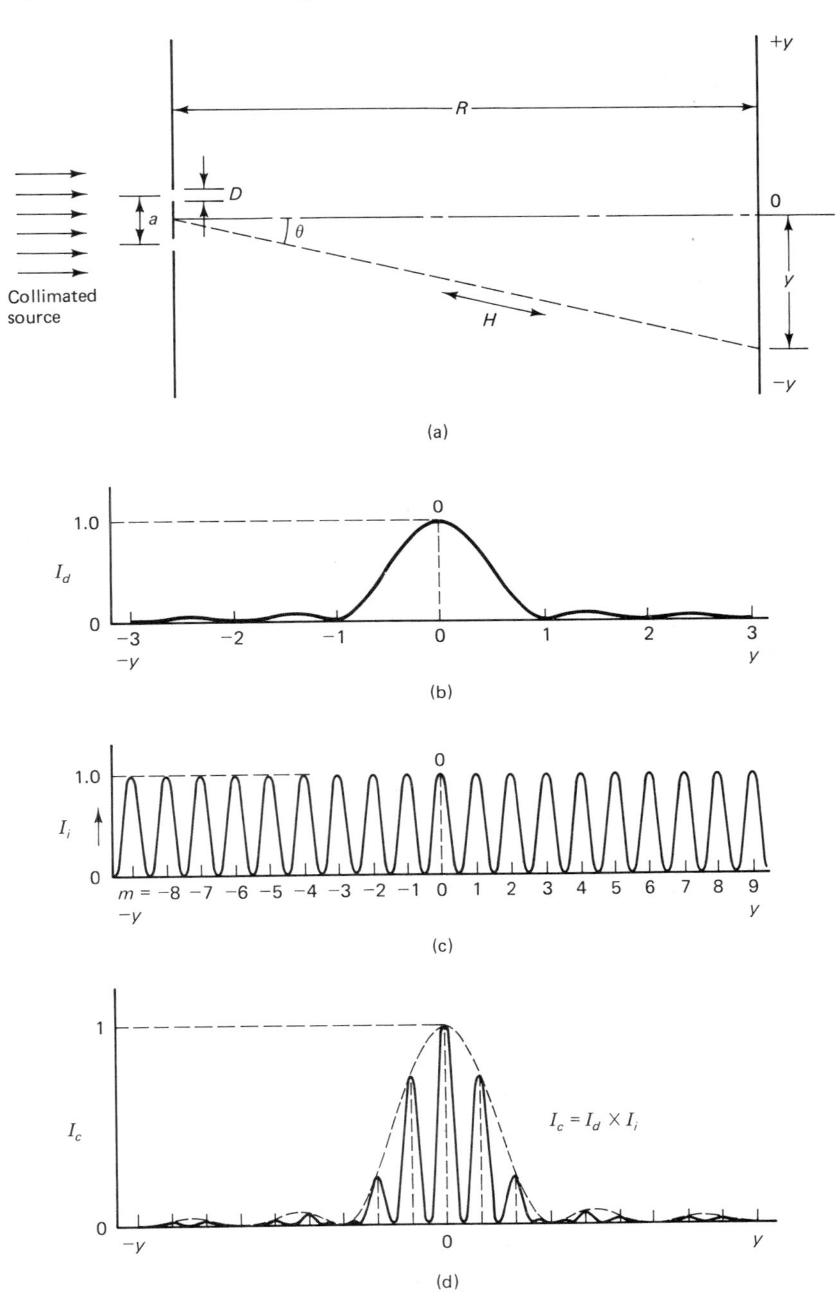

Figure 2-9 Double-slit intensity pattern: (a) top view of test setup; (b) relative intensity caused by diffraction (numbers above the horizontal axis are order numbers); (c) relative intensity caused by interference; (d) composite, actual, relative intensity.

equation (2-5) for a single slit and can be changed in the same way to a relationship including R and y.

$$D \sin \theta = m\lambda \tag{2-12a}$$

$$D \frac{y}{R} = m\lambda \tag{2-12b}$$

where m is the set of integers.

In summary, the two slits acting together create a diffraction pattern that is similar to that which would be caused by either slit alone except that it is centered between the slits. Diffraction is not the only process taking place. The radiation traveling from the two slits will arrive at the image plane along different path lengths. When the sine waves arriving by these two different paths have the same phase, a peak of illumination will be observed. When the sine waves arriving by these two different paths are 180° out of phase, an illumination null or minimum will be observed. Figure 2-9(b) illustrates the pattern that would be observed if only this interference effect were at work. Notice that the nulls and peaks for the interference pattern are closer together than the nulls and peaks of the diffraction pattern. This is because the location of these nulls and peaks is dependent on the spacing between the slits which must be greater than the slit width. Remember our general rule that the smaller the aperture or object, the larger the pattern, and as in the case here, the larger the aperture or object, the smaller the pattern. The equation for the location of the interference minimum is

$$a \sin \theta = (m + \tfrac{1}{2})\lambda \tag{2-13a}$$

$$a \frac{y}{R} = \left(m + \frac{1}{2}\right)\lambda \tag{2-13b}$$

where a is the center-to-center spacing of the slits
m is the set of integers
θ, y, and R are as shown in Figure 2-9(b)

Equation (2-13) has the same form as the diffraction equation. As was shown in the derivation of equation (2-8), the spacing between adjacent fringes is a constant value Δy and can be solved for in terms of λ, R, and a.

$$\Delta y = \frac{\lambda R}{a} \tag{2-14}$$

Figure 2-9(c) illustrates the actual intensity pattern that would be observed when the two slits are illuminated by a collimated monochromatic source. This pattern is the result of the product of the diffraction pattern and the interference pattern. On an image plane, this pattern would appear as a row of closely spaced lines or dots in clusters (Figure 2-10). The individual closely spaced dots are the result of interference. The spaces between dot clusters are the result of diffraction.

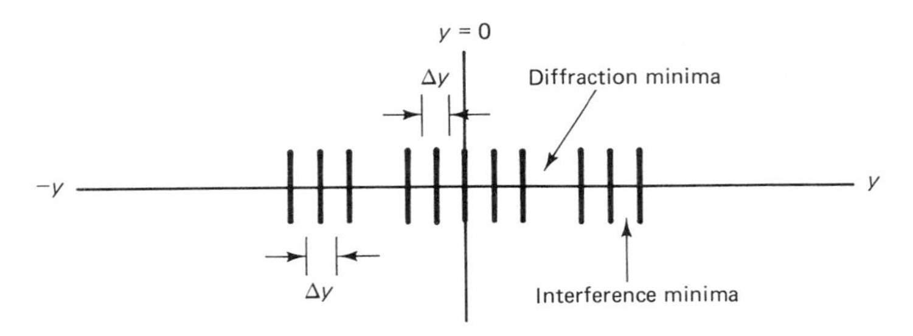

Figure 2-10 Image plane pattern caused by the double-slit intensity pattern illustrated in Figure 2-9(d).

EXAMPLE 2-4

Double-Slit Images

Given: **(1)** $\lambda = 632.8$ nm
(2) $R = 1.5$ m
(3) Δy between dots $= 1.2$ mm
(4) Δy between dot clusters $= 4.8$ mm

Find: **(a)** The slit width.
(b) The slit spacing.

Solution:

(a) $$\Delta y \text{ dots} = \frac{\lambda R}{a}$$

$$a = \frac{\lambda R}{\Delta y \text{ dots}}$$

$$= \frac{(632.8 \text{ nm})(1.5 \text{ m})}{1.2 \text{ mm}}$$

$$= 791\ \mu\text{m}$$

(b) $$D = \frac{\lambda R}{\Delta y \text{ dot clusters}}$$

$$= 198\ \mu\text{m}$$

Example 2-4 illustrates how the image plane pattern caused by a double slit can be used to determine the characteristics of the double slit.

The transmission diffraction grating is a plate with many very thin parallel slits (called *lines*). The transmission grating is illuminated from behind by a collimated source that is normal to the rear surface. If the source is monochromatic and the image plane is a section of a circle centered on the center of the grating, an image will be formed that consists of sharp thin lines separated by arcs of increasing length on the circular image plane. These lines will be arranged symmetrically with respect to the centerline (Figure 2-11).

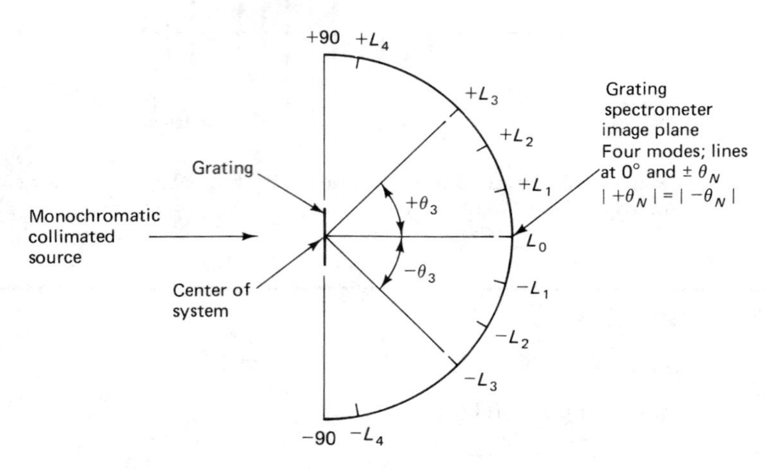

Figure 2-11 Grating spectrometer (top view).

The angular locations of these image lines are determined by the slit spacing and the wavelength of the source.

$$d \sin \theta = m\lambda \tag{2-15a}$$

$$\sin \theta = \frac{m\lambda}{d} \tag{2-15b}$$

$$\theta = \arcsin \frac{m\lambda}{d} \tag{2-15c}$$

where d is the grating slit spacing
θ is the angle measured with respect to the central axis
m is the set of integers
λ is the source wavelength

The now-familiar equation (2-15) describes the operation of the grating. This equation applies when the predicted angle θ is less than 90° and the source beam is normal to the rear surface of the grating. Alignment can be checked by measuring the angular location of lines of the same order with respect to the center. If the measured angles are the same, the system is correctly aligned. Because the equation is restricted to an angular range of $\pm 90°$ the maximum number of lines that will be visible is established by the ratio of d to λ. The justification of equation (2-15) is beyond the scope of this book but can be found in most wave optic texts.

For the maximum number of lines of a fixed wavelength,

$$d \sin \theta = m\lambda \tag{2-15a}$$

at maximum $\theta = 90°$; therefore, $\sin\theta = 1$ and

$$d = m\lambda$$

$$\frac{d}{\lambda} = m \tag{2-16}$$

where m would represent the highest order or the maximum number of lines in one direction, from which

$$n = 2m + 1 \tag{2-17}$$

where n is the total number of lines
$2m$ is the total number of lines in the clockwise and counterclockwise directions
1 accounts for the zero order or centerline

A coarse grating where the grating line spacing d is much larger than λ will yield a large number of projected lines. A very fine grating will yield far fewer projected lines.

EXAMPLE 2-5

Projected Lines Caused by a Transmission Grating

Given: **(1)** $\lambda = 632.8$ nm
(2) s grating specified at 6000 lines/cm

Find: **(a)** The number of lines that will be visible.
(b) Their angular locations.

Solution:

(a) Calculate d, the line spacing, in meters.

$$d = \frac{1 \text{ cm}}{6000 \text{ lines}} = \frac{1 \times 10^{-2} \text{ m}}{6000 \text{ lines}}$$

Calculate the maximum value of m.

$$m = \frac{d}{\lambda} \tag{2-16}$$

$$= \frac{1 \times 10^{-2} \text{ m}}{6000 \text{ lines}} \times \frac{1}{632.8 \times 10^{-9} \text{ m}}$$

$$= 2.63$$

but m only includes integers.

$$m = 2$$

$$n = 2m + 1$$

$n = 5$ total lines: one on axis, two clockwise, and two counterclockwise

(b) Calculate the angular line locations.

$$\theta = \arcsin \frac{m\lambda}{d}$$

$$= \arcsin m \; 632.8 \times 10^{-9} \text{ m} \times \frac{6000 \text{ lines}}{1 \times 10^{-2} \text{ m}}$$

m (± units)	sin θ (± units)	θ (± degrees)
0	0	0
1	0.3797	22.31
2	0.7594	49.41

As illustrated in Example 2-5, the monochromatic helium–neon laser will yield two well-defined lines with this grating. If the beam of light being examined came from a gas lamp, there would be several wavelengths present at one time, and each of these wavelengths would cause a separate set of visible lines of different colors.

Example 2-6 illustrates laboratory data taken with a hydrogen lamp and a fine grating.

EXAMPLE 2-6

Transmission Grating Spectrometer

Given: **(1)** Calibration run: 0 to 75°
Source: 632.8-nm laser

m	θ
0	0
1	22.4
2	49.6

(2) Lamp run: 0 to 75°
Comments: A slow rotation clockwise indicates that four lines are distinguishable between 0 and 30°. The lines begin to repeat at about 29°. These lines are designated by color as encountered: line 1, violet; line 2, violet-blue; line 3, blue; and line 4, red.

LAMP LINE DATA AS A FUNCTION OF ANGLE

θ (deg)	Line Number	Order	Color
0	0	0	0
14.3	1	1	Violet
15.1	2	1	Violet-blue
17.0	3	1	Blue
23.3	4	1	Red
29.6	1	2	Violet
31.5	2	2	Violet-blue
35.8	3	2	Blue
47.8	1	3	Violet
51.6	2	3	Violet-blue
52.2	4	2	Red
61.3	3	3	Blue

Find: **(a)** Use laser data to solve for d.

(b) Use d to solve for λ of each of the four lines 1 to 4.

Solution:

Line	Average Measured (nm)
1	410.2
2	434.0
3	486.1
4	656.3

These results correspond to a typical hydrogen lamp spectrum.

The data of Example 2-6 illustrate some basic characteristics of gratings.

1. Like a prism, a grating will disperse light into a spectrum.
2. A grating can create more than one complete spectral pattern or order. A prism creates only one pattern.
3. At larger angles the spectral patterns may overlap. Notice how the third repetition of the violet line pair occurs before the second occurrence of the red line.
4. When multiple orders are present, closely spaced lines are easier to resolve or separate at higher orders.

Gratings are often used in the manner illustrated in Example 2-6 to resolve a light source into its component spectral lines. For a given separation of spectral lines a grating will be much smaller than the prism required for the same resolution. A prism would have the advantages of superior spectral brightness and no confusing overlapping spectral orders.

In addition to transmission gratings, reflection gratings are also used to separate a beam of light into its spectral components. A reflection grating is made by creating regularly spaced reflecting and nonreflecting lines on a flat surface. The energy reflected from the narrow reflecting lines interferes to produce effects just like transmission gratings. In this section double slits and gratings were discussed. The image created by a double slit illustrates both interference and diffraction effects. If the number of slits is increased and the slit spacing reduced to a few multiples of a wavelength, a grating is formed. The grating can be used to resolve a beam of light into its spectral components.

2-6 COATINGS AND COATED DEVICES

Coatings are layers of material applied to the surface of an optical component to enhance or fix transmission or reflection characteristics. Lenses, mirrors, prisms, polarizers, beamsplitters, and filters are all optical devices that have coatings applied. The effectiveness of a coating will vary with the wavelength of the radiation, the incident angle of the beam, and the polarization of the incident radiation. Important mechanical characteristics of a coating are its thickness and uniformity.

Coatings are applied to the optics by deposition techniques in fractional-wavelength thicknesses. Coatings are fragile and with improper handling can readily be damaged. If it is necessary to clean a coated surface, low-pressure dry gas may be used to blow off loose particles. Some coatings may be cleaned with deionized water and a mild detergent or alcohol. These liquids are applied with a soft lint-free swab or by flow techniques.

Transmission-enhancing coatings are called *antireflection coatings.* These coatings reduce the reflection at the air–glass interface below the typical 4% per surface of the bare glass. Since less of the incident radiation is reflected, more will be transmitted. In addition, the secondary images (ghosts) caused by multiple reflections are also attenuated. This improves image sharpness.

A very common optical coating material is magnesium fluoride (MgF_2). This material, used for the visible-light region, has an index of about 1.38 at 550 nm. Single-layer magnesium fluoride coatings will be applied to a thickness of one-quarter wavelength. Figure 2-12 illustrates a typical single-layer magnesium fluoride antireflection coating. In this kind of coating structure, the ideal index of the material of the coating would be equal to the geometric mean of the transmission medium and the glass. The coating thickness is one-quarter wavelength. Conceptually, under these conditions two potential reflections take place. The first reflection is off the incident surface and the second reflection is off the glass surface. Because the coating is one-quarter wavelength thick, the second reflection will be delayed by one-half wavelength (one round trip) with respect to the first reflection, for small angles of incidence. These two reflections will have almost equal amplitudes, and the half-wavelength difference in phase will cause total destructive inter-

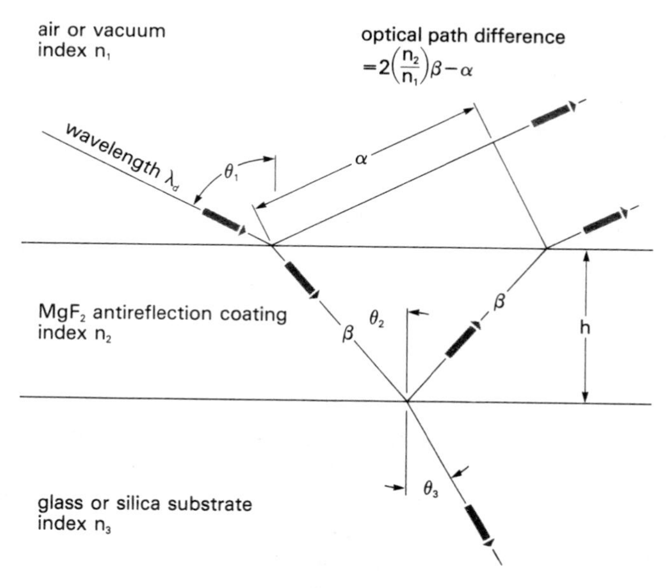

Figure 2-12 Antireflection coating. (*Melles Griot Optics Guide 4,* 1988; courtesy Melles Griot, Irvine, CA.)

ference of the two reflections. In other words, the amplitude of the net reflection will be reduced to almost zero. If the amplitude of the reflection is almost zero, the conservation of energy requires that the amplitude of the transmitted energy be increased to almost 100%.

$$r=\frac{(n_o n_g-n_c^2)^2}{(n_o n_g+n_c^2)^2} \tag{2-18}$$

where n_o is the original medium index
n_g is the glass index
n_c is the coating index
r is the reflection factor

when $n_o=1$ in air,

$$r=\frac{(n_g-n_c^2)^2}{(n_g+n_c^2)^2} \tag{2-19}$$

For $r=0$,

$$0=n_g-n_c^2$$

Therefore,

$$n_c=(n_g)^{1/2} \tag{2-20}$$

Equations (2-18) through (2-20) are the theoretical equations for the reflection constant r at an angle of normal incidence for a surface with a quarter-wavelength coating. Using equation (2-19) and assuming the glass index to be 1.5 and the coating index to be 1.38 (magnesium fluoride), a reflection of 1.4% is predicted. This is, in fact, the reflection percentage realized. A piece of optical glass coated with magnesium fluoride will typically have a reflection of less than 2% for wavelengths in the visible range 400 to 700 nm. As was the case with the uncoated surface, the reflection amplitude will increase as the angle of incidence increases. Measurable increases will be noted for incident angles greater than 40°.

In summary, single-layer quarter-wavelength antireflection coatings enhance transmission by reducing reflection. If a coating glass combination could be found such that the coating would have an index, at the desired wavelength, that was exactly equal to the square root of the glass index, reflection would theoretically be reduced to zero. In practice compromises must be made, but these coatings do result in a significant decrease in reflection compared to the glass alone. The amplitude of the reflection will increase as the wavelength of the incident radiation is varied above or below the design wavelength. The amplitude of the reflection will also increase as the incident angle increases beyond 40°.

In addition to single-layer antireflection coatings, multilayer coatings of various indexes are also available. These coatings typically achieve reflection ratios of less than 0.3%. Multilayer coatings can be designed to operate over a wide range of wavelengths or to maximize transmission at a single specific wavelength. Multilayer coatings are usually specified for a specific family of glass materials and relatively small angles of incidence. Typical specified acceptable angles of incidence would be less than 30° and are sometimes as small as 10°.

Lenses, prisms, and polarizers are examples of components whose surfaces would be coated with transmission-enhancing antireflection coatings. High-reflection coatings are applied to surfaces that we want to behave like mirrors. These coatings may be metallic coatings or dielectric coatings (transparent oxides). Reflection coatings can be placed on the first surface of the mirror or on an internal surface. Metal coatings are soft and are often covered with a half-wavelength dielectric coating. The dielectric coating protects the surface mechanically and prevents oxidation of the metal. Reflectance will be enhanced at the wavelength where the coating is a half-wavelength thick because the reflections off the first and second surfaces of the coating will recombine in phase. Typical metallic coatings are aluminum, silver, and gold. Aluminum quickly forms a low-reflection oxide when exposed to air so is usually overcoated with a dielectric layer. Coated aluminum surfaces can achieve 80 to 95% reflectance over the visible range of wavelengths. With an ultraviolet transmitting coating, aluminum mirrors can also be made for the ultraviolet range of wavelengths.

Silver also oxidizes and tarnishes very rapidly, so it is usually used only on well-protected internal surfaces. Silver surfaces will have reflectances of greater than 95% over much of the visible region.

Gold is tarnish resistant and exhibits excellent constant values of reflectance over the infrared range of wavelengths. Reflectance of better than 98% can be expected for wavelengths longer than 1000 nm. In addition to glass optics, integrating spheres and thermal detectors are often coated with gold. Gold is soft and expensive and needs to be treated with care. Cleaning is usually by a flow wash as rubbing can damage the gold-coated surface.

Mirrors coated with dielectrics only are also available. The reflectance of a surface can be increased by coating it with a material of a higher index of refraction. Alternating layers of high and low indexes can be fabricated to achieve a very high reflectance over a narrow range of wavelengths. Reflectances of greater than 99% are possible with multilayer reflection coatings. Depending on construction, the reflectance of these mirrors will vary with incident angle. Some complicated coating structures are specified over only a $\pm 15°$ angle of incidence.

Layers of dielectrics can be combined to construct devices called *transmission bandpass interference filters.* A typical interference filter would have a construction profile like that illustrated in Figure 2-13. These filters transmit radiation only within a narrow range of wavelengths. The heart of such filters is the half-wavelength-thick cavity. This cavity and the adjacent quarter-wavelength layers are constructed in such a way that internal reflections in the cavity exit the cavity in phase with the transmitted wave at the desired wavelength. Undesired wavelengths exit the cavity out of phase with the transmitted wave and destructively interfere with each other. The other quarter-wavelength layers act to enhance transmission and minimize reflection at the desired wavelength. The filter shown in Figure 2-13 is actually two identical filters connected in series. These filters can be quite selective, with 50% transmission bandwidths of 10 to 15nm. The 50% specification means that the power transmitted through the filter at these band-edge wavelengths will be 50% of the power level transmitted through the filter at the desired center wavelength. Unfortunately, these filters exhibit high loss, with attenuations of the desired wavelength as high as 70% in the visible region and even higher in the ultraviolet. This means that only 30% of the power at the desired wavelength is transmitted and 15% of the power at the band-edge wavelengths.

A variation of the bandpass filter is the *edge filter,* which is designed to yield a very rapid change from transmission to reflection at a specified wavelength. Depending on construction, the edge filters will transmit a fairly wide range of wavelengths above or below the specified edge wavelength. Ideal interference filters

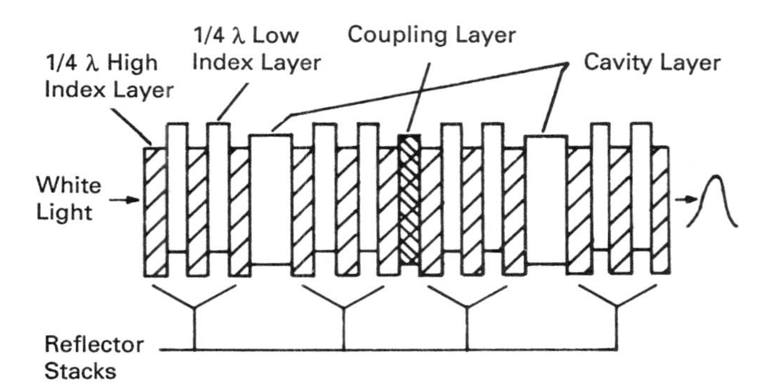

Figure 2-13 Two-cavity filter. (*Oriel Optics and Filters,* Vol. III, 1984; courtesy Oriel Corporation, Stratford, CT.)

would transmit desired wavelengths and reflect undesired wavelengths. Unfortunately, in practice a high level of heat-producing absorption also takes place. Figure 2-14 illustrates some typical interference filter responses. These filter responses are dependent on path length and assume normal angles of incidence.

In addition to filters that employ wave interference to select wavelengths, there are filters that control transmittance by absorbing undesired wavelengths. Colored-glass filters and heat-absorbing filters both employ this principle. Care must be taken with *absorption filters* to keep flux densities low enough to prevent damage to the filter due to overheating from the absorption process. The total absorption is proportional to path length, and the specified absorption assumes normal incidence. Since practical interference filters also exhibit absorption, care must also be taken with them to maintain temperature levels below those that will cause damage or specification changes.

Neutral density filters are a fourth kind of filter used in optical laboratories. A neutral density filter is really a beam attenuator that employs a reflecting surface to control transmission. Neutral density filters are used in visible-light setups and have a relatively constant attenuation characteristic for all visible wavelengths. The attenuation of a neutral density filter is specified in density units D, which are equal to the log base 10 of the reciprocal of the filter's transmission. The function of a neutral density filter is to attenuate the light by a specified amount.

$$D = \log_{10} \frac{1}{T} \tag{2-21}$$

EXAMPLE 2-7

Density of a Filter

Given: A neutral density filter that transmits 40% of the incident energy.

Find: The density of the filter.

Solution:

$$D = \log_{10} \frac{1}{0.400}$$

$$= 0.398$$

Neutral density filters are really partially reflecting mirrors. Therefore, in laboratory bench setups, it is often necessary to mount them at a small angle in order to direct the unwanted reflection out of the system.

In summary: Coating is a process that uses wave optics interference concepts to control the reflection and transmission characteristics of optical devices. Simple single-layer coatings can be used to increase the transmittance of a lens or the reflection of a mirror over a desired range of wavelengths. Complicated multilayer coatings can be used to fabricate narrowband optical filters.

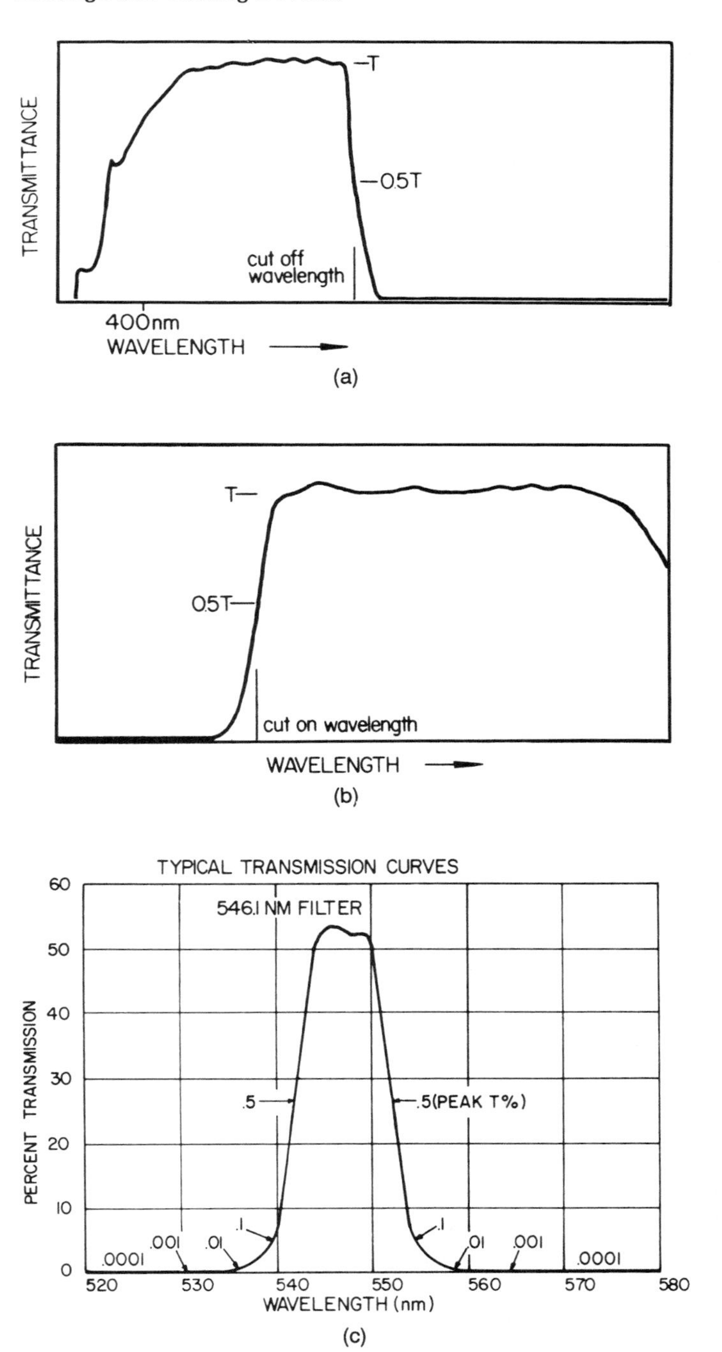

Figure 2-14 Filter characteristics: (a) edge filter short-wavelength passband; (b) edge filter long-wavelength passband; (c) narrowband (line) filter.

2-7 POLARIZATION AND POLARIZERS

In Section 2-1 the concept of polarization was introduced. In this section methods and devices for polarizing waves are discussed. The reflection process is the physical process primarily responsible for the polarization of waves. Reflected beams are really re-radiated beams caused by vibrating electrical charges at the surface of the reflector. These charges are excited by the incident beam's electric field vector. Metals have so many electrons free to move in any direction parallel to the reflecting surface that they re-radiate freely in all directions at visible wavelengths.

We know that glass surfaces exhibit both reflection and transmission. This must mean that some electric-field directions will cause oscillation at the surface and therefore radiate a "reflected" ray, while other directions will not cause oscillation at the surface and therefore achieve transmission. If the electric field is oriented to cause oscillation on the surface, reflection occurs. If the electric field cannot cause oscillation on the surface, transmission occurs. Most electric fields are oriented at angles that cause both reflection and transmission.

The analysis of three cases of linearly polarized waves will be used to illustrate these ideas. Before proceeding to the three cases of linearly polarized waves, we discuss the idea of the plane of polarization. The plane of polarization is illustrated in Figure 2-15. It is the plane formed by the y axis normal to the reflecting surface and the ray velocity vector **v**. This plane can have any angular relationship to the x and z axes.

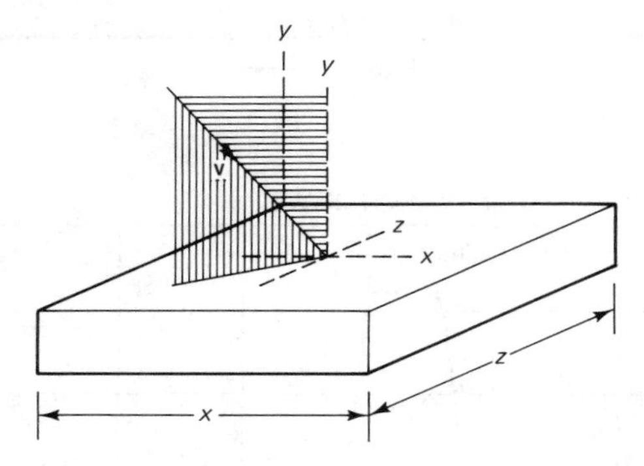

Figure 2-15 Plane of polarization (shaded surface is the plane of polarization).

In Figure 2-16 the special case where the plane of polarization lies on the x axis is shown. Figure 2-16(a) is the special case where the **e** vector is at right angles to the plane of polarization and therefore in this case parallel to the z axis. This is called s polarization. Figure 2-16(b) is the special case where the **e** vector lies in or parallel to the plane of polarization; this is called p polarization. Figure 2-16(c) is the more general case of a linearly polarized wave where the **e** vector is at some angle greater than zero and less than 90° with respect to the plane of polarization. This **e** vector could be resolved into two component vectors: one **p** vector parallel to the plane of polarization and one **s** vector at right angles to the plane of polarization. Remember that the plane of polarization is not the plane of the glass surface but is a plane in the space above the glass surface defined by the ray's velocity vector and the normal to the glass surface.

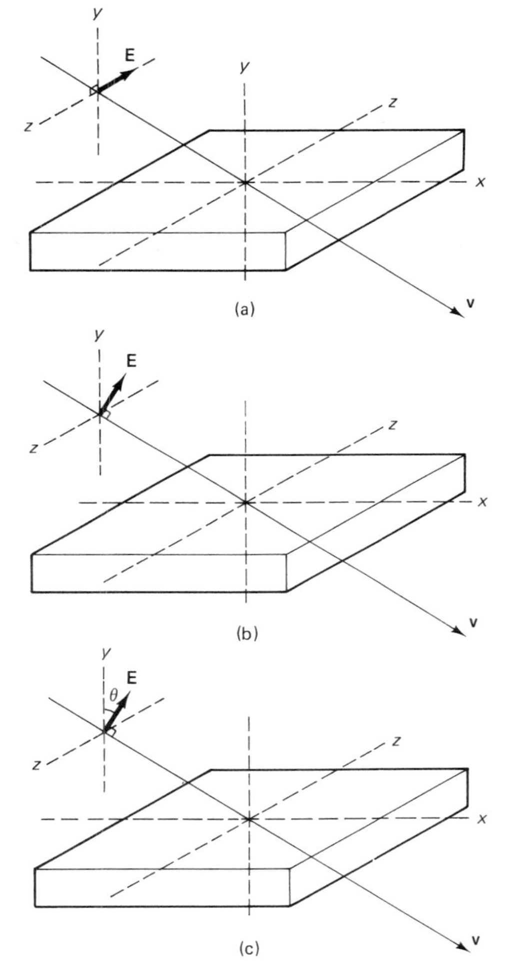

Figure 2-16 Three possible linear polarizations. For this illustration the velocity vector **v** lies on the x,y plane, which makes the x,y plane the plane of polarization. (a) S polarization: The **E** phasor is parallel to the z axis and perpendicular to the plane of polarization. (b) P polarization: The **E** phasor lies on the x,y plane of polarization. (c) Wave containing S and P components: The **E** phasor is at a fixed angle $0 < \theta < 90°$ with respect to the plane of polarization.

The s polarization of Figure 2-16(a) will cause the entire length of the **e** vector to be incident on the surface simultaneously. This will induce maximum vibrations in the surface and give rise to a strong reflection that is s polarized. The p polarization of Figure 2-16(b) causes the various portions of the **e** vector to be incident on the surface at different times. This will induce minimum vibrations in the surface and give rise to a weak reflection and a strong transmission.

The ray of Figure 2-16(c) has one field vector at a fixed angle with respect to the plane of polarization. This **e** field vector can be thought of as having two components. One of these components is an s component that will yield a strong reflection and one is a p component that will yield a strong transmission.

In summary: **s** vectors that are parallel to the glass surface and normal to the plane of polarization will yield maximum **s** vector reflections. **p** vectors that are parallel to the plane of polarization and at an angle with respect to the surface will dig in and be transmitted.

Sir David Brewster investigated the polarization of light by reflection. He discovered that for any given pair of indexes there is an angle of incidence at which the reflection of the *p*-polarized components goes to zero. At this special angle of incidence the reflected ray and the refracted ray will be 90° apart. This special angle of incidence is called the *Brewster angle.* At the Brewster angle the reflected ray will be entirely *s* polarized. If the incident beam is randomly polarized and is incident at the Brewster angle, the reflected beam will be entirely *s* polarized and the transmitted beam will have both *s* and *p* components. For optical glass with an index of 1.5 mounted in air, the Brewster angle is about 57°. If the indexes are known, the Brewster angle can be found using the equation

$$B = \tan^{-1}\frac{n_1}{n_0} \qquad (2\text{-}22)$$

where B is the Brewster angle of incidence
n_1 is the glass index
n_0 is the transmission index, usually air (1)

The Brewster angle is the angle of maximum polarization on reflection, or the maximum transmission of the *p* polarization. Figure 2-17 is a graph of the percent

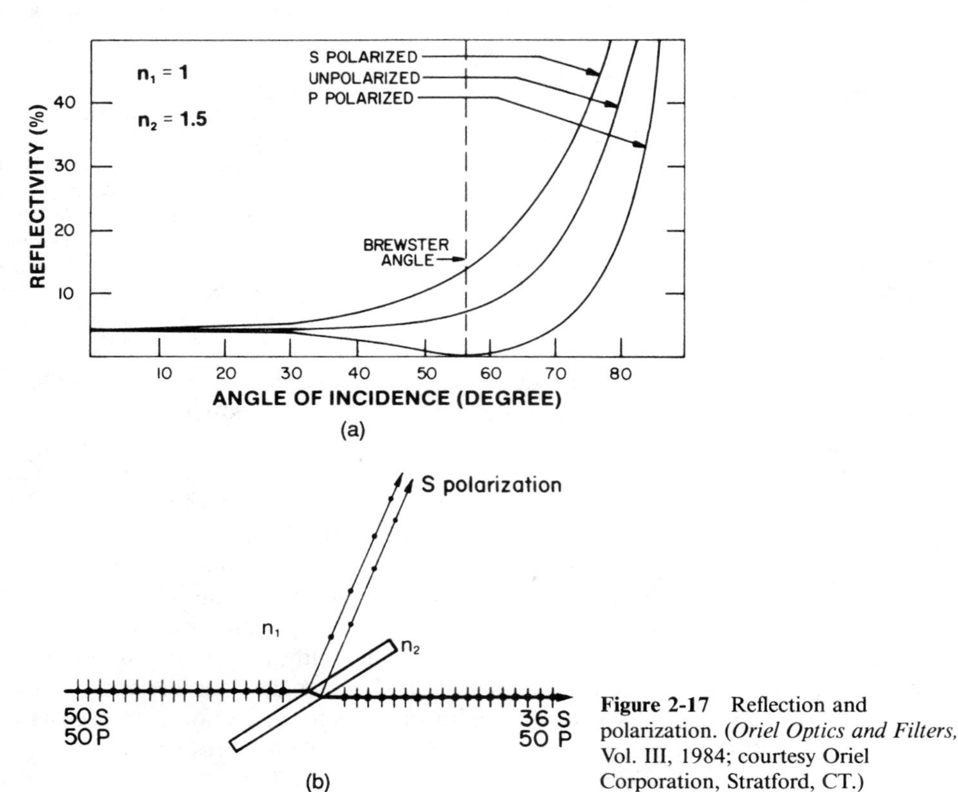

Figure 2-17 Reflection and polarization. (*Oriel Optics and Filters,* Vol. III, 1984; courtesy Oriel Corporation, Stratford, CT.)

reflection from a single flat surface with an index of 1.5 caused by *s*-polarized, *p*-polarized, and randomly polarized radiation.

Because polarization is a reflection phenomenon, the transmission and reflection characteristics of many of the optical devices discussed in Chapter 1 are polarization dependent. For example, a cube beamsplitter with an average transmission of 50% for randomly polarized radiation could have 90% transmission for a *p*-polarized beam or 10% for an *s*-polarized beam. To obtain fixed known wave characteristics, beams are often "polarized" by a class of devices called, logically enough, *polarizers.*

The simplest possible polarizer would be the Brewster window, which is a plane-parallel plate mounted in the beam path at the Brewster angle. At the Brewster angle essentially all of the *p*-polarized vector will be transmitted and about 14% of the *s*-polarized vector will be reflected per surface. A series of plates could be established that would result in a highly *p*-polarized transmitted wave. Table 2-2 illustrates this process.

TABLE 2-2 PERCENT *p* AND *s* POLARIZATION CAUSED BY BREWSTER WINDOWS[a]

Plate	%*p*	%*s*	%*p*	%*s*
	With Respect to Initial Beam Energy		With Respect to Beam Energy at Point of Measurement	
Free space	50	50	50	50
1	50	36	58	42
2	50	26	66	34
3	50	19	72	28
4	50	13	79	21
5	50	10	83	17
6	50	7	88	12

[a] Assumes 14% *s* reflection per surface.

Because reflection is an electrical phenomenon, a grid of wires all running parallel can be used to polarize a beam. The wires must be spaced at small spacings with respect to a wavelength, which means that the wavelengths have to be large. The electrical vector parallel to the wires will be blocked, actually shorted out! The electrical vector at right angles to the wires will be passed.

Polyvinyl sheet polarizers are essentially grid polarizers. In the polyvinyl sheet polarizer, long-chain molecules are arranged in parallel with small spacings. Electrical vectors parallel to these molecule strings are blocked and electrical vectors at right angles to these molecule strings are transmitted. Another common type of device used for polarization is the crystal or birefringence polarizer. The crystals used in such polarizers exhibit the property that *s*- and *p*-polarized vectors will have different velocities of propagation in the crystal. This means that the effective index of refraction for these two orthogonal vectors is different and therefore their angles of refraction are different. As a net result an input beam into the crystal will be broken into two parallel output beams of differing polarizations. By convention, the output ray which undergoes the greatest refraction is called the *ordinary* or *O ray.* The output ray that undergoes the least refraction is called the *extraordinary* or *E* ray. Figure 2-18 illustrates the ray paths through a typical crystal.

Devices similar to cube beamsplitters can be constructed out of these crystal materials which will give greater spatial separation between the *E* and *O* rays than can be obtained with a single crystal. The orientation of these crystals in an optical setup is very important. There will be a particular angle of incidence called the *optical axis* where the *E* and *O* rays will have the same index and so will not be separated. To obtain separation the incident ray must be at some angle with respect to the optical axis.

In summary: Polarizers are optical devices whose output is a wave with the *s* or *p* electrical vector emphasized. Nonmetallic reflection devices will alter the polarization components of the incident beam. The reflection from these devices will be strongly polarized. The terms *s* and *p* refer to two orthogonal electrical vectors that are referenced to the plane of polarization.

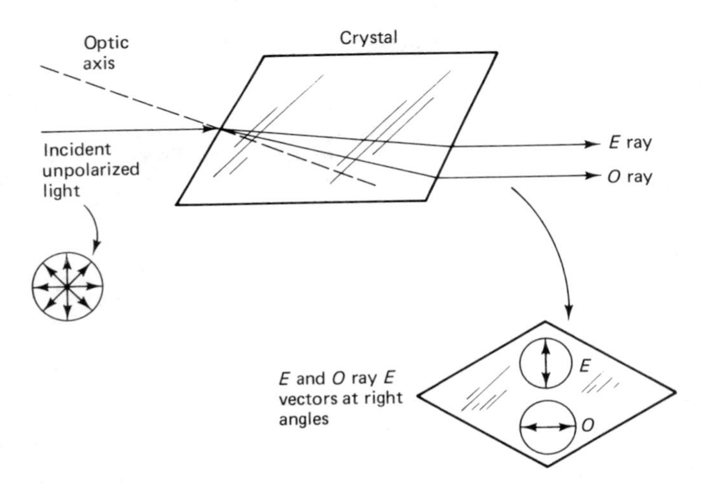

Figure 2-18 Crystal polarizer: refraction and polarization caused by a birefringent crystal.

2-8 BENCH SETUPS

Three common bench setups are the collimated beam, the expanding cone beam, and the spatially filtered beam. The output of a collimated beam setup is a tube or beam of light that has a constant diameter independent of distance from the collimator. An expanding cone beam is a way of simulating the radiation from an ideal point source. The flux density of the radiation on the central axis of the cone beam will obey the inverse-square law. Spatial filtering in the general sense refers to blocking portions of the beam in space, thus causing some desired change in the observed image or radiation pattern. In this section the much more restricted application of "cleaning" up a laser beam is discussed. The resulting "clean beam" can then be collimated and used for interferometric testing or holographic applications.

The components of an expanding cone beam setup are (1) light source (lamp), (2) condenser lens, (3) baffle, (4) pinhole, and (5) lens bench. The condenser lens is positioned so that the filament of the lamp is focused on the pinhole to obtain maximum brightness. The baffle is positioned to block from the setup all light except that which passes through the pinhole. Diffraction effects cause the light emerging from the pinhole to form an expanding cone of light. In this way the pinhole emulates a point source. Figure 2-19 illustrates an expanding cone bench setup. When setting up the cone beam it is important that the axis of the resulting beam be parallel to the lens bench or table axis. To obtain a cone with a circular base, it is important that the foil containing the pinhole and the lens axis both be at right angles to the ray direction.

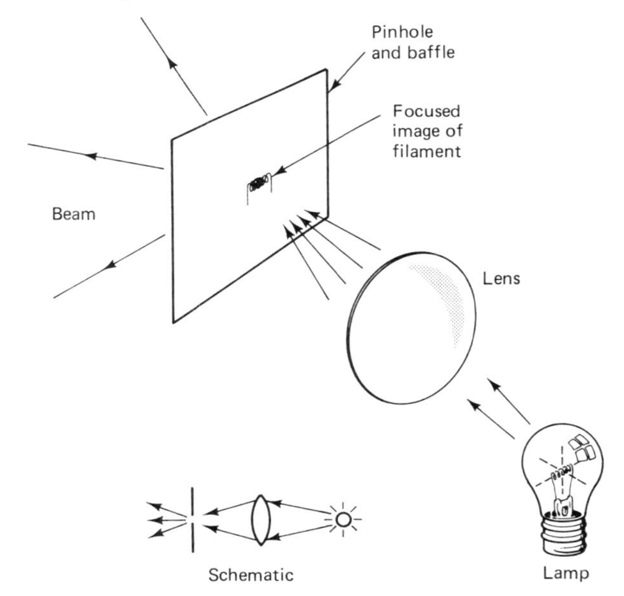

Figure 2-19 Cone beam setup.

In Chapter 1 it was pointed out that if a point source were positioned at the focal point of a condenser lens, a collimated beam would be created. The cone beam setup provides the point source. To obtain a collimated beam, a second lens is placed in the cone of light and positioned so that the pinhole is at the focal point of the lens. Figure 2-20(a) illustrates a collimated beam setup.

Finding the correct location for the second lens (lens 2) is usually done with the aid of a first-surface mirror. The mirror is positioned at right angles to the ray path at a convenient distance from lens 2, as illustrated in Figure 2-20(b). When the beam is collimated the reflection from the mirror will cause a focused spot to be formed in the plane of the pinhole. The physical position of lens 2 is then adjusted to cause this focused spot to fall on the pinhole, at which point the two lenses and the pinhole are locked in position and the mirror is removed. The collimated setup can be checked by positioning a flat target in the beam and sliding it up and down the bench. If the beam stays centered over the bench at a fixed height and exhibits a constant diameter, the setup is correctly aligned.

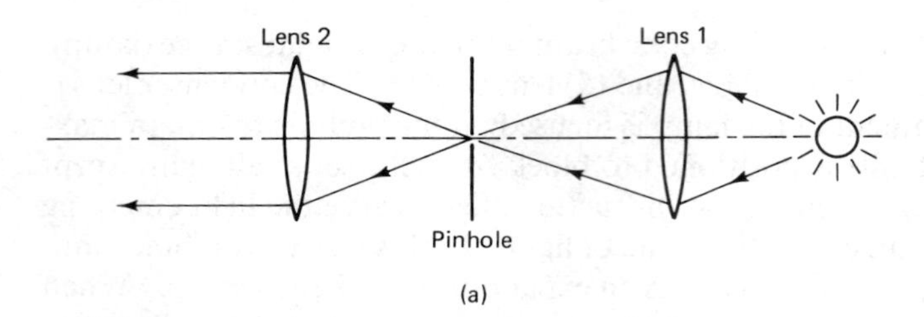

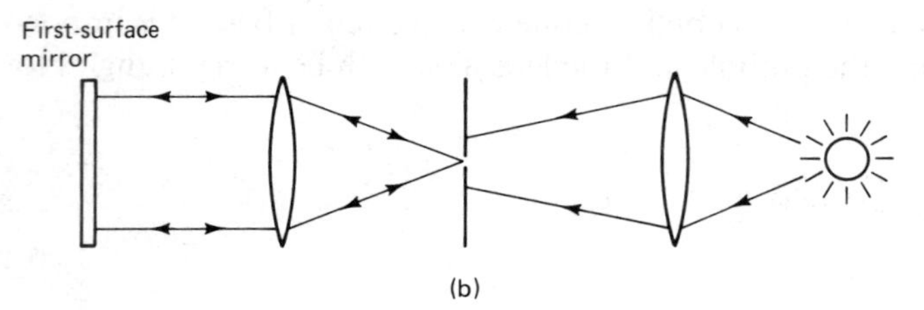

Figure 2-20 Collimated beam setup schematics.

A filter is a device that discriminates against (blocks or stops) unwanted quantities or characteristics and passes desired characteristics. In this chapter we have discussed filters that pass some wavelengths and block others. A *spatial filter* is a system that passes some rays to an image plane and blocks others. The spatial filter operates by blocking the undesired rays and passing the desired rays in space. The basic components of a spatial filter are a condenser lens, masking device, and image plane. The masking device is positioned in the focal plane of the lens, where it intercepts the unwanted rays and passes the desired rays. The image plane displays the filtered image. In designing any filter it is necessary to know what the actual input is and what the desired output is. Once the inputs and outputs are known, a filter can be constructed that removes the unwanted components, leaving only the desired output.

In this section we discuss a method for spatially filtering a laser beam. But first we must define what we want and what we have. The output of an ideal laser is a collimated beam of radiation with a single frequency (wavelength), direction, and polarization that is both spatially and temporally coherent. A laser beam is said to be *coherent* when there is a fixed phase relationship between points on the electromagnetic wave. An ideal monochromatic wave would be coherent everywhere in space. Practical beams are coherent only for finite periods of time over finite distances and areas. The collimated laser beam has a Gaussian transverse irradiance profile (Figure 2-21).

For engineering purposes the radius of the profile (beam radius) is the distance w where the beam irradiance is equal to the peak beam irradiance divided by the Naperian base number ε squared. If this ideal beam is projected onto a smooth diffuse reflecting image plane through an expanding lens, the image will appear to be a perfect circle with a uniformly granular or velvety appearance. When measured, the irradiance profile of this image would be Gaussian. The ideal laser beam image is what we want.

When we project an actual laser beam onto a diffuse reflecting image plane through an expanding lens, what we get is an Airy disk pattern. The primary cen-

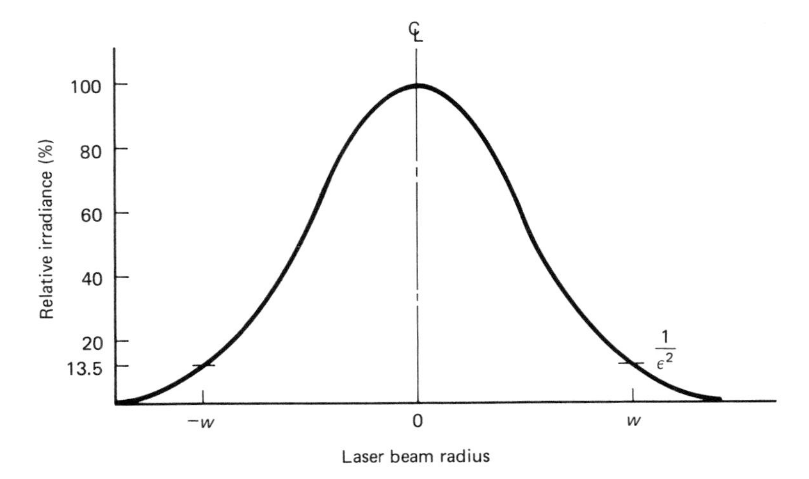

Figure 2-21 Gaussian distribution of relative irradiance as a function of laser beam radius.

tral circle of this image does not have the desired smooth velvety appearance, but has very noticeable large bright and dark regions. These undesired variable-intensity background regions are due to scattering of the laser energy from the ideal single beam path.

To achieve the desired smooth beam irradiance pattern of the perfect laser, the spatial filter must block the scattered beams and pass the undirectional beam. A condensing lens and a pinhole will accomplish this. A laser spatial filter is just a fancy cone beam setup. The desired rays will be focused to a very tight spot by the lens. A pinhole is placed in the focal plane around this focal spot. The scattered rays will not be focused to the spot and will be blocked by the material surrounding the pinhole. Thus the rays emerging from the pinhole will be spatially filtered. These rays can then be collimated for use in the optical setup.

This process is much easier to describe than to accomplish. Typical laser beam diameters are on the order of 1 mm and pinholes that will effectively filter the beam will be from 6 to 40 μm in diameter. Focusing a laser beam on a pinhole this small can be a tedious and time-consuming process. Typically, the condensing lens used is a microscope objective, and this lens and the pinhole are mounted in a mechanical jig with fine micrometer adjustments.

For every combination of laser beam diameter, wavelength, and lens focal length, there is an optimum pinhole size. The pinhole size can be calculated as

$$D_h > 4\left(\frac{\lambda F}{\pi w_d}\right) \tag{2-23}$$

where D_h is the pinhole diameter
λ is the laser wavelength
F is the focal length of the focusing optics
w_d is the $1/\varepsilon^2$ beam diameter

EXAMPLE 2-8

Calculation of a Spatial Filter Pinhole Diameter

Given: **(1)** $\lambda = 632.8$-nm (laser wavelength)
(2) $F = 4.0$-mm (objective focal length)
(3) $w_d = 0.6$-mm (specified laser beam diameter)

Find: D_h, the pinhole diameter.

Solution:

$$D_h = 4\left(\frac{\lambda F}{\pi w_d}\right)$$

$$= 4\left[\frac{(632.8 \times 10^{-9}\text{ m})\,(4.0 \times 10^{-3}\text{ m})}{n\,(0.6 \times 10^{-3}\text{ m})}\right]$$

$$= 5.37 \times 10^{-6}\text{ m}$$

Comment: As a practical matter, a 6- or 8-μm pinhole would be used.

In this section we have seen that a simple pinhole and condenser lens combination can be used to create three very useful bench setups. In all three cases careful attention to the details of axial and planar alignment are necessary to achieve the desired results. In a spatial filtering setup, because of the very small dimensions involved, the alignment process can most honestly be described as tedious and time consuming.

2-9 SUMMARY

In this chapter, concepts and components that emphasize the wave characteristics of radiation have been discussed. Important concepts discussed include (1) wavelength, (2) dispersion, (3) diffraction, (4) interference, (5) polarization, and (6) spatial filtering. Important devices and processes discussed are (1) dispersing prisms, (2) diffraction gratings, (3) pinholes, (4) coatings, (5) filters, (6) polarizers, (7) collimators, and (8) spatial filters.

It is the purpose of this chapter and Chapter 1 to give the technician a functional grasp of the optical principles and devices that will be encountered in an electro-optics lab. Entire books are available on these topics. The brief treatment presented here is only intended as introductory and does not pretend to be a thorough coverage of these topics.

PROBLEMS

1. A sine wave has a frequency of 0.55×10^{15} Hz. What is its period?
2. An electromagnetic wave traveling in a vacuum has a frequency of 0.55×10^{15} Hz. What is its wavelength?

3. An electromagnetic wave traveling in free space has a wavelength of 400 nm.
   (a) What will the wavelength be when the wave is traveling in glass with an index of 1.53?
   (b) What will the wave's velocity be when traveling in glass?
   (c) What will the wave's frequency be when traveling in free space?
   (d) What will the wave's frequency be when traveling in glass?
4. A mercury lamp violet line ($\lambda = 404.7$ nm) and a hydrogen lamp red line ($\lambda = 656.3$ nm) are used to test a piece of Schott glass. Both beams are incident at an angle of 20°. The violet line is refracted 10.80°; the red line is refracted 11.10°.
   (a) What is the glass index at 404.7 nm?
   (b) What is the glass index at 656.3 nm?
   (c) Estimate the white light (550 nm) index.
5. It is desired to fabricate a single lens with minimum chromatic aberration. Three materials are available. Which material will cause the least aberration?

λ (nm)	Material A	Material B	Material C	
435.8	1.534	1.527	1.825	Mercury blue
656.3	1.520	1.514	1.776	Hydrogen red

6. A single slit is illuminated by a laser with $\lambda = 632.8$ nm. A fringe pattern is formed on an image plane 2 from the slit. The dark fringes are 6.3 apart. What is the slit width?
7. A single slit is illuminated by a laser with $\lambda = 632.8$ nm. The image pattern is formed on an image plane 1.5 m from the slit. The dark fringes are 6.2 mm apart. An unknown blue source is used to illuminate the same slit and the dark fringes are 4.3 mm apart. What is the unknown wavelength?
8. A pinhole is illuminated by a laser with $\lambda = 632.8$ nm. The diameter of the fourth dark ring is 26.8 mm. The pinhole is 2 m from the image plane. What is the pinhole diameter?
9. A hydrogen red lamp 656.3 nm will be used to illuminate a system containing a pinhole and a lens. The lens will be placed about 10 ± 0.5 cm from the pinhole. It is desired to just fill the lens with the central maximum of the Airy disk pattern. The lens diameter is 4 cm. What size pinhole is required?
10. A double slit with a slit width of 10 μm and a slit spacing of 200 μm is illuminated by a laser with $\lambda = 632.8$ nm. The image plane is 2 m from the slit.
    (a) How far apart from the diffraction minima?
    (b) How far apart are the interference minima?
11. A transmission grating has 6000 lines/cm. The source is a laser with a wavelength of 632.8 nm.
    (a) How many off-axis lines will be formed?
    (b) At what angles will the off-axis lines be?
12. A laser of 632.8-nm wavelength used with a transmission grating causes the first line at 25.00°. An unknown source is then used to illuminate the grating. The first line occurs at 27.50°. What is the wavelength of the unknown source?
13. A manufacturer offers three different coated lenses. All three lenses are transmisson coated with magnesium fluoride, which has an index of 1.33. The index for each is:

fused silica, 1.46; Schott glass A, 1.52; Schott glass B, 1.79. Which lens will have the smallest reflection?

14. A reflecting surface is coated with gold. Over what range of wavelengths is it designed to operate?

15. Two lenses are combined to minimize chromatic aberration. What is the combination called?

16. A lens exhibits chromatic aberration.
   (a) Will the red or blue focal point be closer to the lens surface?
   (b) When lateral color is measured will the red or the blue line be farther from the center?

17. How thick will be a single-layer reflecting coating be?

18. A setup is developed to measure the transmission characteristics of a multilayer interference filter. Do the following data seem reasonable?

λ (nm)	Transmission (%)	λ (nm)	Transmission (%)
540	10	548	52
542	20	549	52
543	30	550	50
544	40	551	40
545	50	552	30
546	55	553	20
547	54	554	10

19. A neutral density filter transmits 70% of the incident light. What is the density?

20. A laser spatial filter is being assembled. The wavelength of the laser is 611.8 nm. A microscope objective with a focal length of 4.0 mm is used. The laser beam diameter is 0.5 mm. What pinhole size is required?

3

Radiation and Radiometry

3-1 INTRODUCTION

Sources of radiation are described in terms of both their electrical and their radiation characteristics. Optical technicians must understand source electrical specifications such as voltage, current, and power. Important optical specifications are source spectra, color temperature, intensity, and spatial distribution. Finally, mechanical specifications relating to size and expected useful life must be considered.

The measurement units for the electrical specifications are well standardized and lend themselves to comparison. The optical units are unfortunately not as easy to deal with. There are two types of radiation measurement systems: *radiometric,* which relates to all radiation, and *photometric,* which confines itself to the visible region of radiation. Each of these systems has its own vocabulary and units. In most cases there is more than one name and unit for the same type of parameter.

In this chapter we describe the most common concepts and units. The radiometric system is discussed in this chapter, and the photometric system is discussed in detail in Chapter 4. We discuss radiometric measurements in more detail in Chapter 8.

3-2 TYPES OF SOURCES

Sources can be categorized into three large classes based on the nature of the radiation they emit:

1. *Monochromatic sources* ideally emit a single wavelength of radiation. Practically, they emit a very narrow band of wavelengths.

2. *Continuous spectra sources* emit radiation over a wide range of wavelengths, with a smooth transition between the amplitudes of adjacent wavelengths.
3. *Line spectra sources* also emit radiation over a wide range of wavelengths, with very sharp peaks of energy (lines) at particular wavelengths.

Lasers and light-emitting diodes are examples of sources that are nearly monochromatic. Incandescent lamps are continuous spectra sources. Arc lamps, which are gas filled, are sources of line spectra. Carbon arc lamps also emit line spectra. Figure 3-1 illustrates the spectra of a light-emitting diode (LED), an incandescent lamp, and a xenon arc lamp. The xenon arc lamp is an example of a line spectra source.

Incandescent lamps, LEDs, and arc lamps are often called *incoherent* or *noise sources* because there is no fixed phase relationship between the waves being emitted. Monochromatic gas lasers and semiconductor lasers are often classed as *coherent sources.* These sources are discussed in detail in Chapters 6 and 13.

A line spectra source emits radiation when an arc is established in the gas between a lamp's electrodes. In operation a high starting voltage is applied across the lamp, which causes an arc. As the flow of ions is established in the arc, the voltage across the lamp drops but the arc is maintained. The exact nature of the emitted spectra is a function of the gas in the lamp. As the arc current passes through the gas, it causes changes in the energy levels of the electrons in the individual gas ions. Whenever an electron changes energy level in an atom, a photon of radiation is emitted or absorbed. The wavelength of this energy is given by the equation

$$\lambda = \frac{hc}{\Delta E} \tag{3-1}$$

where λ is the wavelength emitted or absorbed
h is Planck's constant
c is the velocity of light
ΔE is the difference in energy between permitted energy levels

For atoms of a specific element, electrons can exist only at permitted energy levels. For this reason the values that ΔE can assume are fixed for each element. The wavelengths of radiation emitted by each element are therefore fixed. Energy levels are usually measured in electron volts and wavelengths in nanometers; the terms h and c are constants. To use these units, equation (3-1) can be rewritten

$$\lambda = \frac{1.24 \times 10^3 \text{ eV} \cdot \text{nm}}{\Delta E} \tag{3-2}$$

A special case of the arc lamp is the fluorescent lamp. In a fluorescent lamp the glass case of the lamp tube is coated with a fluorescent powder consisting pri-

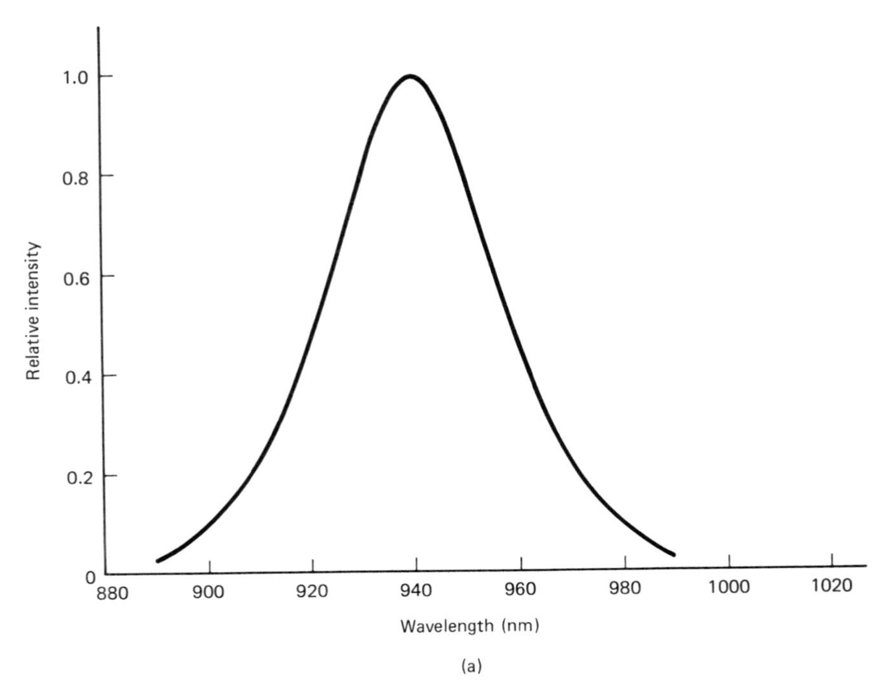

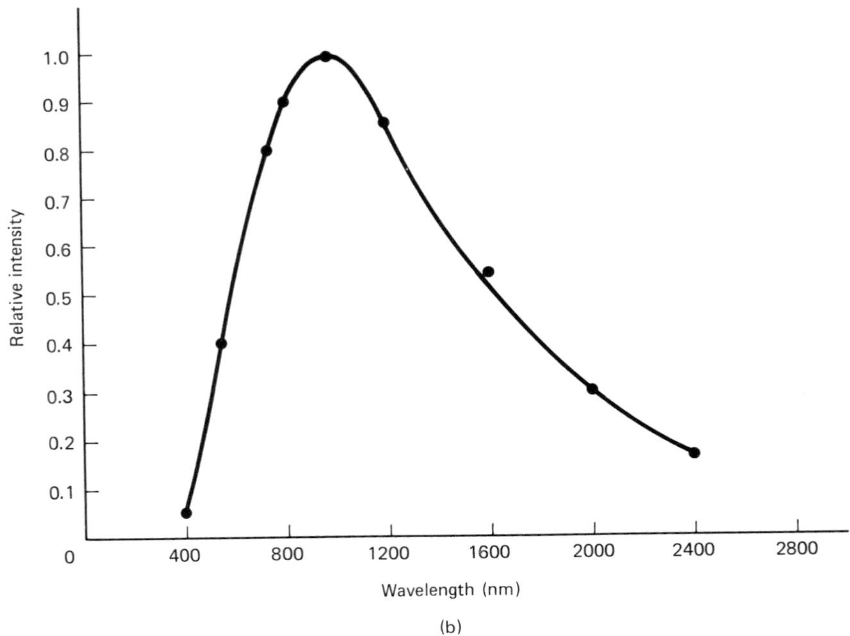

Figure 3-1 Spectral intensity graphs: (a) IRED; (b) tungsten lamp, 2856K; (c) xenon arc lamp.

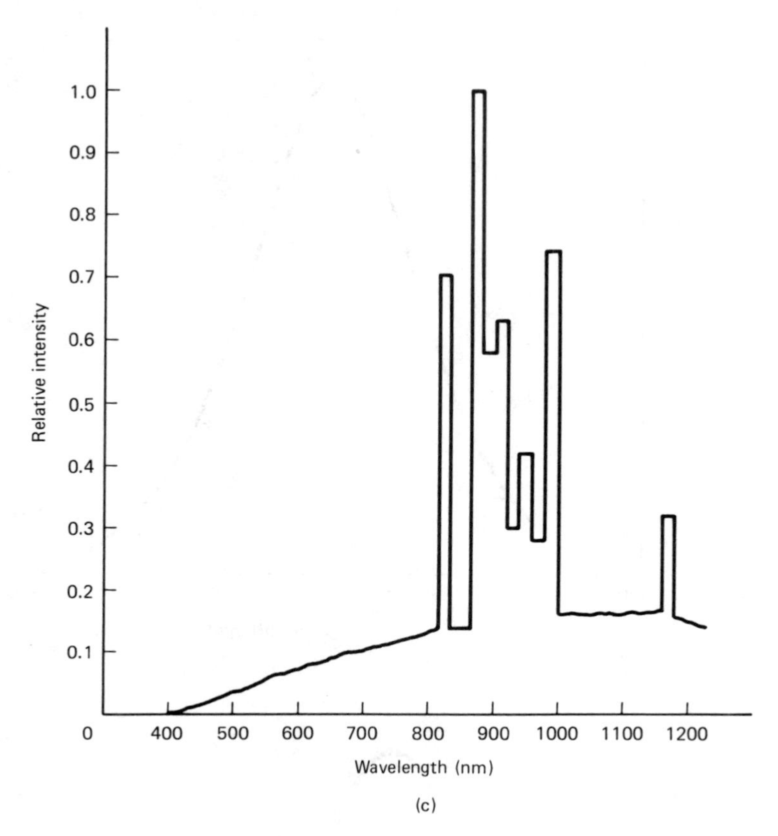

Figure 3-1 *(cont'd)*

marily of phosphors. Fluorescence is the property of a material to reemit radiation in the visible range of wavelengths when irradiated by energy at wavelengths outside the visible range. In a fluorescent lamp an arc is created in a mercury vapor. The mercury vapor emits photons at visible and ultraviolet wavelengths. These ultraviolet (253.7-nm) photons cause the lamp coating to fluoresce. Figure 3-2 illustrates the spectra of a fluorescent lamp. The fluorescence caused by the interaction of the ultraviolet photons and the lamp coating results in a concentration of the emitted energy in the visible region. For a given watt of input electrical power, a fluorescent lamp will have more emitted energy in the visible spectra than an incandescent lamp. The input power not radiated as visible light is given off as infrared radiation and conducted heat. Gas lamps are discussed in greater detail in Chapter 5.

Sources are also grouped together according to the range of wavelengths that they radiate. The optical range of wavelengths extends from 10 nm to 10^6 nm. This range is subdivided into the regions listed in Table 3-1.

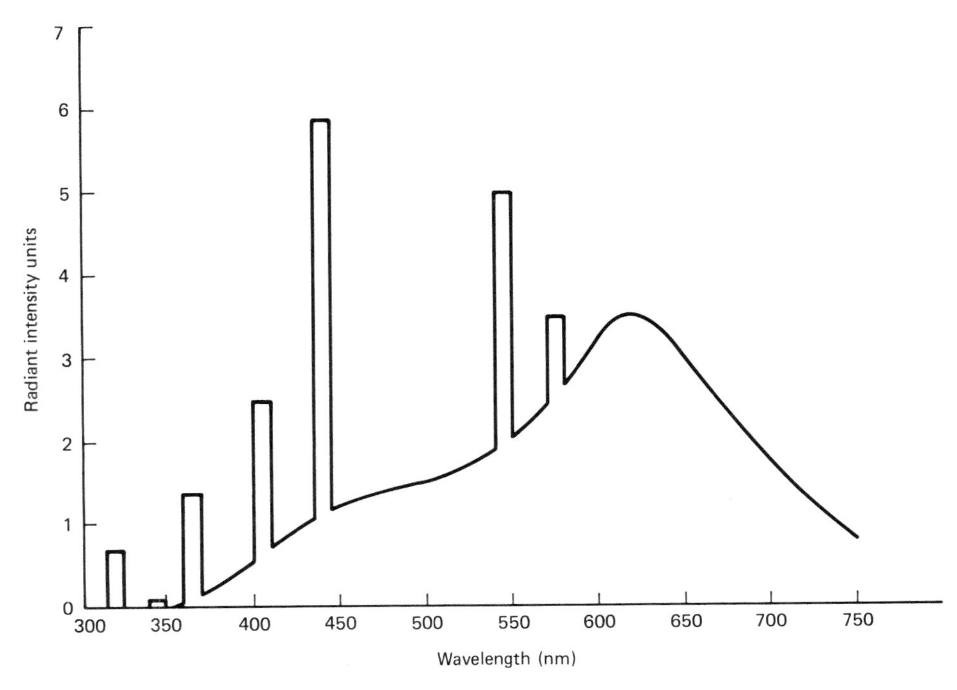

Figure 3-2 Spectral intensity fluorescent lamp.

TABLE 3-1 NOMENCLATURE ASSIGNED TO WAVELENGTH RANGES

Name	Range (nm)
Extreme UV (ultraviolet)	10–200
Far UV	200–300
Near UV	300–380
Visible	380–770[a]
Near IR (infrared)	770–1500
Middle IR	1500–6000
Far IR	6000–40,000
Far-far IR	40,000–1,000,000

[a] The visible range is often given as 400 to 700 nm.

3-3 RADIOMETRIC SPECIFICATIONS

When solids or liquids have their temperatures raised, they radiate energy over a continuous spectra. If the temperature is high enough, some of the radiation will be in the visible spectra. Incandescence is the emission of light by thermal radiation.

The common light bulb or incandescent lamp is an example of this process. Lamps of this type are used in a variety of optical applications. Important from the technician's point of view is the use of this type of lamp as a calibrated source. Many data sheets of opto-electronic devices list specifications that are based on measurements that use incandescent lamps as the illuminating source.

Lamps are specified in terms of their electrical and radiation specifications. Power, voltage, and current are the primary electrical specifications. Color temperature, or graphs of lamp spectra, irradiance, and intensity, are the usual optical specifications.

The vocabulary of radiation units and parameters can be confusing. In the interest of clarity we have selected a restricted vocabulary and system of units. Table 3-2 lists the radiometric concepts and units that are used in this chapter. Before getting quantitative about radiation specifications it is important that we develop a qualitative understanding of the processes and units involved. A solid has its temperature elevated and the solid radiates. The radiation can be thought of as energy exiting from the solid. The energy can be described in terms of joules (a unit of energy), or if the process is continuous, as joules per second. When the radiation is described only in terms of energy, we call it the *radiant energy* and measure it in joules. More commonly, the radiation is measured over a period of time and the rate of flow of energy is measured. The rate of flow of energy is power and radiant power is called *radiant flux,* measured in watts. A power level of 1 watt is the same as an energy flow rate of 1 joule per second.

Stop for a moment and envision the measurement of the radiant flux. You, the observer, are at some distance R from the radiant source Q_e. You have a power detector of area A. Essentially, your detector area A will form a small part of the surface area of a sphere of radius R around the source. This is illustrated in Figure 3-3. The actual measurement being made is the total power being intercepted by the area of your detector. This measurement of power per unit area is power density, called in radiometric terms the *flux density.* The flux density is the *incident*

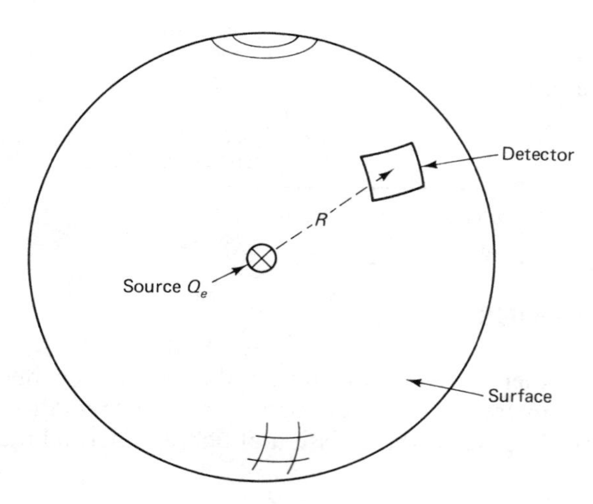

Figure 3-3 Source, detector, and spherical surface.

TABLE 3-2 RADIOMETRIC AND SPECTRAL QUANTITIES SYMBOLS AND UNITS

Radiometric Units			Spectral Radiometric Units		
Quantity	Symbol	Unit	Quantity	Symbol	Unit
Radiant energy	Q_e	joule (J)			
Radiant flux: power across a surface	$\phi_e = \frac{dQ_e}{dt}$	watt (W)	Spectral radiant power; radiant power (flux) per unit wavelength	$\phi_\lambda = \frac{dQ_e}{d\lambda}$	W/nm
Flux density: (incident flux) (irradiance) flux per unit area	$H_e = \frac{d\phi_e}{dA}$	watt per meter squared (W/m^2)			
Radiant emittance: exiting flux per unit area	$M_e = \frac{d\phi_e}{dA}$	W/m^2	Spectral emittance	$W_\lambda = \frac{dM_e}{d\lambda}$	$\frac{W}{m^2 \cdot nm}$
Solid angle: area on surface of sphere divided by radius of sphere squared	$\omega = \frac{dA}{R^2}$	Steradian (sr)			
Radiant intensity: flux per solid angle	$I_e = \frac{d\phi_e}{d\omega}$	W/sr	Spectral radiant intensity	$I_\lambda = \frac{dI_e}{d\lambda}$	$\frac{W}{sr \cdot nm}$
Radiance: radiant power per unit solid angle and unit projected area	$L_e = \frac{d\phi e}{d\omega\, dA \cos\theta}$ $L_e = \frac{dJ_e}{dA \cos\theta}$	$\frac{W}{sr \cdot m^2}$	Spectral radiance	$L_\lambda = \frac{dI_e}{d\lambda}$	$\frac{W}{sr \cdot m^2 \cdot nm}$

flux on the detector, sometimes called the *irradiance.* In terms of the source under test the *exiting flux density* being measured is often called the *radiant emittance.* In a rigorous sense, exiting flux density exists only at the surface of the source.

The numerical magnitude of the flux density will decrease as the distance from the source increases. This is most easily understood by considering a *point source,* a source that radiates power equally in all directions. Imagine a point source of small diameter that radiates a power ϕ_e of 10 W. The power ϕ_e will be evenly distributed over a spherical surface area of $4\pi R^2$. The flux density at any radius is found by simple calculations.

$$H_e = \frac{\phi_e}{4\pi R^2} \tag{3-3}$$

where H_e is the radiometric flux density
ϕ_e is the radiated power
R is the distance from the source to the detector

Table 3-3 lists the radiant flux density at various radial distances given by equation (3-3) for a 10-W point source.

TABLE 3-3 VALUES FOR A 10-W SOURCE POINT

R (cm)	H_e (mW/cm^2)
10	7.96
20	1.99
50	0.32
100	0.08

It is true that actual sources are not usually point sources, but the same reasoning holds if they radiate so that the beam size expands. If you pick a radius line going through the center of the imaginary sphere and made measurements at different distances along that line with a detector, the magnitude of the measured flux density will decrease as the inverse square of the distance. Equation (3-4) summarizes the proportionality that exists between flux density and radial distance. You may hear this proportionality referred to as the *inverse-square law.*

$$H_e \propto \frac{1}{R^2} \tag{3-4}$$

Now imagine a different situation, one where the area of the detector could be altered. Suppose that you could alter the area of the detector so that you always intercepted the same percentage of the spherical surface. In other words,

$$\frac{A}{4\pi R^2} = \text{constant}$$

In this case the total power being intercepted by the detector would be constant even though the flux density decreased.

Solid angles are a way of describing the percentage of the spherical surface intercepted. The solid angle ω is measured in steradians (sr). The solid angle ω is the ratio of the intercepted area to the square of the radius of the imaginary sphere:

$$\omega = \frac{A}{R^2} \tag{3-5}$$

There are 4π steradians on any spherical surface, such as the one illustrated by Figure 3-4.

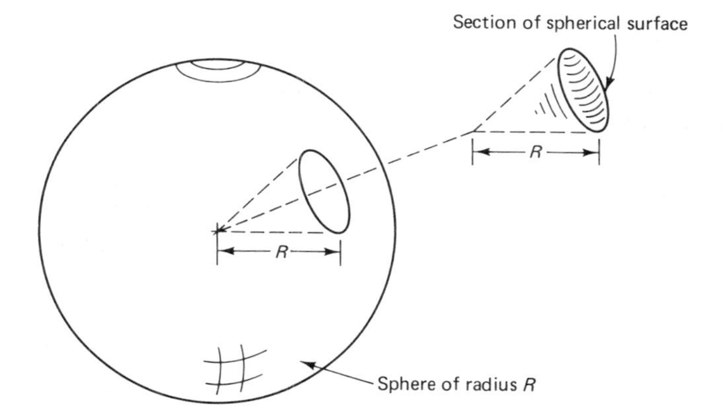

Let: A be the area of the section
R the radius of sphere
ω the solid angle, in steradians
$\omega = \frac{R}{R^2}$

Figure 3-4 Solid angle intercepted by a spherical surface section with an area of A.

The flux per unit solid angle is called the radiant intensity I_e and is expressed in watts per steradian.

$$I_e = \frac{\phi_{ei}}{\omega} \tag{3-6}$$

where I_e is the radiant intensity
ϕ_{ei} is the power intercepted by the detector
ω is the solid angle formed by the detector

Philosophically, intensity is a number that describes the source output; flux density describes the power per unit area at some location in the space surrounding the detector.

Table 3-4 lists the resulting measured flux density and calculated radiant intensity for various radial distances. As shown in Table 3-3, the flux density decreases according to the inverse-square law. Table 3-4 illustrates that the radiant

TABLE 3-4 MEASURED FLUX DENSITY AND CALCULATED RADIANT INTENSITY FOR A 10-W POINT SOURCE AND A 1-CM DETECTOR AT VARIOUS DISTANCES

Measured Data		Calculated Quantities		
R (cm)	H_e (irradiance; mW/cm^2)	$\phi_{ei} = H_e A$ (intercepted power; mW)	ω (solid angle; sr)	I_e (radiant/intensity; mW/sr)
10	7.96	7.96	0.0100	796
20	1.99	1.99	0.0025	796
50	0.32	0.32	0.0004	800
100	0.08	0.08	0.0001	800

intensity, on the other hand, has a constant value that is independent of distance. This is consistent with the idea that intensity is a characteristic of the source itself and not of detector location.

The algebra below summarizes the relationships that exist between power, flux density, and radiant intensity for an isotropic point source at any radius R where ϕ_{ei} is the intercepted power and ω is the solid angle formed by the detector of area A at radius R.

$$\phi_{ei} = H_e A$$

$$H_e = \frac{\phi_e}{4\pi R^2} \quad \text{(for an isotropic point source)} \tag{3-3}$$

$$\omega = \frac{A}{R^2}$$

$$I_e = \frac{\phi_{ei}}{\omega}$$

$$I_e = \frac{\phi_e}{4\pi R^2} \frac{A}{A/R^2}$$

$$I_e = \frac{\phi_e}{4\pi} \quad \text{(for an isotropic point source)} \tag{3-7}$$

Finally, H_e, the irradiance, can be converted to the radiant intensity, I_e.

$$I_e = \frac{\phi_{ei}}{\omega}$$

$$I_e = \frac{H_e A}{A/R^2}$$

$$I_e = H_e R^2 \tag{3-8}$$

or

$$H_e = \frac{I_e}{R^2} \tag{3-9}$$

Examples 3-1 and 3-2 illustrate the use of equations (3-3) and (3-8).

EXAMPLE 3-1 *Calculation of Radiant Intensity from Flux Density*

Given: On its alignment axis a carbon-filament calibration lamp yields an irradiance of 45.9 $\mu W/cm^2$, measured at 2 m.

Find: The radiant intensity at the 2-m location caused by the lamp.

Solution:

$$H_e = 45.9\ \mu W/cm^2$$

$$I_e = H_e R^2$$

$$= 45.9\ \mu W/cm^2 \times (200\ cm)^2$$

$$= 1.836\ W/sr$$

EXAMPLE 3-2 *Calculation of Flux Density for a Point Source*

Given: When the carbon filament lamp is delivering 45.9 $\mu W/cm^2$, it has an electrical power input of 23.8 W. The source is located 2 m from the detector.

Find: Does the measured value of flux density have a value similar to what you would expect from an ideal point source?

Solution: For a point source,

$$H_e = \frac{\phi_e}{4\pi R^2} \qquad \phi_e \simeq \text{electrical power in}$$

$$= \frac{23.8\ W}{4\pi (200\ cm)^2}$$

$$= 47.4\ \mu W/cm^2$$

Conclusion: Along this axis the measured value of H_e is close to that which would be caused by a uniform radiator or point source. This would not be the case for every possible axis of measurement. We must not forget that the input electrical power is being used to heat the lamp and to cause useful radiation.

3-4 FLUX DENSITY BENCH MEASUREMENTS

In Section 3-3 an ideal point source was used as a basis for deriving equations for flux density and intensity. When making measurements on an optical bench, it is more likely that we would be working with a collimated beam of radiation, or an expanding beam emitted from a pinhole or other small aperture. In this section we examine these two common test radiation conditions.

By definition, a collimated beam has a constant diameter and therefore a constant cross-sectional area. It can be envisioned as a tube of radiation extending from the collimating lens to infinity. Over the short distances involved in lens benchwork, there are no detectable power-loss mechanisms. The average power measured across the entire beam cross section is therefore independent of the distance from the collimating lens. If your detector has a surface area that is equal to or larger than the cross-sectional area of the beam, and if it is positioned so that it intercepts the entire beam area at right angles to the beam path, the instrument readings of power or flux density will be constant and independent of distance from the collimating lens.

There are two possible alignment errors that can occur with this measurement. In each case the incorrect positioning of the detector causes a portion of the beam's radiation to bypass the detector, which results in a low reading. If the detector is rotated so that its surface is no longer at right angles to the beam path, some of the beam energy will be lost and a low reading of average power will be obtained. You can illustrate this to yourself with a simple experiment. Hold a quarter directly in your line of sight and observe how much of the field of view is blocked. Now rotate the quarter slowly and observe how a smaller segment of your field of view is blocked.

In addition to rotational position errors it is possible to have translational position errors. A *translational error* will occur if the detector is not centered on the beam axis. If we assume a normal x, y positioning mechanism, the detector may be too high or too low or too far to one side of the beam. When working with visible light you can usually "eyeball" these position errors. With invisible radiation, such as infrared, greater care and more sophisticated methods of alignment verification will be required.

Additional problems arise when the detector diameter is smaller than the beam diameter. If you want to measure the power of the entire beam cross section, a lens will be used to focus the beam down to the detector diameter and compensations will be made for lens reflections and losses. If you want to make measurements of relative magnitude at different distances the detector must always be located at the same transversal position with respect to the beam axis. The beam has a transversal flux density profile that has a peak value at the center and falls off to zero at the beam edges. If you made one measurement at the beam center and another near the beam edge, you would obtain significantly different values. If you position the detector near the beam center for each measurement location and then "fine tune" the detector's position with minor x, y adjustments to obtain a peak reading, you will obtain consistent results.

Small-diameter beams with high flux densities can also cause problems. The output of a bench laser would be an example of such a beam. The localized flux density may cause a "hot" spot on the detector that will damage the detector. A more subtle problem arises from the calibration and sensitivity characteristic of the detector. The detector is calibrated with its detection surface fully illuminated.

This calibration process averages out any variations in device responsivity over its surface. A small beam that hits only a small segment of detector surface area may yield an incorrect measured value. When you measure small-diameter beams, they should be expanded to fill the detector surface. An alternative approach would be to use the small-diameter beam to illuminate the interior of an integrating sphere and then place your detector in a sphere port. The radiation is diffused uniformly over the interior of the sphere by reflection. Your detector then measures a portion of this diffused radiation at the surface of the sphere. The sphere manufacturer can provide appropriate equations for calculating the input beam power from the measured value.

The diffraction of radiation caused by a pinhole was discussed in Chapter 2. The radiation pattern observed on a plane in front of the pinhole was described as a large centered circular disk (Airy disk) and much smaller low-intensity rings concentric with the disk. For simplicity, in this chapter we assume that only the radiation that causes the Airy disk is of interest. Given this simplification, we can envision a beam that expands like a cone from the pinhole. The flux density is most intense at the center of the beam and decreases to a value of zero at the beam edge.

Figure 3-5 illustrates the flux density profile at two different locations on the z axis. Two features are noticeable. At the greater distance z_2, the magnitude of the central peak is lower than it is at location z_1; also, the peak tends to be flatter at the greater distance. In the discussion that follows, assume that a circular detector is placed exactly centered on the beam and exactly at right angles to the beam direction. Three regions of flux density measurement are of interest: long distances

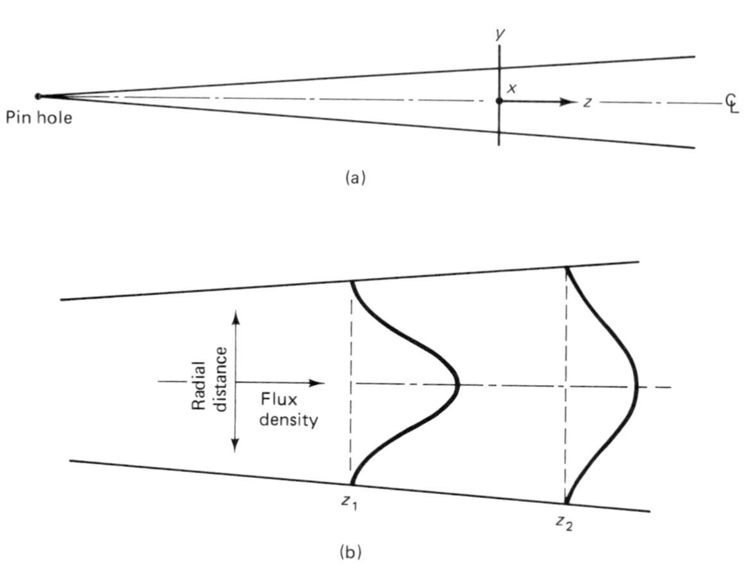

Figure 3-5 Expanding beam: (a) side view illustrating axis definitions; (b) flux density profiles at two locations: z_1 and z_2.

where the diameter of the detector is small with respect to the beam diameter, short distances where the detector diameter is equal to or greater than the beam diameter, and the intermediate distances between long and short distances.

At long distances the measured flux density will appear to obey the inverse-square law. This is called the *far field;* in this region of measurement the centered detector is intercepting flux density profiles that appear to be uniform across the detector's diameter. The magnitude of these flux density peak regions decrease with increasing distance according to the inverse-square law.

At short distances where the beam diameter is less than or equal to the detector diameter, the measured flux density is essentially constant; this is called the *near field.* Figure 3-6 illustrates measurements of the flux density of a cone beam radiating from a pinhole. The horizontal line represents the constant flux density measured when the detector is larger than the beam diameter. The sloped line is drawn through values of flux density that decrease in accordance with the inverse-square law. The two lines intersect where the beam diameter is equal to the detector diameter.

There is an intermediate range of distances where the data points fall below both of these straight lines. In this intermediate region, the flux density varies noticeably across the diameter of the detector and the detector area intercepts a sig-

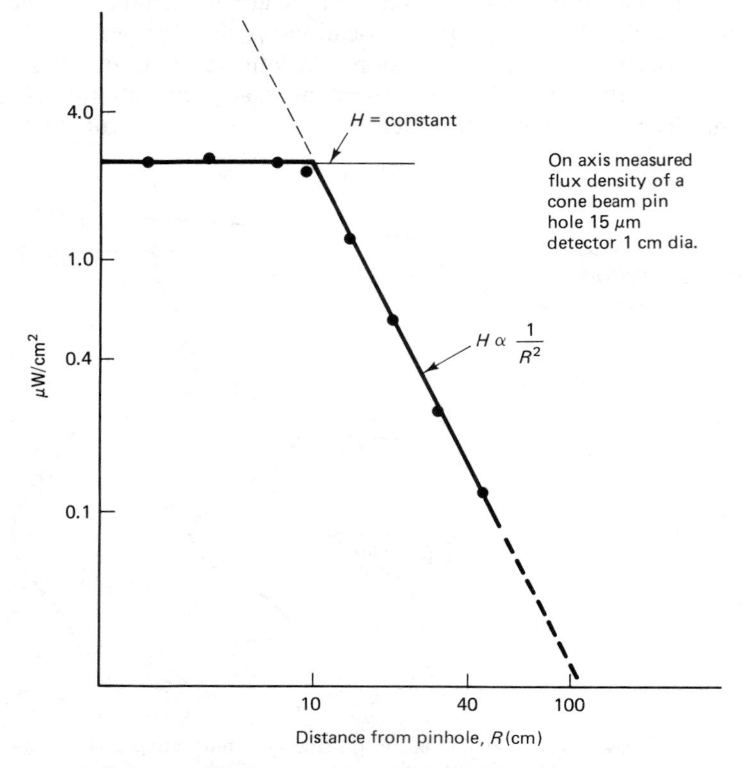

Figure 3-6 Measure flux density of a cone beam radiating from a 15μm (diameter) pinhole and detected on axis by a 1-cm (diameter) detector.

nificant percentage of the beam area. For the data illustrated on this graph, the first data point that falls on the inverse-square-law line is at a distance of 24 cm from the pinhole. At this distance the detector area covers about 16% of the beam area. For the data plotted here the intermediate region of measurement appears to lie between about 7 and 24 cm. The distance corresponding to the point where the horizontal and sloped lines on Figure 3-6 intersect is called the *critical distance.* You are generally safe in assuming that near-field conditions exist when your measurement distance is less than 75% of the critical distance. Far-field conditions are assumed for distances that are longer than 130% of the critical distance. For the data plotted in Figure 3-6, the critical distance appears to be 10 cm, so near-field distances would be less than 7.5 cm and far-field distances would be greater than 13 cm, which is consistent with our general observations above.

3-5 INCANDESCENT AND BLACKBODIES

Solids and liquids radiate visible light at temperatures of 500°C and above. A surface that absorbed all incident radiant energy would appear black. Such a surface would also be a complete radiator. Such idealized surfaces are called *blackbodies.* A practical incandescent radiator will exhibit behavior similar to that of a blackbody. The nature of the radiation from blackbodies was analyzed in some detail by Max Planck. He found that the energy radiated by complete radiators is distributed over a range of wavelengths in a mathematically predictable pattern. As the temperature of the body is altered, the distribution of energy is also altered.

Figure 3-7 illustrates the distribution of energy from a blackbody as a function of wavelength at a fixed temperature. This graph can be thought of as a relative plot of energy at each wavelength. Qualitatively, there will be much more energy at 3748 nm than there will be at 2000 nm. The total radiant emittance, M_e, of a surface is the area under the curve. The radiant emittance between two wavelengths would be the area under the curve between the two wavelengths. A new term, *spectral emittance,* W_λ, is used for the vertical axis of the curve. Spectral emittance W_λ is the radiant emittance, M_e, per unit wavelength and would have the units watts per meter cubed.

$$W_\lambda = \frac{C_1 \lambda^{-5}}{e^{C_2/\lambda T} - 1} \tag{3-10}$$

where λ is the wavelength, in meters

$C_1 = 3.740 \times 10^{-16}$ W · m^2

$C_2 = 1.4385 \times 10^{-2}$ m · K

T is the temperature, in Kelvin (temperature °C + 273°)

A variation of the algebra expression for equation (3-10) would be

$$W_\lambda = \frac{C_1 \lambda^{-5}}{\exp(C_2/\lambda T)} - 1$$

We shall use this form in calculations.

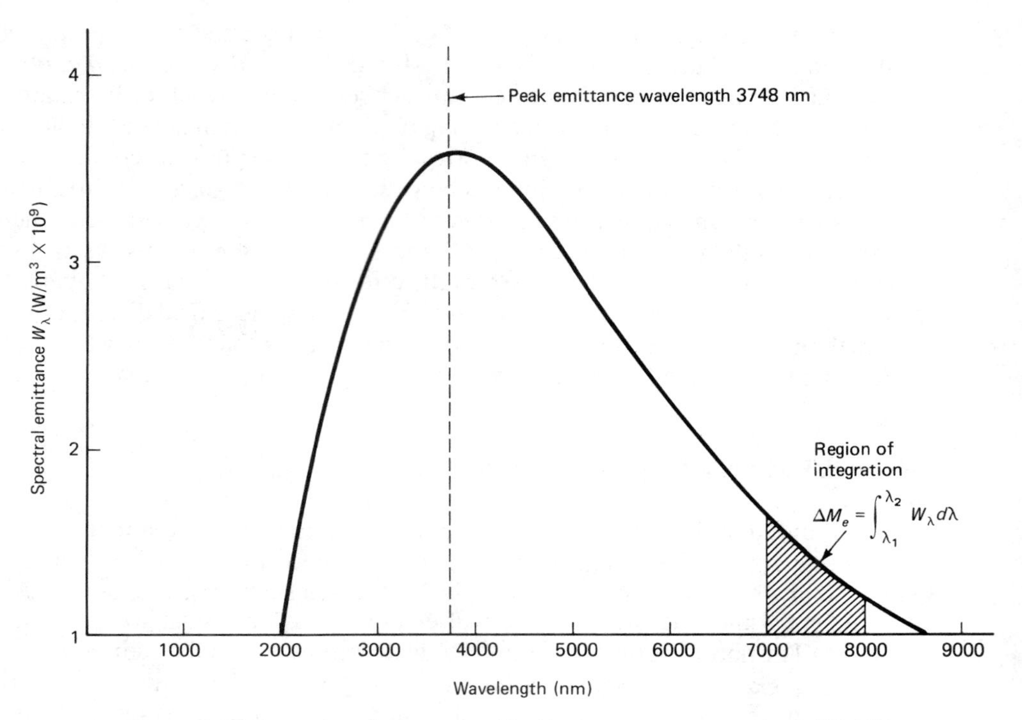

Figure 3-7 Spectral emittance of a blackbody at a temperature of 773.15K (500°C).

The radiant emittance between any two lengths is then given by

$$\Delta M_e = \int_{\lambda_2}^{\lambda_1} W_\lambda d\lambda \tag{3-11}$$

Equation (3-11) can be used to calculate the area under the curve between λ_1 and λ_2. The total radiant emittance is given by

$$M_e = \int_0^\infty W_\lambda \, d\lambda \tag{3-12}$$

Equation (3-11) can be solved using numerical techniques. Equation (3-12) can be solved using calculus. The result obtained by solving equation (3-12) is

$$M_e = \sigma T^4 \tag{3-13}$$

where $\sigma = 5.672 \times 10^{-8}$ W/m^2 · K^4
T is the temperature of the source, in kelvin

This means that if the temperature is known, the radiant emittance is known, and if the radiant emittance is known, the temperature is known. The wavelength at which the spectral emittance is at a peak can be calculated using

$$\lambda_{\max} = \frac{2.8971 \times 10^6}{T} \; nm \cdot \text{K} \tag{3-14}$$

Figure 3-8 illustrates the manner in which the wavelength at which the spectral emittance peak occurs changes with temperature. Example 3-3 illustrates the use of equations (3-13) and (3-14).

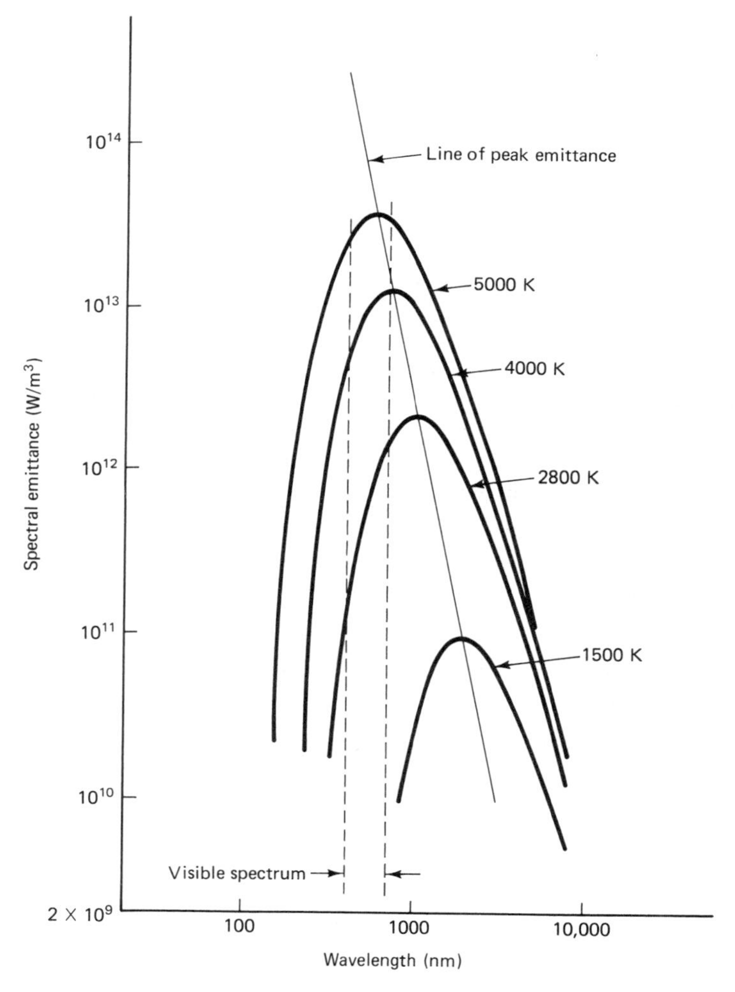

Figure 3-8 Spectral emittance of a blackbody at various temperatures.

EXAMPLE 3-3

Calculation of Total Radiant Emittance and Peak Wavelength for a Blackbody of a Fixed Temperature

Given: A blackbody with a temperature of 500°C.

Find: The total radiant emittance M_e and the wavelength at which spectral emittance has a peak value.

Solution: Recall that the temperature in kelvin is obtained by adding 273° to temperature in °C.

$$M_e = \sigma T^4$$

$$= 5.672 \times 10^{-8}\ \mathrm{W/(m^2 \cdot K^4)} * [(500 + 273)\ \mathrm{K}]^4$$

$$= 20{,}251\ \mathrm{W/m^2}$$

$$= 2025\ \frac{\mathrm{mW}}{\mathrm{cm^2}} \text{ at the radiating surface}$$

$$\lambda_{max} = \frac{2.8971 \times 10^6}{T}\ \mathrm{nm \cdot K}$$

$$\lambda_{max} = \frac{2.8971 \times 10^6}{773\ \mathrm{K}}\ \mathrm{nm \cdot K}$$

$$\lambda_{max} = 3.748\ \mathrm{nm}$$

Example 3-4 illustrates how the radiant emittance between two wavelengths can be estimated using geometric techniques. This value of radiant emittance for a limited range of wavelengths is then compared to the total radiant emittance.

EXAMPLE 3-4

Calculation of the Radiant Emittance Emitted Over a Limited Range of Wavelengths

Given: A blackbody with a temperature of 500°C.

Find: The percentage of the total radiated power that will be radiated between the wavelengths 7000 and 8000 nm.

Solution: Find the approximate area under the W_λ curve between 7000 and 8000 nm and divide this value by the value of M_e from Example 3-3. A good approximation for the area would be a rectangle and a triangle. This geometric region is illustrated in Figure 3-9.

Equation (3-10) is used to evaluate the magnitude of the spectral emittance W_λ at the 8000- and 7000-nm wavelengths.

$$W_\lambda = \frac{C_1 \lambda^{-5}}{\exp(C_2/\lambda T) - 1}$$

At 8000 nm: $C_1 \lambda^{-5} = 3.740 \times 10^{-16}\ \mathrm{W \cdot m^2}\ (8000 \times 10^{-9}\ \mathrm{m})^{-5}$

$$C_1 \lambda^{-5} = 11.414 \times 10^9\ \mathrm{W/m^3}$$

At 8000 nm: $\exp\left(\frac{C_2}{\lambda T}\right) = \exp\left[\frac{1.4385 \times 10^{-2}\ \mathrm{m \cdot K}}{(8000 \times 10^{-9}\ \mathrm{m})(773\ \mathrm{K})}\right] = 10.239$

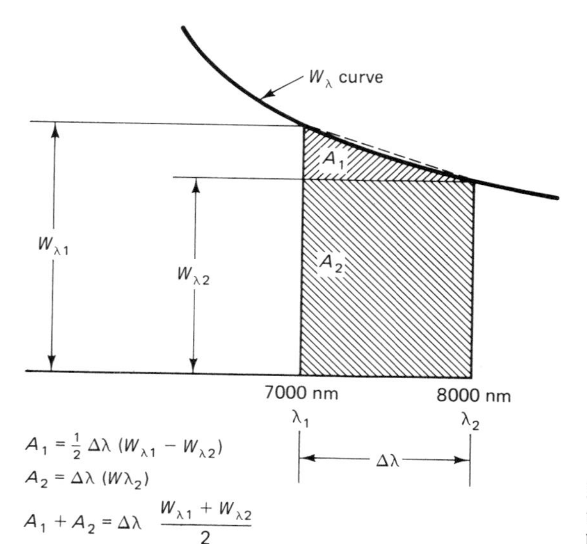

Figure 3-9 Approximation of the area under the spectral emittance curve between two wavelengths.

At 8000 nm: $W_{\lambda 2} = 1235 \times 10^6 \text{ W/m}^3$

At 7000 nm: $W_{\lambda 1} = 1676 \times 10^6 \text{ W/m}^3$

$$\begin{aligned}
\text{height of the triangle} &= W_{\lambda 1} - W_{\lambda 2} \\
&= 441 \times 10^6 \text{ W/m}^3 \\
A_1 \text{ (area of triangle)} &= \tfrac{1}{2} \text{ base} \times \text{height} \\
&= \tfrac{1}{2}(1000 \text{ nm})(441 \times 10^6 \text{ W/m}^3) \\
&= 221 \text{ W/m}^2 \\
A_2 \text{ (area of rectangle)} &= W_{\lambda 2}\,\Delta\lambda \\
&= 1235 \text{ W/m}^2 \\
A_T \text{ (total area)} &= A_1 + A_2 \\
&= 1456 \text{ W/m}^2
\end{aligned}$$

The total area, A_T, is the radiant emittance that occurs between 7000 and 8000 nm. This radiant emittance is given the symbol ΔM_e.

$$\begin{aligned}
\Delta M_e &= A_T \\
&= 1456 \text{ W/m}^2
\end{aligned}$$

The ratio obtained by dividing ΔM_e by the total radiant emittance M_e will be the same as the ratio obtained by dividing the power radiated between the two wavelengths by the total radiated power or flux. For this example we shall define this ratio as N.

$$N = \frac{\Delta M_e}{M_e} = \frac{\Delta \phi_e}{\phi_e}$$

For this example,

$$N = \frac{\Delta M_e}{M_e}$$

$$= \frac{1456 \text{ W/m}^2}{20{,}251 \text{ W/m}^2}$$

$$= 0.072 \quad \text{or} \quad 7.2\%$$

Examples 3-3 and 3-4 illustrate that the total radiant emittance is dependent on source temperature and that the total radiant emittance can be found only if we consider all wavelengths being emitted. Radiant emittance is defined in terms of power per unit area. This means that a source of a fixed area, such as a lamp filament, would have a specific output power per unit area. This idea is summarized by

$$M_e = \frac{\phi_e}{A} \tag{3-15}$$

where ϕ_e is the radiated power or flux A is the area of the source.

If we assume that we are dealing only with radiation (conduction negligible), the conservation of energy requires that the output flux be equal to the electrical input power P.

$$\phi_e = P \tag{3-16}$$

Measurements of M_e, radiant emittance, should be directly proportional to input power. Even if conduction cannot be neglected, it is often a fixed percentage of the input power, and therefore ϕ_e and P would be directly proportional. If you use a detector that responds equally to all wavelengths being emitted, this proportionality will be observed. If, on the other hand, you use a detector that responds only to a narrow range of wavelengths, you will not observe a linear relationship.

Example 3-4 indicates that 7.2% of the total radiant emittance would be detected between the wavelengths of 7000 and 8000 nm when the source is at 500°C. If the source temperature were increased by the application of more input power, P, the peak of the energy distribution would shift to a shorter wavelength and the amplitude of the peak will increase. A recalculation of the radiant emittance between the wavelengths 7000 and 8000 nm would yield a larger value of ΔM_e but a smaller percentage of the total M_e. This is illustrated in Table 3-5, which lists values of λ_{max}, M_e, ΔM, and N as a function of T.

A graph illustrating the general way that λ_{max}, M_e, and ΔM_e vary can be obtained by plotting the relative values of λ_{max}, M_e and ΔM_e versus relative input power. The three parameters λ_{max}, M_e, and ΔM_e can be converted to relative values by dividing the maximum value in each column of Table 3-5 by the maximum value in that column. The relative input power can be obtained by recalling

TABLE 3-5 λ_{max}, M_e, λM, AND N FOR THE WAVELENGTH RANGE 7000 TO 8000 NM AS A FUNCTION OF BODY TEMPERATURE

T (K)	λ_{max} (nm)	M_e (W/m²)	ΔM_e (W/m²)	N (%)
773	3748	20,251	1,456	7.19
1000	2897	56,720	2,774	4.89
1500	1931	287,145	6,260	2.18
2000	1419	907,520	10,073	1.11
2500	1159	2,215,625	14,180	0.64
2900	999	4,011,699	17,250	0.43
3000	966	4,594,320	18,377	0.40

that M_e and input power are directly proportional and therefore relative M_e will be equal to relative input power. Table 3-6 lists the parameters of Table 3-5 converted to relative values.

TABLE 3-6 RELATIVE VALUES OF RADIANT EMITTANCE FOR VARIOUS BODY TEMPERATURES

T (K)	λ_{max} (%)	M_e (%)	ΔM_e (%)
773	100	0.44	7.92
1000	77	1.23	15.10
1500	52	6.25	34.10
2000	39	19.75	54.81
2500	31	48.23	77.16
2900	27	87.32	93.87
3000	26	100.00	100.00

The values of relative λ_{max} and relative M_e are plotted against relative input power in Figure 3-10. Figure 3-10 illustrates that as input power to a radiating body is increased, the magnitude of the peak wavelength decreases while the emittance measured over a limited wavelength range, ΔM_e increases in a nonlinear manner. A plot of total output power or total radiant emittance versus input power would be linear.

You will be happy to know that technicians do not usually have to make calculations of blackbody characteristics or of the energy between wavelengths. These ideas are presented to illustrate what happens when source temperature is changed or when the wavelength range being considered is restricted. It is interesting to note that in this discussion the area under the curve, rather than the points on the curve, is the item of interest. There are a number of other occasions in the study of electro-optics, where the area under the curve or under portions of the curve is the item of greatest interest.

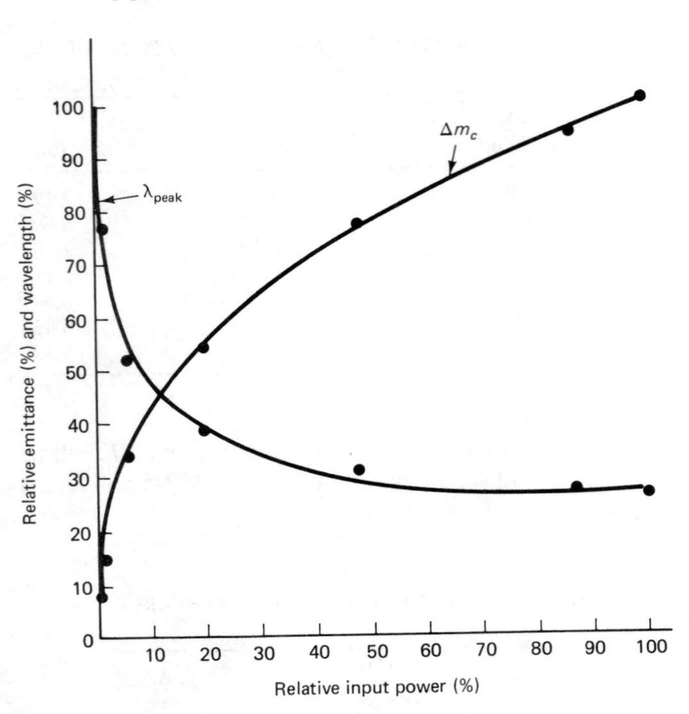

Figure 3-10 Relative peak wavelength (%) and relative emittance over a limited wavelength range (%) versus relative input power (%).

3-6 INCANDESCENCE OF REAL BODIES

Real bodies do not emit as much radiant flux as a blackbody at the same temperature. The ratio of the actual radiant emittance to that of a perfect blackbody is called the *emissivity, e.* It turns out that the ratio of the absorbed radiant flux of a body to that of a perfect blackbody will have the same numerical value, called the absorptance, a. The reflectance of a surface R is equal to 1 minus the absorptance. These ideas are summarized as follows:

$$e = a \tag{3-17}$$

$$R = 1 - a \tag{3-18}$$

$$M_e = e\sigma T^4 \ (\sigma \text{ a constant}) \tag{3-19}$$

$$\frac{M_e}{\sigma T^4} = e \qquad \text{definition of } e \tag{3-20}$$

The power, radiant flux, absorbed from incident flux would be given by

$$\phi_e = aHA \tag{3-21}$$

Typical radiating bodies are mounted in evacuated containers, an example being a lamp filament. Figure 3-11 is a simplified drawing representing a radiating body mounted in a container.

Some power P is supplied to the body, causing it to radiate. A portion of the

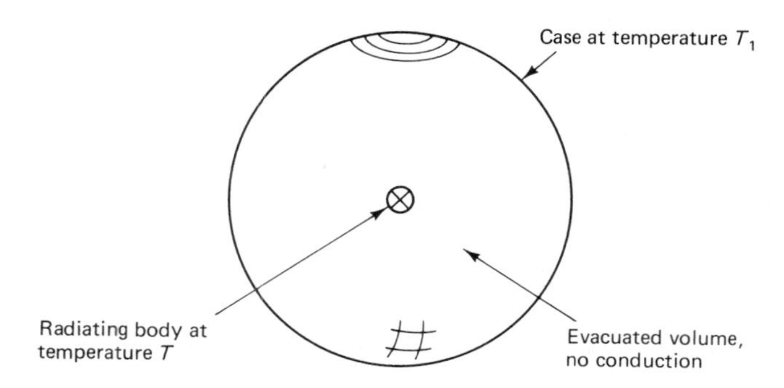

Figure 3-11 Enclosed radiating body.

radiated power is intercepted by the enclosure and reflected back to the body and the remainder escapes the system. When a steady-state operating temperature has been reached, the escaping power must equal the supplied power. These relationships are illustrated in Figure 3-12.

Equations (3-22) through (3-26) summarizes the relationships that exist between the various values of power illustrated in Figure 3-12, taking into account the emissivity of the source.

$$P = \phi_e - P_a \tag{3-22}$$

$$\phi_e = Ae\sigma T^4 \tag{3-23}$$

$$P_a = AaH \tag{3-24}$$

$$P_a = Ae\sigma T_1^4 \tag{3-25}$$

$$P = Ae\sigma(T^4 - T_1^4) \tag{3-26}$$

The symbols used above relate to Figure 3-12 and are defined as follows:

P: Applied electrical power and total radiated power. This assumes that conduction loss is zero.

ϕ_e: Power emitted by the enclosed source at steady-state temperature T.

P_a: The power absorbed by the enclosed source. This absorbed power is due to reflection of energy from the case.

T: The steady-state temperature of the source when power P is applied to the system.

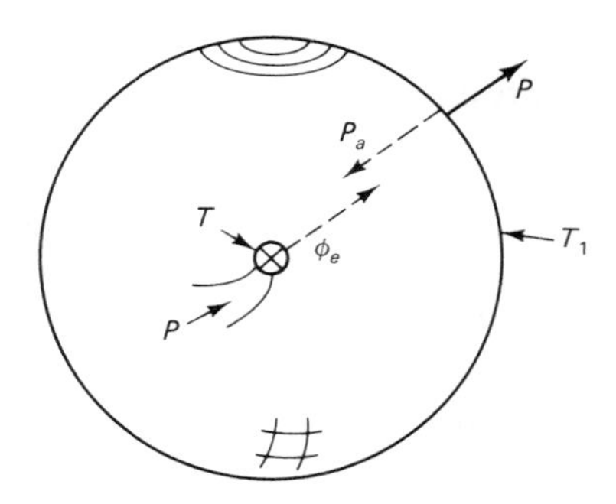

Figure 3-12 Steady-state power relationships for an enclosed radiating body.

A: The surface area of the source.
e: The emissivity of the source.
a: The absorptance of the source $a = e$
σ: A physical constant from equation (3-13); this physical constant has the value $5.672 \times 10^{-8}\ \mathrm{W/(m^2 \cdot K)^4}$.

Example 3-5 illustrates the use of Equation (3-26).

EXAMPLE 3-5

Calculation of Input Power for an Enclosed Radiating Source

Given: **(1)** Filament with a 0.1-cm^2 surface area ($0.1 \times 10^{-4}\ \mathrm{m^2}$)
(2) $e = 0.35$
(3) Filament temperature of 2700 K
(4) Case temperature of 100°C (383 K)

Find: The required input power P.

Solution:

$$P = Ae\sigma(T^4 - T_1^4)$$
$$= (0.1 \times 10^{-4})(0.35)(5.672 \times 10^{-8})\ [(2700)^4 - (383)^4]\ \mathrm{W}$$
$$= 10.5\ \mathrm{W}$$

The power required to keep the small filament at the desired temperature is a reasonable 10.5 W. If we desired to build a lamp with a greater flux output at the same temperature, both power and filament area would have to be increased.

Equation (3-26) and Example 3-5 use a single value of emissivity for the radiating body. It would be nice if nature provided each material with a single fixed value of emissivity, but unfortunately, this is not the case. For any given material and temperature, the emissivity varies as a function of wavelength. In addition, for a fixed wavelength and material, the emissivity varies as a function of temperature.

In Figures 3-13 and 3-14 the emissivity of tungsten, a common filament material, is plotted versus temperature and wavelength. For tungsten, the blue end of the visible spectrum (400 nm) has a higher emissivity than the red end (700 nm). This means that at a fixed filament temperature the relative magnitude of the red end of the spectrum with respect to the blue end of the spectrum will be smaller than that of a blackbody of the same temperature. The emissivity acts like a filter that enhances the blue output with respect to the red output.

As blackbody temperature increases, the relative magnitude of the red end of the visible spectrum decreases with respect to the blue end of the spectrum. For a blackbody to have the same red-to-blue balance as a tungsten filament operating at

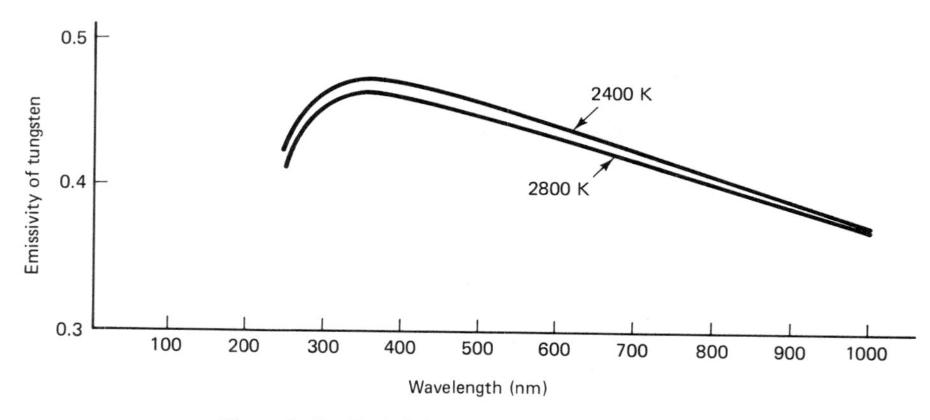

Figure 3-13 Emissivity of tungsten versus wavelength.

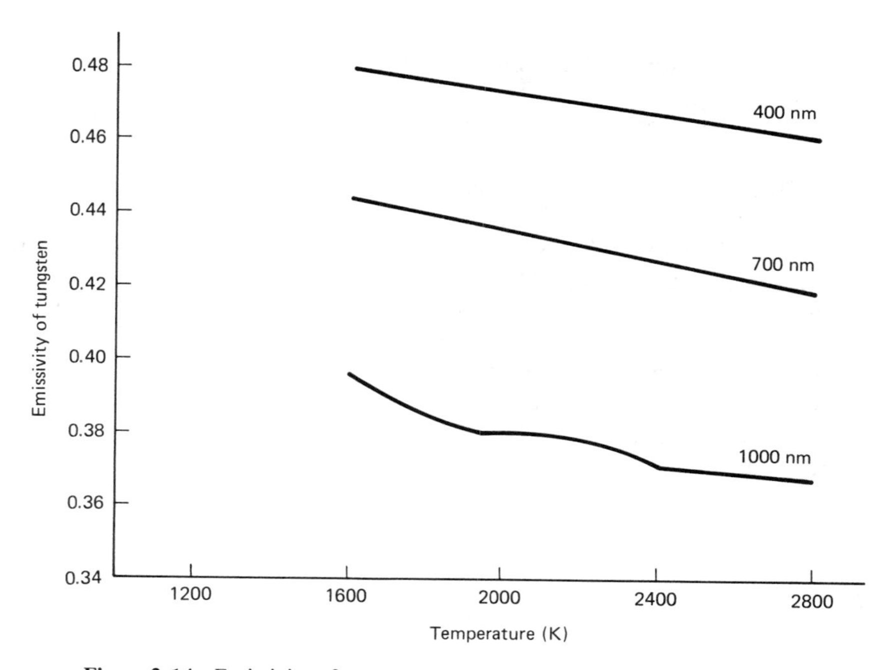

Figure 3-14 Emissivity of tungsten at three wavelengths versus temperature.

2800 K, the blackbody would have to be operated at a temperature higher than 2800 K. As a general rule, we can conclude that an actual incandescent source will have a red–blue color balance that is different from that of a blackbody operating at the same physical temperature. This characteristic of actual sources has led to the establishment of a lamp specification called *color temperature.*

Color Temperature (Light Source) The absolute temperature at which a blackbody radiator must be operated to have a chromaticity equal to that of the light source.

(*Source:* ANSI/IEEE Standard 100-1984. Reprinted by permission of the IEEE Standards Department.)

Emissivity variation as a function of wavelength is one process that can establish the color temperature of a source. The transmission characteristics of the filament enclosure will also alter color temperature. For example, by using a blue-coated bulb around a tungsten filament, a color temperature of 5000 K can be achieved. This high color temperature is much greater than the 3653 K melting point of tungsten. A typical operating temperature tungsten filament would be 2800 K. There would be a small increase in color temperature to about 2876 K, due to the emissivity characteristics of the tungsten. The remaining increase in color temperature to 5000 K is caused by the filtering effects of the blue bulb.

3-7 ELECTRICAL INPUT POWER AND RADIATED POWER

A constant-value voltage source is called a direct-current source. Batteries are an example of a constant-value voltage source. Unfortunately, the power supply used to provide power to many lamps is not a constant-value voltage power supply. The electrical outlets in homes, labs, and theaters supply a continuously varying voltage, dubbed by Edison in his conflicts with George Westinghouse, the "deadly alternating current."

In the United States the typical home or laboratory electrical outlet provides a voltage that varies continuously in a sinusoidal manner from zero to ± 170 V peak. This voltage produces 60 complete cycles per second, 60 hertz, where 1 hertz (Hz) is equal to 1 cycle per second. The current through a resistive load changes value instantaneously with the applied voltage. Figure 3-15 illustrates the voltage waveform of an ac source.

Equations (3-27) and (3-28) summarize the current and voltage relationships in an ac system:

$$i(t) = \frac{e(t)}{R} \tag{3-27}$$

$$i(t) = \frac{E_p \sin \omega t}{R} \tag{3-28}$$

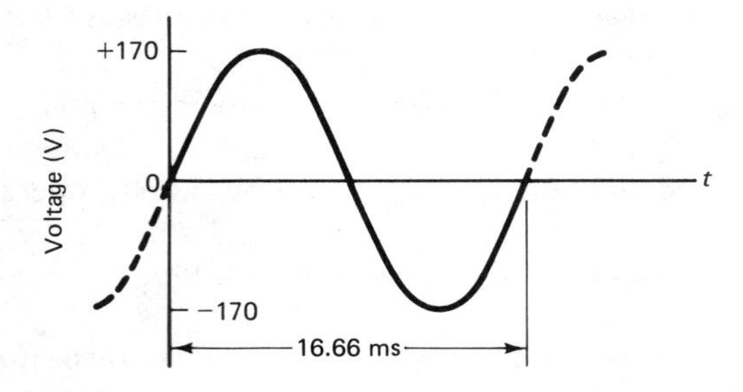

Figure 3-15 AC line voltage, 60 Hz and 120 V.

The symbols used in equations (3-28) represent E_p, the peak value of the voltage; ω, the radian frequency, which is equal to $2\pi f$; and t, the elapsed time. The frequency f of the standard power line voltage in the United States in 60 Hz. In Europe the standard frequency is 50 Hz, and in most mobile and airborne applications, 400 Hz is the standard frequency. Figure 3-16 illustrates the resulting current and voltage waveforms for a resistive load connected to an alternating voltage source.

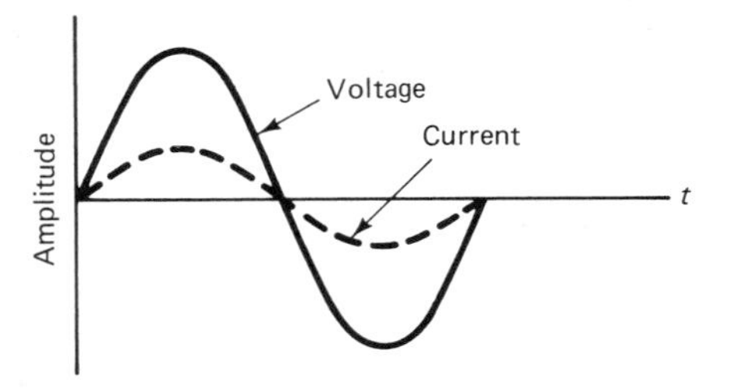

Figure 3-16 AC voltage and resulting current caused by a resistance.

When the voltage polarity across the load is positive the current flows in one direction, and when the polarity is negative the current flows in the other direction. This idea is illustrated in Figure 3-17. Regardless of the direction of current flow, energy is converted to radiation by the resistive load. Power is delivered to the load during both half-cycles. The frequency of the power peaks is twice the line frequency. This can also be shown graphically by calculating and plotting the instantaneous values of power for each cycle. For a resistive load the instantaneous power values are always positive or zero, which means that power flows only from the source to the load. This is illustrated by the graph of instantaneous power illustrated in Figure 3-18.

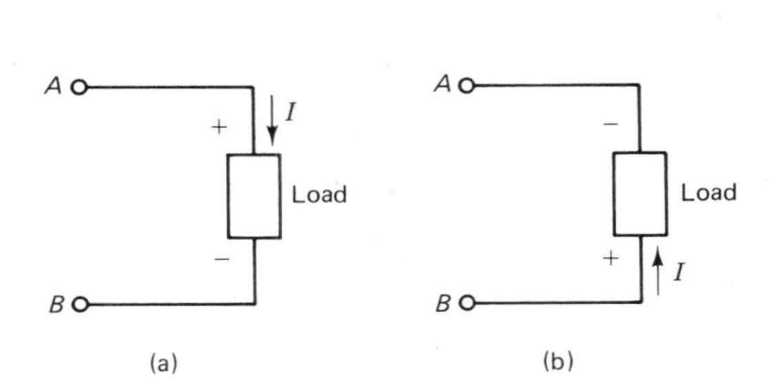

Figure 3-17 Polarity and current direction.

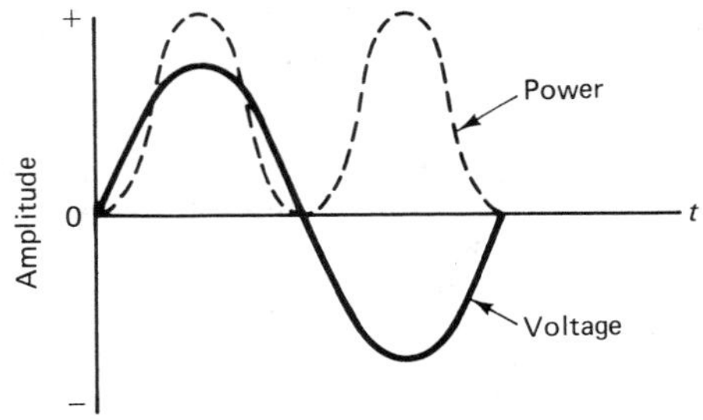

Figure 3-18 Voltage and power waveforms.

Instantaneous power is obtained by calculating the product of instantaneous voltage and instantaneous current at each instant of time. The following equations summarize these relationships:

$$P(t) = E(t)\, I(t) \tag{3-29}$$

$$P(t) = E_p \sin \omega t \; I_p \sin \omega t \tag{3-30}$$

$$P(t) = E_p I_p \sin^2 \omega t \tag{3-31}$$

As this power is applied to the load, the load heats to some steady-state average temperature. You can demonstrate this with any common electrical heating element or lamp. After the ac power is applied to the heating element, the element takes a few moments to rise to its operating temperature, and when power is removed, the element takes some time to cool down again. The rate of heat rise and cooldown is due to a number of factors, of which mass and surface area are major components. Massive objects heat and cool slowly, while low-mass objects respond more quickly to changes in input power. At steady state a plot of temperature versus time for a lamp filament in response to ac power would look like Figure 3-19.

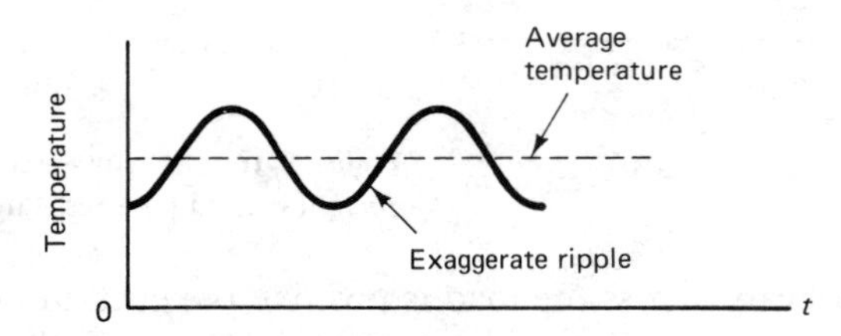

Figure 3-19 Steady-state temperature of a lamp filament caused by an AC voltage source.

The lamp filament will maintain some average operating temperature with a small periodic variation (*ripple*) about the average value. The ripple is caused by the peaks and nulls of the power waveform. The amplitude of the temperature ripple depends on the lamp's physical characteristics, but it is usually very small compared to the average temperature. The frequency of the temperature ripple will be twice the line frequency because the frequency of the power peaks that cause the ripple is twice the line frequency.

A plot of the change in average temperature from zero, the time when power is applied, until the average temperature stabilizes would look like Figure 3-20. The time required for the lamp's average temperature to go from zero to steady state can be thought of as a system *rise time.* The longer a lamp's temperature rise time is, the smaller its temperature ripple will be at steady state. A longer rise time also means that it will take longer for the lamp temperature to change if the supply voltage changes.

The idea of *thermal rise time* is analogous to that of electrical rise time. A long rise time means that a large time constant is present in the system. The larger the time constant, the more steady-state fluctuations are attenuated. A thermal

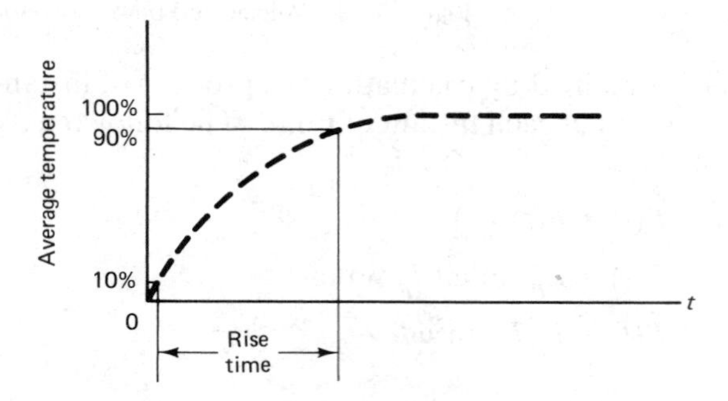

Figure 3-20 Thermal rise time.

system exhibits thermal resistance and capacitance. Where thermal resistance opposes the flow of heat, thermal capacitance stores heat and opposes a change in temperature.

The final average temperature is directly related to the average power of the ac waveform. The average power–temperature relationship was developed in Section 3-6 and is restated here:

$$\phi_e = Ae\sigma T^4 \tag{3-23}$$

The average power of an ac waveform will cause the same average operating temperature as an equivalent amount of dc power. For sine-wave systems the average power is given by

$$P = \frac{I_p E_p}{2} \tag{3-32}$$

where I_p and E_p are the peak instantaneous values of voltage and current.

The peak voltage value E_p divided by the square root of 2 is called the *root-mean-square value* of the waveform. The rms voltage value of the waveform when squared and divided by the device resistance R will yield the average power. Appliances, lamps, and heating elements designed to be operated from ac (sinusoidal) power lines are rated in terms of rms voltage:

$$P = \frac{E^2}{R} \tag{3-33}$$

where E is the rms voltage
P is the average power
R is the device resistance

When "120 V" is printed on a standard light bulb, it is 120 V rms that is meant. Because rms values are so widely used, they are not subscripted or identified in any special way. Other values of the sine wave, such as the peak value, are generally subscripted. There is also an rms value of current. The rms values of voltage and current can be used just like dc values to calculate average power. This is illustrated by the equations

$$I = \frac{I_p}{2^{1/2}} \tag{3-34}$$

$$E = \frac{E_p}{2^{1/2}} \tag{3-35}$$

$$P = IE \tag{3-36}$$

$$P = I^2 R \tag{3-37}$$

Rms values are numbers assigned to waveforms to enable us to predict the average power a waveform will deliver to a resistive load. A dc voltage source with a constant output voltage equal to the rms value of a sine-wave source will deliver the same average power to a resistive load. A lamp connected to either a 120-V rms line or a 120-V dc source would run at the same average temperature and deliver the same average radiant emittance in each case.

In the preceding discussion the resistance of the device to which the voltage was applied was presumed to be constant. Resistance is constant only when temperature changes are small. Lamp filaments have much higher resistances when they are hot than when they are at room temperature because of the typical 2500 K change in operating temperature. For this reason lamp input power and voltage relationships are not adequately described by equations involving constant resistance values. An equation relating lamp voltage and lamp power will be presented in Chapter 4.

3-8 SUMMARY

Electrical specifications of voltage current and power are related to radiometric specifications of flux, radiant emittance, and color temperature. Rms values of voltage and current can be used in the same way as dc voltage and current to calculate average power by taking their product. Changes in average electrical input power will yield linear changes in measured radiant emittance if the detector responds equally to all wavelengths emitted. There will be a nonlinear relationship between changes in input power and measured emittance if the detector does not respond to all wavelengths equally. The behavior of blackbodies provides a useful model for predicting how incandescent sources will behave. Spectral radiometric units are used to describe blackbody radiation. The total radiated flux density between two wavelength values is a function of the area under the spectral emittance curve between the wavelengths of interest.

PROBLEMS

1. The persistent lines of mercury in an arc in order of magnitude occur at the wavelengths listed below. What energy difference is associated with each line? Which lines are visible?

λ (nm)	ΔE (eV)
435.835	
253.652	
366.328	
365.015	
404.656	

2. What range of energy levels in electron-volts corresponds to the visible region 400 to 700 nm?

3. A spherical source radiates a total of 100 W. The diameter of the source is 1 cm.
   (a) What is the flux density at the surface of the source?
   (b) What is the flux density 10 m from the source?

4. **(a)** A measurement of radiant emittance at a point in space with a 1-cm^2 detector is 10 $\mu W/cm^2$. A second measurement 2 m from the first is 15 $\mu W/cm^2$. How far away is the point source from the second position of measurement?
   Both positions of measurement lie on the same axis and the same side of the source.
   (b) Estimate how much power the point source is radiating.
5. A detector of 1-cm^2 area is used on an optical bench 1 m from the source.
   (a) What is the solid angle in steradians formed by the detector?
   (b) What percentage of the spherical surface of measurement does the detector intercept?
6. A source delivers a total 100 mW uniformly in all directions. What is the source's radiant intensity I?
7. A source has radiant intensity of I of 10 mW/sr. What is the irradiance caused by the source:
   (a) At 1 m?
   (b) At 10 m?
8. On an optical bench 1 m from the source a fluxmeter with a 1-cm^2 detector indicates 2 μW.
   (a) What is the irradiance along this axis?
   (b) What is the radiant intensity along this axis?
   (c) Assuming a point source, what is the source output power?
9. Calculate the total radiant emittance of a 2870-K source.
10. Calculate the wavelength of peak output of a 2870-K source.
11. Estimate the radiant emittance of a 2870-K source in the visible range of wavelengths. Refer to Example 3-4.
12. Determine the radiant emittance of a 2870-K source in the ultraviolet (300 to 400 nm) range of wavelengths. Refer to Example 3-4.
13. **(a)** Using the results of Problems 9 and 12, what must the IR radiant emittance be?
    (b) What percentage of the total radiant emittance is in the infrared?
    (c) What percent do the visible and UV represent?
    (*HINT:* Refer to Example 3-4.)
14. For a tungsten filament with an emissivity of 0.43 and a temperature of 2800 K, what is the total radiant emittance?
15. Carbon has an emissivity of 0.82 and tungsten of 0.43. Equal-size filaments of both materials are placed in identical envelopes and driven at steady state by equal power inputs.
    (a) Which filament will be hotter?
    (b) By what percentage?
16. For a carbon filament of emissivity of 0.82, surface area 0.1 cm^2, filament temperature 2800 K, and case temperature 400 K (neglect conduction), what is the required input power?
17. Given a 1-cm^2 filament at a temperature of 3000 K in a case at 400 K, what would the required input power be for a tungsten filament of emissivity 0.43?
18. What happens to the output radiant intensity and filament temperature if input power is halved?
19. It is desired to have a maximum spectral intensity at 1000 nm. The filament has a surface area of 2 cm^2 and an enclosure temperature of 400 K. What input power is required? Assume that emissivity = 1.

4

Photometry and Incandescent Lamps

4-1 INTRODUCTION

Light is the range of wavelengths from 400 to 700 nm. Human eyes are the fundamental detectors of light and the light measurement system is based on simulation of the eye's response. *Photometry* is the measurement of quantities associated with light.

The graph of Figure 4-1 illustrates the relative response of radiometric and photometric detectors as a function of the power applied to an incandescent source. The vertical axis of the graph is a relative axis for the measurements made with each of the three instruments, and the horizontal axis is relative input power to the lamp. Photometric measurements are the least linearly related to lamp power, and the very wide wavelength range radiometric instrument yields the most linear response to increasing lamp power. A radiometric detector with a flat but narrower wavelength response falls in between these two linearity extremes. Because of this and other differences that will be discussed, photometric quantities are different in nomenclature although similar in concept to radiometric quantities. Photometric nomenclature and units are discussed in this chapter, and photometric, radiometric, and electrical units are used to describe sources and interpret data sheets in the remainder of the book.

4-2 THE PHOTOMETRIC SYSTEM

The photometric system is based on the use of detectors that respond to radiated energy in the same way that the human eye responds. Large amounts of statistical data were accumulated to create an empirical standard curve describing the spectral

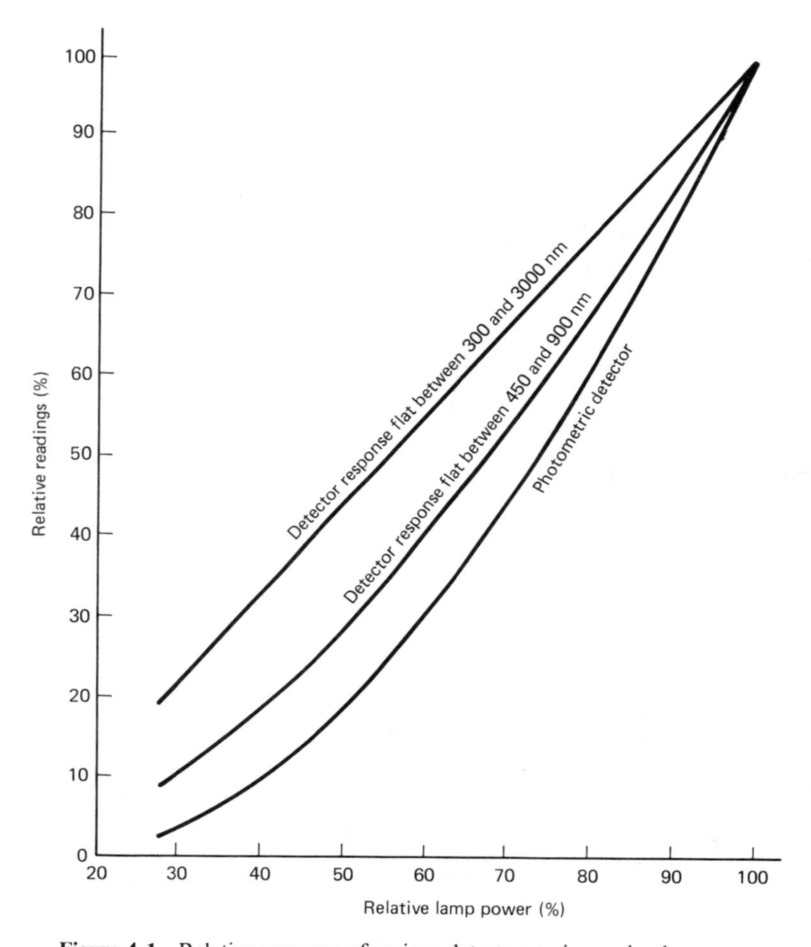

Figure 4-1 Relative response of various detectors to increasing lamp power.

response of the eye. This curve is called the *luminosity curve for the standard observer* or, more simply, the *standard observer curve.* It represents the average result from many trials of testing with numerous subjects. Figure 4-2 is a graph of the standard observer curve. Photometric detectors are constructed to duplicate this spectral response. This curve was established using the methods described below.

A brightness scale running from black, minimum brightness, through gray to white, maximum brightness, was the basic measurement element of the standard observer curve. Human observers were asked to rank by brightness various wavelengths of light of known flux density. The averaging of these responses results in the luminosity curve for the standard observer, sometimes called the *CIE curve.* The letters CIE stand for Commission Internationale de l'Eclairage, the French name for the International Commission on Illumination, which sets standards for illumination and color.

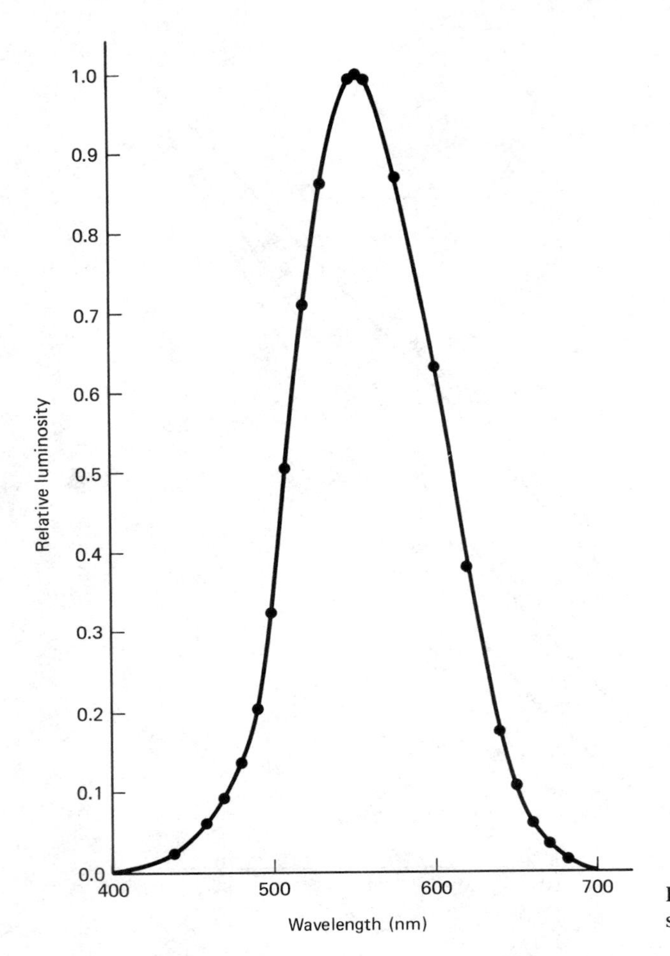

Figure 4-2 Luminosity curve for the standard observer.

The curve itself can be interpreted as follows: A wavelength of 555 nm will appear brighter than any other wavelength of the same radiometric power. A source that can radiate the same radiometric energy at 555 nm and 610 nm will appear only half as bright when operated at 610 nm as it does when operated at 555 nm. This relative brightness is called the *relative luminosity.* Table 4-1 lists numerical values of relative luminosity as a function of wavelength.

Radiometric flux is measured in watts and is assigned to symbol ϕ_e. Photometric flux is measured in lumens and is assigned the symbol F_v. The lumen is similar in nature to a unit of power except that it applies only to the visible range of wavelengths. Within the visible range of wavelengths the flux in lumens has a weighted value based on the CIE curve. The following equation summarizes the relationship that exists between radiometric flux and photometric flux:

$$F_v = \phi_e \times 683 \frac{\text{lm}}{\text{W}} \times \eta \tag{4-1}$$

TABLE 4-1 RELATIVE LUMINOSITY η[a]

Wavelength (nm)	Relative Luminosity	Wavelength (nm)	Relative Luminosity
410	0.001	570	0.952
420	0.004	580	0.870
430	0.012	590	0.757
440	0.023	600	0.631
450	0.038	610	0.503
460	0.060	620	0.381
470	0.091	630	0.265
480	0.139	640	0.175
490	0.208	650	0.107
500	0.323	660	0.061
510	0.503	670	0.032
520	0.710	680	0.017
530	0.862	690	0.008
540	0.954	700	0.004
550	0.995	710	0.002
560	0.995	720	0.001

[a] $\eta = 1$ at 555 nm.

where F_v is the luminous flux, in lumens
ϕ_e is the radiometric flux, in watts
683 lm/W is a physical constant
η is the relative luminosity for the wavelength under consideration

Example 4-1 illustrates the use of equation (4-1) and the relative luminosity data of Table 4-1.

EXAMPLE 4-1

Calculation of Required Radiometric Flux Density Using Relative Luminosity Data

Given: Two monochromatic sources, one radiating at 555 nm and the other source radiating at 490 nm. The 555-nm source has a flux density of 1 mW/cm^2.

Find: The radiant emittance that would be required from the 490-nm source to achieve the same level of brightness that the 555-nm source exhibits.

Solution:

λ (nm)	Relative Luminosity η
555	1.000
490	0.208

Let

F_0 be the luminous flux of a 555-nm source
F_1 be the luminous flux of a 490-nm source
η_0 be the relative luminosity of 555 nm (1)
η_1 be the relative luminosity of 490 nm (0.208)
H_0 and H_1 be the two radiometric flux densities

Then

$$\eta_0 = \frac{F_0}{F_0} = 1$$

$$\eta_1 = \frac{F_1}{F_0} = 0.208$$

In general,

$$F_0 \propto H_0 \eta_0 \qquad \text{refer to equation (4-1)}$$

$$F_1 \propto H_1 \eta_1$$

For this problem it is desired that

$$\mathrm{F}_1 = F_0$$

Therefore,

$$H_1 \eta_1 = H_0 \eta_0$$

$$H_1 = H_0 \frac{\eta_0}{\eta_1}$$

$$H_1 = \frac{1 \text{ mW/cm}^2}{0.208} = 4.81 \text{ mW/cm}$$

The 490-nm source will have to deliver 4.81 mW/cm^2 of irradiance (flux density) to obtain the same sensation of brightness and the same number of lumens.

Incandescent sources radiate energy throughout the visible and infrared ranges of wavelengths. To convert this spectral response to a luminous response would require multiplying the spectral amplitude of small ranges of wavelengths by the luminous efficiency for a wavelength in that range and generating a new curve. The area under the new curve would be the *luminous exitance.* Without going through this exercise it should be obvious that the number obtained will be significantly different from that obtained from just the radiant spectral response. In general we can expect radiometric and photometric measurements to differ significantly when the spectral content changes from measurement to measurement.

Table 4-2 illustrates which radiometric and photometric units utilize the same methods of measurement. No equality is implied by similar quantities of measure-

TABLE 4-2 COMPARISON OF RADIOMETRIC AND PHOTOMETRIC QUANTITIES, SYMBOLS, AND UNITS

Radiometric Units			Photometric Units		
Quantity	Symbol	Unit	Quantity	Symbol	Unit
Radiant energy	Q_e	joule (J)	Luminous energy	Q_v	lumen-second ($\text{lm} \cdot \text{s}$)
Radiant flux power across a surface	$\phi_e = \frac{dQ_e}{dt}$	watt (W)	Luminous flux power weighted to wavelength by the CIE curve	$F_v = \frac{dQ_v}{dt}$	lumen (lm)
Flux density: (incident flux) (irradiance) flux per unit area	$H_e = \frac{d\phi_e}{dA}$	watt per meter squared (W/m^2)	Illuminance luminous flux density	$E_v = \frac{dF_v}{dA}$	lumen per meter squared (lm/m^2)
Radiant exitance: flux per unit area	$M_e = \frac{d\phi_e}{dA}$	W/m^2	Luminous exitance	$M_v = \frac{dF_v}{dA}$	lm/m^2
Solid Angle: area on surface of sphere divided by radius of sphere squared	$\omega = \frac{dA}{A^2}$	steradian (sr)			
Radiant intensity: flux per solid angle	$I_e = \frac{d\phi_e}{d\omega}$	W/sr	Luminous intensity	$I_v = \frac{dF_v}{d\omega}$	lumen per steradian (lm/sr) [cd (candela)]
Radiance: radiant power per unit solid angle and unit projected area	$L_e = \frac{d\phi e}{d\omega\, dA \cos\theta}$ $L_e = \frac{dI_e}{dA \cos\theta}$	$\frac{W}{sr \cdot m^2}$	Luminance (brightness)	$L_v = \frac{dF_v}{d\omega\, dA \cos\theta}$	$\frac{lm}{sr \cdot m^2}$

ment because of the wavelength limitations and the relative luminosity weighting factors inherent in the photometric system.

Figure 4-3 is a plot of luminous efficiency (measured in lm/W) for a blackbody at various temperatures. Typical incandescent lamps operate at color temperatures in the range 2500 to 2900K. The luminous efficiency in this range would be no greater than 20 lm/W, and a typical value would be 11 lm/W. On the other hand, a monochromatic source with a 555-nm wavelength would have a luminous efficiency of 683 lm/W. The great difference in magnitude between the luminous efficiency of the blackbody and the luminous efficiency of the monochromatic source occurs because the blackbody distributes the energy of the 1-W input over visible and nonvisible wavelengths. If 1 W is applied to a source, causing it to reach a color temperature of 2870K, only 0.074 W of what power causes radiation in the visible range of wavelengths. Even if the entire 0.074 W were concentrated at 555 nm, an output of only 51 lm would be obtained. Because incandescent lamps distribute energy over a broad range of wavelengths, the luminous efficiency of incandescent sources is very low.

It is usually not possible to easily convert radiometric flux density measured in W/m^2 to luminous flux density measured in units of lm/m^2. The only time the conversion is relatively simple is when the source is monochromatic and the wavelength is known, as was illustrated in Example 4-1. In general, results can be obtained faster, easier, and more accurately by direct measurement than by mathematical conversion. If you want to know a source's photometric flux density, meas-

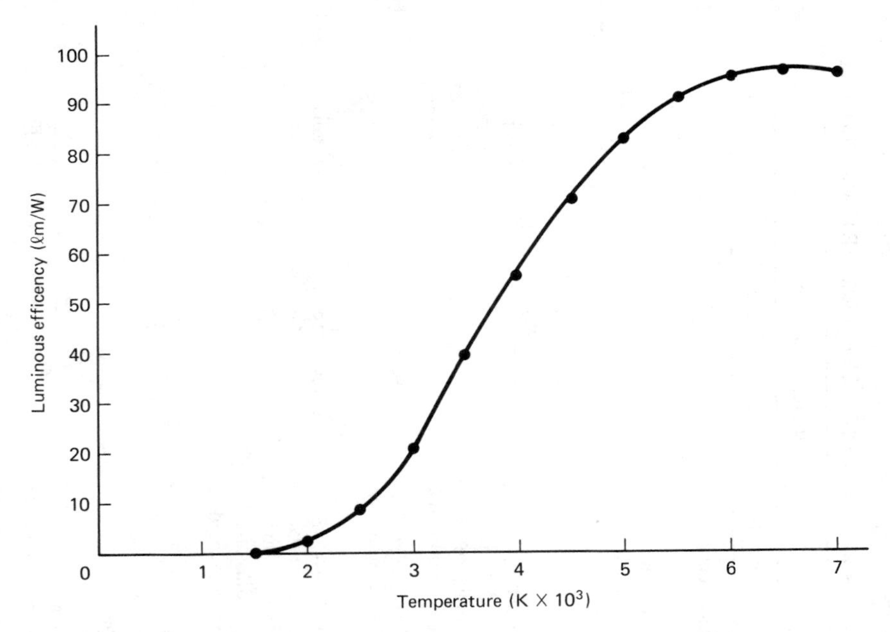

Figure 4-3 Luminous efficiency of a blackbody as a function of temperature. Values obtained by numerical integration. $\Delta\lambda = 10$ nm.

ure it with a photometric detector. If you want to know its radiometric flux density, measure it with a radiometric detector. The preferred set of units for luminous flux density are metric. That is, illuminance is expressed in lm/cm^2 or lm/m^2. Unfortunately, many references and data sheets still use the footcandle (fc) as the unit of illuminance. Figure 4-4 illustrates the relationship that exists between these three units of measure of illuminance caused by a 1-lm/sr source. As was the case with radiant emittance and radiant intensity, luminous exitance and luminous intensity are related mathematically as follows:

$$E_v = \frac{I_v}{R^2} \tag{4-2}$$

where I_v is luminous intensity of the source
E_v is luminous flux density
R is the distance from the source to the detector

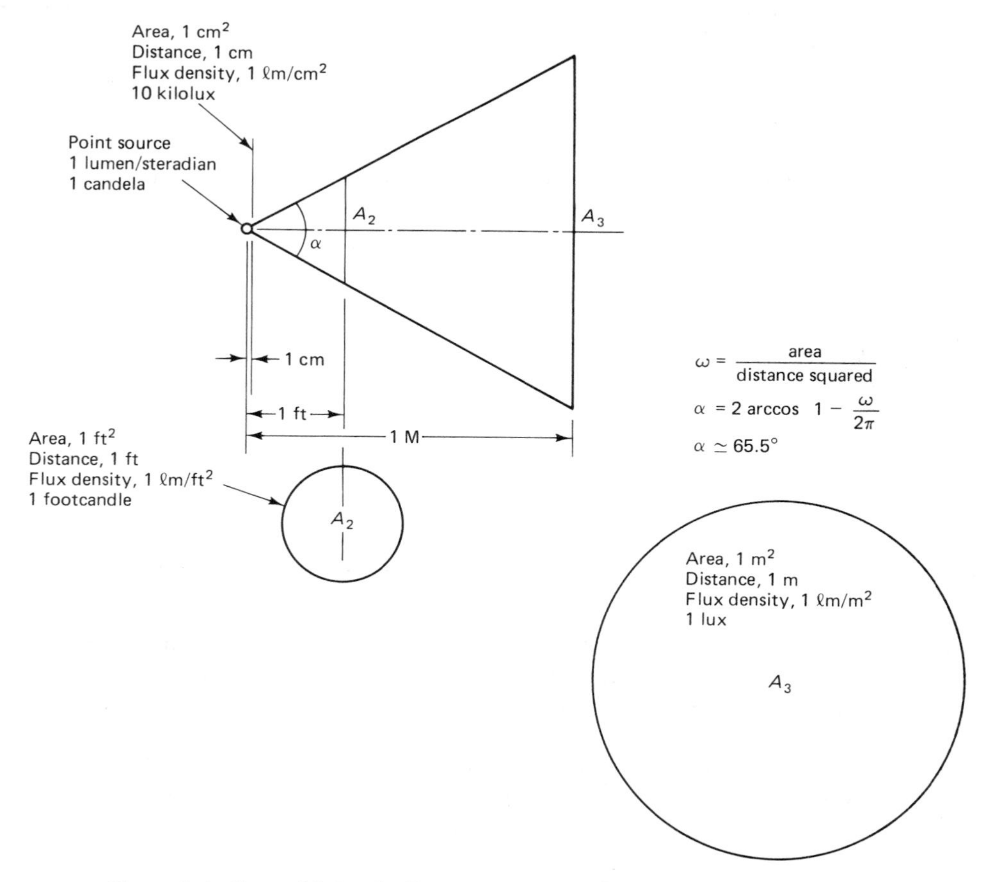

Figure 4-4 Cone of light of solid angle 1 steradian, illustrating flux density units and intercepted areas. Approximately to scale.

It is sometimes necessary to convert from illuminance in footcandles to one of the metric units. This is accomplished simply by applying the appropriate conversion factor. Table 4-3 lists the required conversion factors. To convert photometric units, multiply by the number in the table.

TABLE 4-3 CONVERSION FACTORS

To:	*From:* fc	lux	phot
fc *or* lm/ft^2	1	10.7639	1.08×10^{-3}
lux *or* lm/m^2	0.0929	1	1×10^{-4}
phot *or* lm/cm^2	929	$1 \times 10^{+4}$	1

EXAMPLE 4-2

Conversion of Units of Luminous Flux

Given: A measurement of 5 fc. $E_v = 5\ lm/ft^2$.

Find: The illuminance at this location in (a) lm/cm^2 and (b) lm/m^2.

Solution:

$$\text{(a)}\quad E = \frac{F}{A}$$

$$= \frac{5\ lm}{ft^2} \times \frac{1.08 \times 10^{-3}\ ft^2}{cm^2}$$

$$= 5.4 \times 10^{-3}\ lm/cm^2$$

$$\text{(b)}\quad E = \frac{5\ lm}{ft^2} \times \frac{10.7639\ ft^2}{m^2}$$

$$= 53.82\ lm/m^2 \text{ or } 53.82\ lux$$

Hopefully, the first part of this section has made the point that the magnitude of photometric flux is dependent not only on the radiometric energy present but also on the wavelength of measurement. The two basic measurements that can be made, however, are still flux density and intensity. When flux density is measured at a location removed from the source, it is called *illuminance.* Luminous intensity is a specification that is used to characterize a source. Equation (4-2), which relates flux density to intensity, presumes a point source with a detector of area A at a distance R from the point source. This is illustrated in Figure 4-5.

A cone is projected back from the detector to the point source, which defines a Cartesian angle, α, and a solid angle, ω. The power intercepted by the detector is

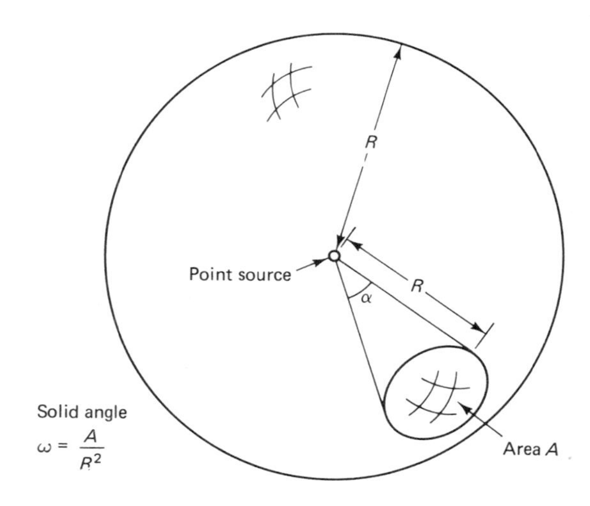

Figure 4-5 Point source at the center of an imaging sphere with a radius R and a detector with an area A on the surface of the sphere.

divided by the solid angle, ω, to determine the intensity, I_v. Two different assumptions can be made about the shape of the detector. The detector can be envisioned as having a surface curved to conform to the surface of the sphere as illustrated in Figure 4-6, or as a flat surface-mounted tangent to the surface of the sphere, as illustrated in Figure 4-7. These two different assumptions lead to different "magic" equations relating the Cartesian angle, ∝, and the solid angle, ω. Short derivations of these two equations are given below.

The surface area of a curved detector is given by

$$A = 2\pi Rr \tag{4-3}$$

where A is the area of the curved detector
R is the distance to the source
r is the "saggital depth" of the detector as illustrated in Figure 4-3

Referring to Figure 4-6 and employing trigonometry yields

$$R - r = R \cos \frac{\propto}{2} \tag{4-4}$$

Therefore,

$$r = R - R \cos \frac{\propto}{2}$$

Substitution into (4-3) yields

$$A = 2\pi R\left(1 - \cos \frac{\propto}{2}\right)R$$

$$A = 2\pi R^2 \left(1 - \cos \frac{\propto}{2}\right) \tag{4-5}$$

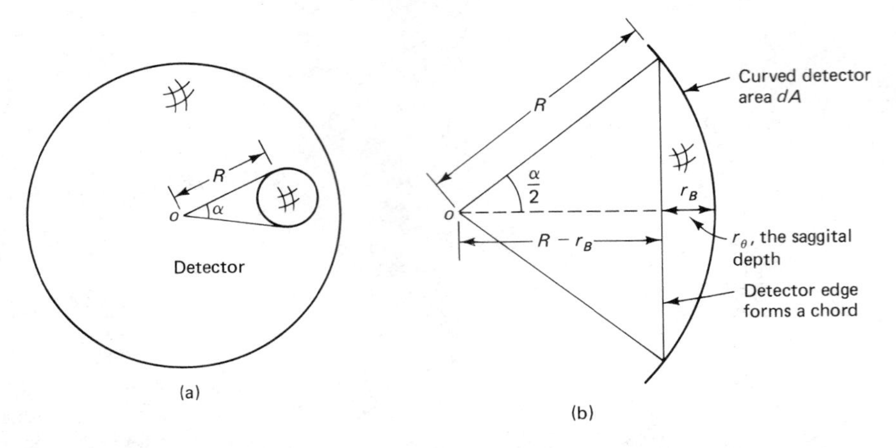

Figure 4-6 Detector with a curvature that conforms to the surface of a sphere: (a) detector on spherical surface; (b) expanded view of detector that conforms to the surface of the sphere.

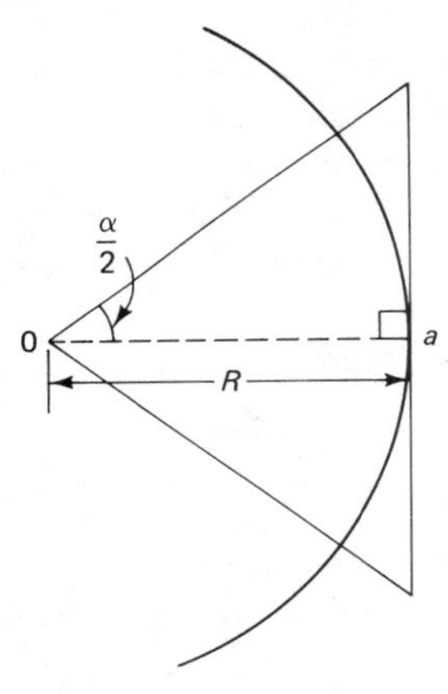

Figure 4-7 Flat circular detector mounted tangent to the surface of a sphere.

Recalling that the solid angle ω is defined as the detector area divided by the distance squared yields the final result:

$$\omega_1 = 2\pi\left(1 - \cos\frac{\alpha}{2}\right) \tag{4-6}$$

where ω_1 is the solid angle formed by a curved detector that intercepts the Cartesian angle α as illustrated in Figure 4-6.

A more practical assumption is that a flat circular detector is located at radius R, tangent to the surface of the imaginary sphere. In this case the area of the detector is given by

$$A = \pi r_D^2 \tag{4-7}$$

where A is the detector area
r_D is the detector radius

The tangent of one-half of the Cartesian angle is given by

$$\tan \frac{\propto}{2} = \frac{r_D}{R} \tag{4-8}$$

Combining equations (4-7) and (4-8) yields the following results:

$$A = \pi R^2 \tan^2 \frac{\propto}{2}$$

$$\frac{A}{R^2} = \pi \tan^2 \frac{\propto}{2}$$

$$\omega_2 = \pi \tan^2 \frac{\propto}{2} \tag{4-9}$$

where ω_2 is the solid angle
$\propto$ is the Cartesian angle

Table 4-4 compares solid-angle values obtained when these two equations are used.

TABLE 4-4 VALUES OF THE SOLID ANGLES

$\propto$ (deg)	ω_1 (sr)	ω_2 (sr)	$\omega_1 - \omega_2$ (sr)
5	0.005980	0.005989	0.000009
10	0.023909	0.024047	0.000138
15	0.053754	0.054451	0.000697
20	0.095456	0.097676	0.002220
25	0.148937	0.154405	0.005468
30	0.214094	0.225556	0.011462
35	0.290805	0.312316	0.021511
40	0.378922	0.416180	0.037258

Equation (4-5) conforms nicely to the rigorous definition of a steradian but envisions an unrealistic detector shape. Equation (4-9) envisions a realistic detector but departs from the exact definition of a steradian. Fortunately, for Cartesian angles smaller than 30° there is a difference of less than 5% between values calculated by the two methods. When making computations with small angles the radian measure of the angle may be substituted for the tangent function or the sine function and the arc length and the chord may be assumed to be equal. In general, angles that are smaller than 30° (0.524 rad) can be assumed to be small angles. I prefer equation (4-9) because it is simpler to calculate. Happily, you usually know detector area and the distance to the source, in which case the solid angle is simply detector area divided by distance squared. This approach corresponds to the conditions of Figure 4-7. These derivations have been presented to demystify the magic equations you often see on data sheets and application notes.

4-3 LUMINANCE AND RADIANCE

Luminance (formerly called *brightness*) is a term used to describe the visible radiation from a surface that is of a significant size when compared to the distance of observation and the size of the detector. Measurements made close to a source like this are called near-field measurements. In Chapter 8 mathematical techniques are developed that define the boundaries of the near-field region in terms of the dimensions of the source and detector and the distance between them. A large surface fills a big portion of the total space being viewed. A projection screen is an example of this type of surface. When the radiometric characteristics of the radiation are being described, the term *luminance* is replaced by the term *radiance.* The luminance or radiance of the source will cause an intensity, $I(\theta)$, at the detector's location that is a function of the angle, θ, of observation (Figure 4-8).

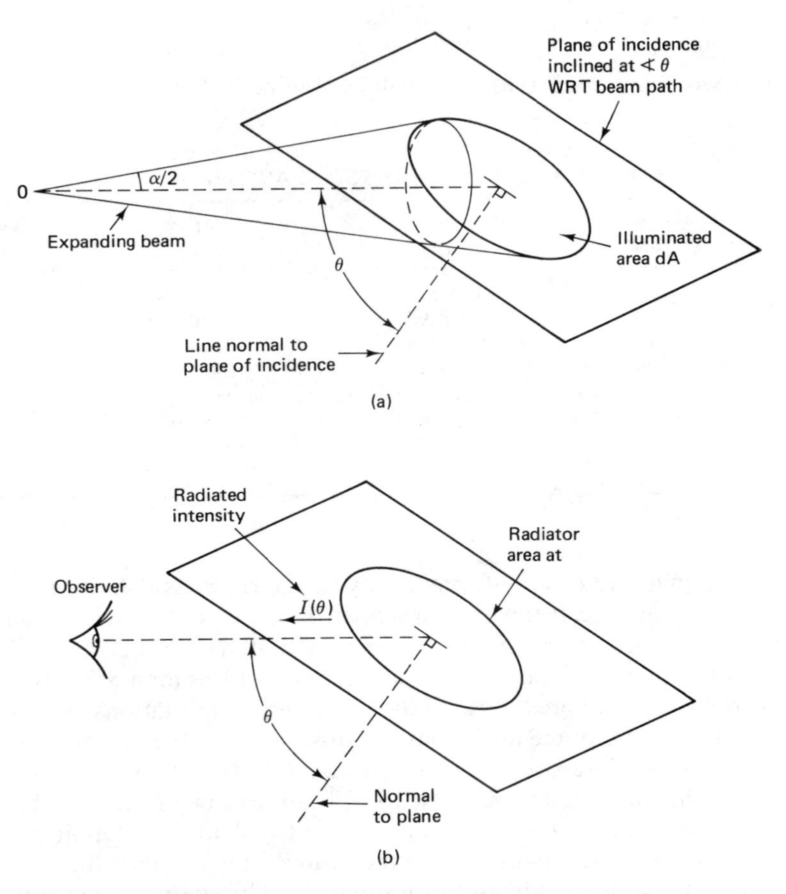

Figure 4-8 Illustrations of luminance: (a) illumination of a surface by a cone beam; (b) illumination radiating from a large surface with an area a_t.

$$L_v = \frac{I(\theta)}{a_t \cos\theta} \tag{4-10}$$

where $I(\theta)$ is intensity as a function of angle θ
a_t is the area of the radiator
θ is the angle measured with respect to the normal to the surface
L_v is the luminance

For the special case where $I(\theta)$ is equal to the product of the intensity along the normal, I_n, and the cosine of θ, the source's luminance (or in the radiometric case, radiance) is equal to a constant,

$$L_v = \frac{I_n \cos\theta}{a_t \cos\theta} = \frac{I_n}{a_t} \tag{4-11}$$

where I_n is the intensity along the normal to the radiating surface area a_t. This type of surface has a constant luminance or radiance regardless of viewing angle and is called a *Lambertian surface.* A Lambertian surface is said to be perfectly diffuse.

In the photometric system intensity is sometimes specified in candela rather than lm/sr. This makes the unit of luminance cd/m^2. If you have one of these perfectly diffuse Lambertian sources, you can calculate the flux density of the surface by using

$$E_v = \pi L_v \tag{4-12}$$

which means that a Lambertian surface with a luminance of 1 cd/m^2 has a luminous flux of π lm/m^2. This relationship leads to a set of units for luminance called lamberts.

$$1 \text{ lambert} = \frac{1}{\pi} \times \frac{1 \text{ cd}}{\text{cm}^2}$$

If you are not confused yet, wait until you see Table 4-5, which lists all the various units of measure that we have avoided using up to this point. It is sad but true that the difficult concepts of radiometry and photometry are made more confusing by the many different units of measure used. I am sure I must have missed someone's favorite unit of measure, but the table covers those you are most likely to encounter. I think it is an elegant example of Murphy's law that the most confusing concept, luminance, has the greatest number of different units!

Now, to close out this section, a few additional details about the CIE curve will be presented. The CIE curve introduced in section 4-2 is a curve that models *photopic* or *day vision.* It is commonplace to assume that this curve peaks at 555 nm. The standard that defines photometric terms defines a frequency of peak response

TABLE 4-5 RADIOMETRIC AND PHOTOMETRIC UNITS

Radiometric		Photometric
	Energy	
Watt-second (W · s) joule (J) erg = J/10		lumen-second (lm · s) talbot
	Flux	
watt (W)		lumen (lm)
	Flux Density	
W/m^2		lux *or* lm/m^2 phot *or* lm/cm^2 footcandle *or* lm/ft^2
	Intensity	
W/sr		candela (cd) (lm/sr)
Radiance		Luminance
$W/m^2 \cdot sr$		$lm/m^2 \cdot sr = cd/m2$ nit *or* cd/m^2 stilb *or* cd/cm^2 lambert *or* $1/\pi$ stilb millilambert = apostilb footlambert *or* $1/\pi\ cd/ft^2$

rather than a wavelength. One candela is defined as the luminous intensity of a source at a monochromatic frequency of 540×10^{12} Hz, with a radiant intensity of 1683 W/sr.

With low background illumination the eyes' characteristic curve changes. This is called *scotopic* or *night vision.* The general shape of the curve is the same, but it peaks at a lower wavelength of 507 nm. Night vision is measured in scotopic lumens. There are 1754 scotopic lm/W. Except for some exotic night vision applications, it is very unusual to encounter scotopic lumens in specifications.

One closing note of caution about lumens. The conversion factor from watts to lumens has had several values during different periods of time, going from a low of 621 lm/W to a high of 685 lm/W. When reading literature on photometry it is helpful to know which conversion factor the author is using.

An illustrative summary of this discussion of units and the relationships that exist between illuminance, intensity, and luminance is provided in Example 4-3. Table 4-6 presents the luminance conversions.

TABLE 4-6 LUMINANCE CONVERSION FACTORS[a]

	stilb *or* cd/cm²	nit *or* cd/m²	lambert	foot lambert
stilb	1	0.0001	$1/\pi$	0.00034
nit	10,000	1	$10{,}000/\pi$	3.426
lambert	π	$\pi/10{,}000$	1	0.00012
footlambert	2919	0.2919	9294	1

[a] The units in the column headings multiplied by the numbers in the table yields the number of units in the first column.

EXAMPLE 4-3

Determination of Lamp Output (in Lamberts)

Description of test setup:

1. The source is an integrating sphere with an exiting port 5 cm in diameter. The average internal photometric reflectance of the sphere is 0.8
2. A photometric detector is mounted 150 cm from the port of the sphere. The detector is mounted on a line normal to the port and centered on the port.

Given: Measured flux density $\left(230 \times 10^{-6}\right) \dfrac{\text{lm}}{\text{cm}^2}$ (phot).

Find: The luminance of the lamp mounted in the sphere in lamberts.

Solution:

1. Assume that the far-field relationships between flux density and intensity are valid because the distance from the detector to the sphere port is 30 times larger than the port diameter.
2. Solve for intensity.

$$I_n = E_v R^2$$

$$= 230 \times 10^{-6} \frac{\text{lm}}{\text{cm}^2} \times (150 \text{ cm})^2$$

$$= 5.18 \frac{\text{lm}}{\text{sr}} \textit{ or } \text{cd}$$

3. Solve for illuminance.

$$L_v = \frac{I_n}{a_t}$$

$$= 5.18\ \frac{\text{lm}}{\text{sr}} \times \frac{1}{\pi(2.5\ \text{cm})^2}$$

$$= 264 \times 10^{-3}\ \frac{\text{lm}}{\text{sr} \cdot \text{cm}^2}\ \textit{or}\ \frac{\text{cd}}{\text{cm}^2}\ \textit{or}\ \text{stilb}$$

4. Convert luminance to lamberts.

$$L_v = \pi \cdot 264 \times 10^{-3}\ \text{lambert}$$
$$= 829 \times 10^{-3}\ \text{lambert}$$

5. Estimate the luminance of the lamp L_{vt}.

$$L_v = L_{vt} \times 0.8$$

$$L_{vt} = \frac{829 \times 10^{-3}}{0.8}\ \text{lambert} = 1.04\ \text{lamberts}$$

4-4 INCANDESCENT LAMPS

Incandescent lamps are used for calibration, illumination, and image projection, and in flash applications. Tungsten, tungsten-halogen, and carbon filament lamps are the primary incandescent lamp types. High-temperature tungsten-halogen lamps have gases introduced into the lamp envelope, that enable these lamps to maintain over 90% of their initial light output over the lamp life. Due to deposition of tungsten on the glass and to aging of the filament, conventional tungsten lamps exhibit a light output decrease of about 50% of its original value over the lamp life. Carbon filament lamps are used primarily as radiation or illumination standards.

The spectral distribution of an incandescent filament is essentially the same as that of a blackbody's at the same color temperature as the filament. Color temperature is the physical temperature that a standard blackbody would have when it has the same distribution of energy in the visible range (that is, the same chromaticity), as the filament. Color temperature was discussed and defined in Chapter 3.

Table 4-7 illustrates the variation of color temperature and resistivity with changes in the actual temperature of tungsten. The resistance of a filament is directly proportional to its resistivity ρ when its length and cross-sectional area are constant.

$$R = \rho\,\frac{l}{A} \tag{4-13}$$

where R is the filament resistance
ρ is the resistivity
l is the filament wire length
A is the cross-sectional area of the filament wire

TABLE 4-7 PHYSICAL TEMPERATURE–RESISTIVITY–COLOR TEMPERATURE OF TUNGSTEN

Physical Temperature		Color Temperature (K)	Resistivity ($\mu\Omega \cdot$ cm)
K	°C		
300	26.85		5.65
500	226.85		10.56
700	426.85		16.09
1000	726.85	1007	24.93
1200	926.85	1210	30.98
1400	1126.85	1414	37.19
1600	1326.85	1619	43.55
1800	1526.85	1825	50.05
2000	1726.85	2032	56.67
2200	1926.85	2241	63.48
2400	2126.85	2451	70.39
2600	2326.85	2662	77.49
2800	2526.85	2876	84.70
3000	2726.85	3092	94.04
3200	2926.85	3310	99.54
3400	3126.85	3530	107.20
3600	3326.85	3754	115.00
3653.15	3380.00	3800	Tungsten melts

The resistivity and therefore the resistance of a lamp filament change dramatically as the lamp temperature changes from room temperature to the operating temperature near 2800 K. The bulb material will affect the lamp's ultraviolet output. Quartz bulbs essentially pass all the ultraviolet while glass lamps attenuate wavelengths shorter than 320 nm. The emissivity of the filaments themselves decreases the longer infrared wave outputs relative to those of an ideal blackbody. Even with these differences, the general trends of blackbodies still provide an excellent model for predicting incandescent lamp behavior. The power flowing into a lamp is directly proportional to the area of the filament and the fourth power of lamp temperature, just like the power flowing into an ideal blackbody.

$$P \propto AT^4 \tag{4-14}$$

where P is the electrical power into the lamp
A is the radiating surface area
T is the color temperature

Calibration lamps can be purchased to perform irradiance, illuminance, or spectral irradiance calibration. The data of Table 4-8 lists typical characteristics for a carbon filament standard lamp. The listed irradiance at 2 m is for a specified aperture and background condition at the currents and voltages shown.

The data sheet of Figure 4-9 lists irradiance and illuminance for a specific mounting configuration at a distance of 50 cm. This data sheet describes three

TABLE 4-8 CARBON LAMP CHARACTERISTICS

Current (mA)	Voltage (V)	Irradiance at 2 m (μw/cm^2)
300	79.4	45.9
350	89.8	62.3
400	99.4	81.3

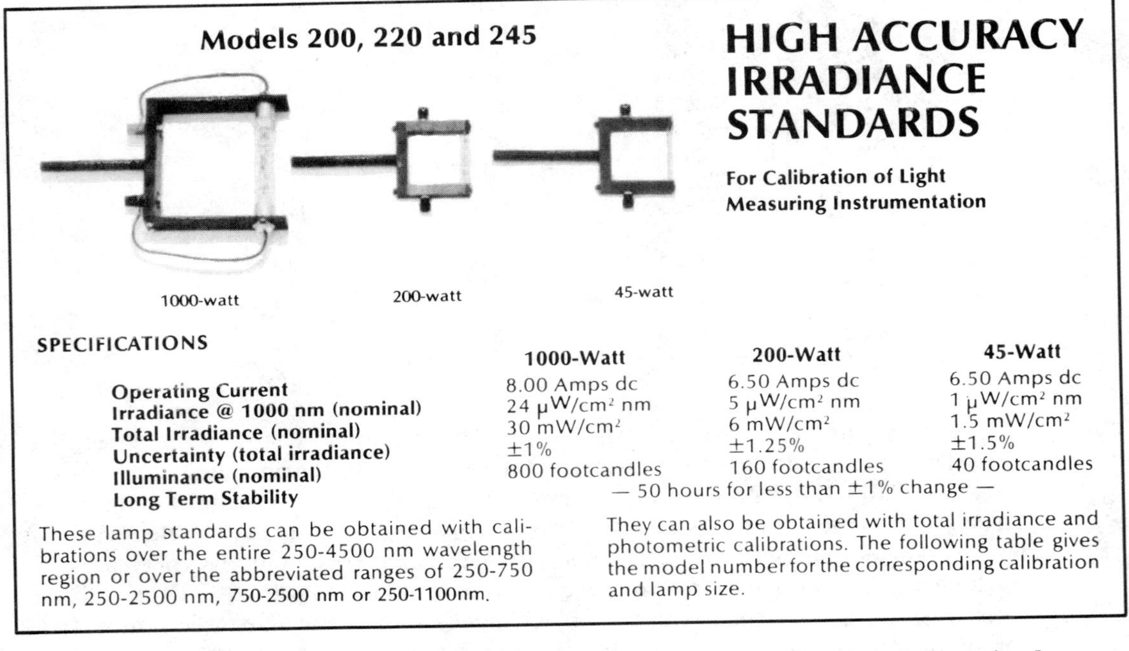

Figure 4-9 High-accuracy irradiance standards. (Courtesy Optronic Laboratories, Inc., Orlando, FL.)

lamp sizes with five choices of calibration over different ranges of wavelengths. In addition, the lamps may be calibrated in photometric or radiometric units. To obtain the calibrated values of irradiance, these lamps must be operated with a specified value of dc current.

The lamp described in Figure 4-10 is a special deuterium ultraviolet source with a window in the side. Again, specific electrical and alignment conditions must be met to obtain the output specified.

Practical lamps do not radiate equally in all directions. The orientation of a lamp in a calibration setup or an illumination application will strongly affect the system output. Specified output is also dependent on the power applied to the lamp. In a calibration setup, the accuracy of the power supply settings and system alignment will determine the accuracy of the radiation output.

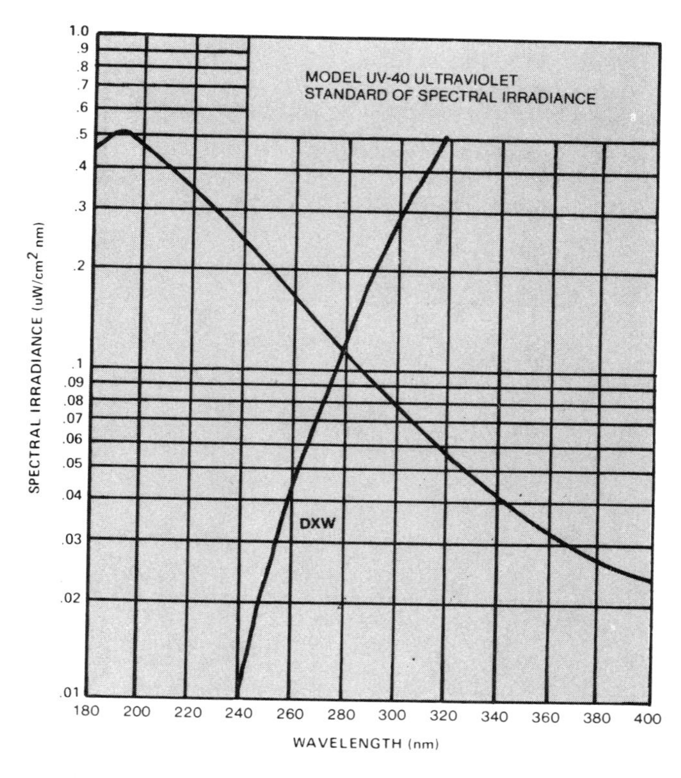

SPECIFICATIONS

Lamp	Deuterium (40 watts)
Wavelength Range	200 to 400 nm
Operating Current	500 ma
Irradiance @ 250 nm (30 cm)	0.2 μW/cm²nm (typ.)
Uncertainty	±3 to 10%
Long Term Stability	50 hours for less than ±2% change

Figure 4-10 Ultraviolet irradiance standard. (Courtesy Optronic Laboratories, Inc., Orlando, FL.)

Looking closely at the units used for spectral irradiance in Figures 4-9 and 4-10, we find that they are $\mu W/cm^2 \cdot nm$ rather than W/m^3 as was used for the blackbody curve in Chapter 3. These units are used because the measured radiant flux will probably have units of $\mu W/cm^2$ and the wavelength of interest will most likely be expressed in nanometers. The spectral emittance units of $\mu W/cm^2 \cdot nm$ are a practical engineering unit.

Example 4-4 illustrates how the spectral emittance graph of Figure 4-10 can be used to predict the radiant emittance from the lamp described by the curve for a given range of wavelengths. The procedure used is identical to that used in Example 3-4 for determining the radiant emittance of a blackbody.

EXAMPLE 4-4

Flux Density Calibration Using a Lamp's Spectral Emittance Curve

Given: The UV-40 lamp described in Figure 4-10.

Find: The approximate radiant emittance between 250 and 340 nm.

Solution: From the graph of Figure 4-10:

λ (nm)	$W\lambda$ ($\mu W/cm^2 \cdot nm$)
250	0.2
340	0.04

Thus

$$\Delta\lambda = 90 \text{ nm}$$

$$M_e = \text{total area} \simeq \frac{(0.04\ \mu W/cm^2 \cdot nm + 0.2\ \mu W/cm^2 \cdot nm)(90\ nm)}{2}$$

$$\simeq 11\ \mu W/cm^2$$

In addition to the spectral content and intensity of radiation, the spatial distribution of radiation from a source is an important consideration. The typical polar pattern of radiation for a standard light bulb would look like Figure 4-11. This plot shows that the maximum luminous intensity will be radiated from the side of the bulb. A luminous intensity equal to about 40% of the maximum value will be radiated from the top of the bulb, and zero radiation will be radiated from the bottom. On data sheets for lamps it is common to find the specification MSCP, or *mean spherical candlepower,* which is the average of the luminous intensity measured in all directions. The total flux in lumens being radiated by a lamp can be found by multiplying the MSCP by the constant 4π:

$$F_v = \text{MSCP} \times 4\pi \tag{4-15}$$

where F_v is the luminous flux, in lumens
MSCP is the mean spherical candle power

If the radial polar axis of the spatial distribution graph is given in absolute units of lm/sr or candela rather than relative output, the MSCP can be found by constructing a circle, or in the case of the plot of Figure 4-11, a half-circle, which has an enclosed area equal to the enclosed area of the polar plot. Do not assume that all lamps will have an identical polar plot; lamps designed for other applications may have most of the light emitted from the top or one side.

The voltage applied to an incandescent lamp is by far the easiest parameter to

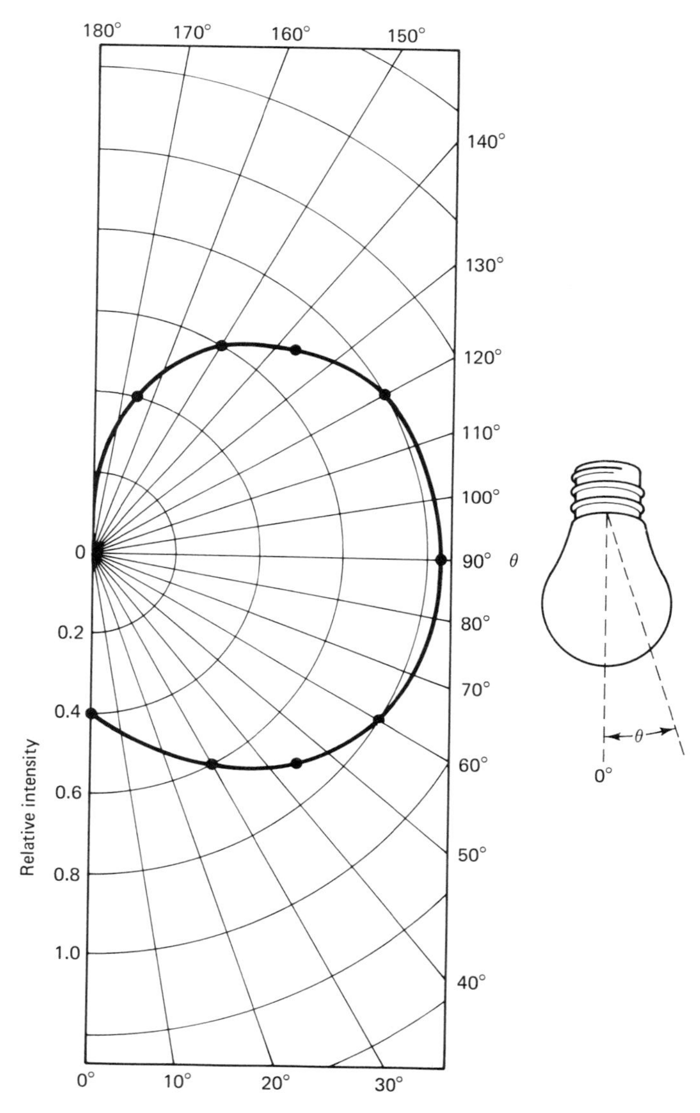

Figure 4-11 Spatial variation of intensity for an incandescent lamp.

measure. If you know the applied voltage and the rated specifications of a lamp, you can make numerical estimates of the lamp's operating characteristics at other voltage levels.

Rated Specifications	Operating Conditions
V_0 rated voltage	V_N new voltage
I_0 rated current	I_N new current
$(MSCP)_0$ rated MSCP	$(MSCP)_N$ new MSCP

$$I_N = I_0 \left(\frac{V_N}{V_0} \right)^{0.55} \tag{4-16}$$

$$(\text{MSCP})_N = (\text{MSCP})_0 \left(\frac{V_N}{V_0} \right)^{3.5} \tag{4-17}$$

Figure 4-12 illustrates these relationships. If color temperature were plotted on the same graph, it would lie almost on top of the current line. The life line is a prediction of how long the average lamp of a given type will last. The *life* is a function of the physical design of the lamp, operating temperature, and environmental condi-

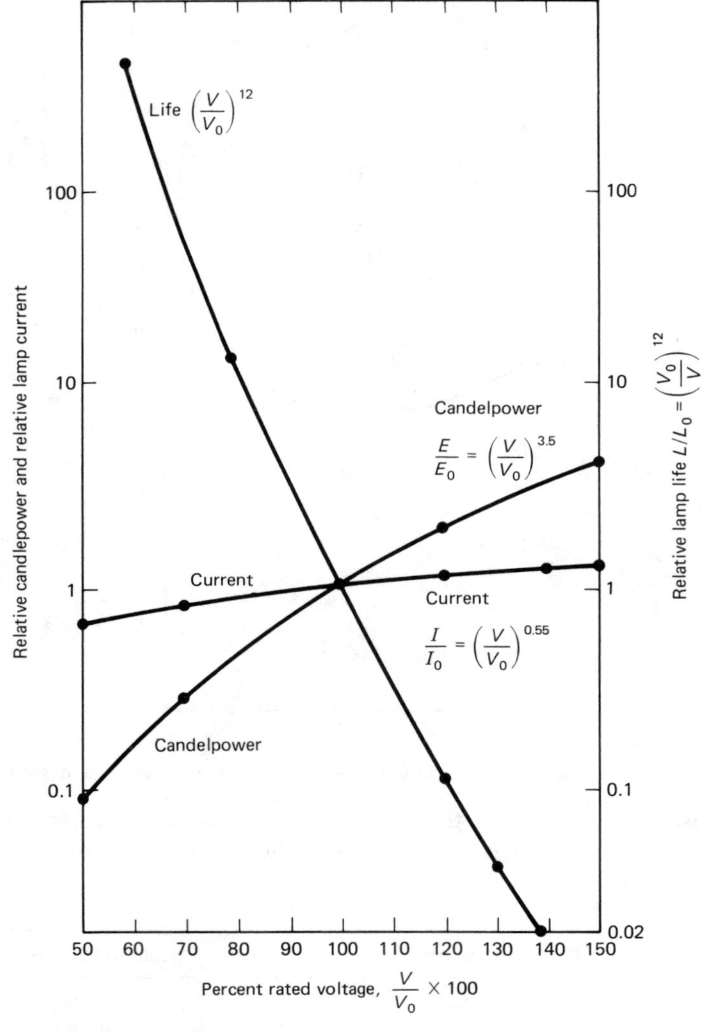

Figure 4-12 Relative operating characteristics of a tungsten incandescent lamp as a function of relative lamp voltage. V_0 rated voltage; V, operating voltage.

tions. Above the rated voltage a lamp will have minimum life because of the high temperatures. As the temperature approaches 3653 K, the melting point of tungsten, the life goes toward zero. As the temperatures decrease, the life increases.

$$\text{life}_N = \text{life}_0\left(\frac{V_0}{V_N}\right)^{12} \tag{4-18}$$

where life_N is the new life

life_0 is the rated life

The use of equations (4-16), (4-17), and (4-18) is illustrated in Example 4-5.

EXAMPLE 4-5

Lamp Calculations

Given: The following lamp data:

Voltage (V)	Current (A)	MSCP	Life (hours)
120	0.025	0.320	10,000 (1.14 years)

Find: The new characteristics at a voltage of 84 V.

Solution:

$$I_N = I_0\left(\frac{V_N}{V_0}\right)^{0.55}$$

$$= 0.25\ \text{A}\left(\frac{84}{120}\right)^{0.55}$$

$$= 0.021\ \text{A}$$

$$(\text{MSCP})_N = 0.320\ \frac{\text{lm}}{\text{sr}}\left(\frac{84}{120}\right)^{3.5}$$

$$= 0.0921\ \frac{\text{lm}}{\text{sr}} \text{ or } 0.0921\ \text{cd}$$

$$\text{life}_N = 10^4\ \text{hours}\left(\frac{120}{84}\right)^{12}$$

$$= 722{,}000\ \text{hours (82 years)}$$

Luminous flux density and luminous intensity are directly proportional. Flux density is directly proportional to luminous flux, which equation (4-15) shows is directly proportional to MSCP. These relationships allow equation (4-17) to be used to predict luminous flux density or luminous intensity simply by substituting these variables for MSCP. I am aware that using I for both current and intensity might be confusing. Unfortunately, that is the way it is: The electrical world uses I for current and the optics world uses I for intensity.

Lamp input power as a function of lamp voltage can also be calculated. A general algebraic equation can be developed by using equation (4-16) and the equation for electrical power. This derivation is illustrated as follows. Let

$$P_0 = \text{rated power}$$
$$V_0 = \text{rated voltage}$$
$$I_0 = \text{rated current}$$
$$P_N = \text{new power}$$
$$V_N = \text{new voltage}$$
$$I_N = \text{new current}$$
$$P_0 = V_0 I_0$$
$$P_N = V_N I_N$$

$$I_N = I_0 \left(\frac{V_N}{V_0} \right)^{0.55} \tag{4-16}$$

$$\frac{P_N}{P_0} = \frac{V_N I_N}{V_0 I_0} = \frac{V_N}{V_0} \frac{I_N}{I_0} \left(\frac{V_N}{V_0} \right)^{0.55}$$

$$\frac{P_N}{P_0} = \left(\frac{V_N}{V_0} \right)^{1.55}$$

$$P_N = P_0 \left(\frac{V_N}{V_0} \right)^{1.55} \tag{4-19}$$

A similar approach can be used to derive an equation for color temperature by combining equations (4-19) and (3-25) and assuming that emissivity remains constant. This is left as an exercise for the reader in Problem 13.

Table 4-9 is an example of a gas-filled incandescent lamp data sheet. It lists all of the major specifications discussed in this section. These lamps are filled with a high-pressure inert gas such as argon, krypton, or an active halogen. A halogen gas minimizes tungsten filament evaporation, raises color temperature, and extends lamp life. The vacuum lamps discussed previously will have a lower case surface temperature than will an equivalent gas-filled lamp.

TABLE 4-9 GAS-FILLED LAMPS: HIGH BRIGHTNESS, HIGH COLOR TEMPERATURES

Lamp	Voltage (V)	Current (A)	MSCP (cd)	Flux (lm)	Color Temp. (K)	Life (hours)
A	2.5	0.550	1.14	14	2950	15
B	3.5	0.610	1.27	16	2930	20
C	3.5	0.750	2.55	30	3030	60
D	4.0	0.290	0.56	7	2680	200

As Problems 14 and 15 will illustrate, the equations presented in this section predict the behavior of gas-filled incandescent lamps almost as well as they predict the behavior of evacuated incandescent lamps. The only exception to the statement above will be with respect to lamp life. The equation for lamp life presented in this section is applicable to vacuum lamps; it is not a good predictor of lamp life for gas-filled incandescent lamps. To obtain life data for gas-filled incandescent lamps, consult the manufacturers' data sheets.

4-5 FLASHLAMPS

Flashlamps are used extensively in photographic applications. Unlike the lamps we have looked at so far, these lamps are not radiating continuously. These lamps are turned on and off repeatedly. Flashbulbs are flashlamps which have filaments that melt during use. Figure 4-13 is a typical graph of flashlamp luminous output with respect to time. A new specification called *light output* is used to describe flashlamps. This specification is measured in lumen-seconds. The lumen-second is the area under the luminous output versus time curve and is a measure of the available luminous energy.

Referring back to radiometric concepts for a moment, the unit of radiometric power is the watt, which is defined as a joule per second. A joule is the fundamental unit of energy and power is the time rate of change of energy. The product of power and elapsed time is the total energy input to the system over the elapsed time inter-

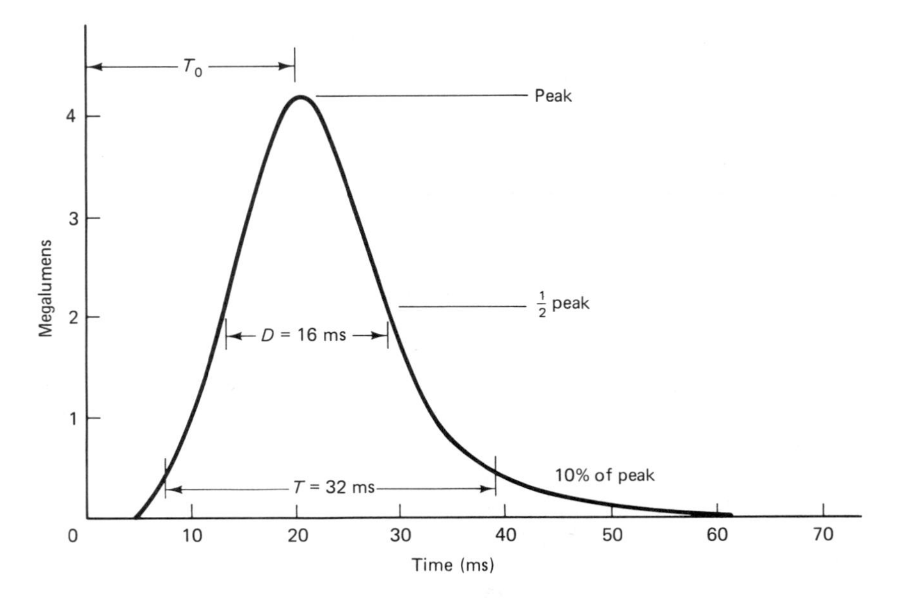

Figure 4-13 Typical flashlamp luminous output as a function of time illustrating time specifications. T_0, time to peak; T, pulse width at 10% of peak; D, duration of pulse between half-peak points.

val. The lumen is the photometric power-like unit. The product of photometric "power" in lumens and elapsed time is the luminous energy. The lumen-second is the unit of luminous energy. The area under a curve of luminous flux versus time is given by the product of lumens and time and so is a measure of luminous energy.

The curve of luminous flux in megalumens lumens versus time illustrated in Figure 4-13 would require a relatively complex exponential function to describe it. Fortunately, a triangle is a "good enough" approximation to the shape of a curve. Assuming that the curve is a triangle simplifies the determination of the area under the curve. Example 4-6 illustrates this calculation for the graph shown in Figure 4-13. To compute the area, the base of the triangle, which is the elapsed time, and the height of the triangle, which is the peak value in lumens, must be determined. The height is read off the curve; in this case it would be about 4.2 Mlm. Determination of elapsed time is a little more difficult because as the curve illustrates, the change of light output with respect to time is very nonlinear near the beginning of the interval. One method that seems to work fairly well is to use the time interval between light levels equal to 10% of the peak. In this case the pulse width at the 10% light level is 32 ms. One-half the base of the triangle is therefore 16 ms.

EXAMPLE 4-6

Calculation of Luminous Energy

Given: The graph of Figure 4-13.

Find: The lumen-second specification.

Solution: Approximate the area under the curve as a triangle by using one-half the elapsed time $\times$ the peak.

$$\begin{aligned} \text{Area} &= 16 \times 10^{-3}\ \text{s} \times 4.2 \times 10^{6}\ \text{lm} \\ &= 67{,}200\ \text{lm} \cdot \text{s} \end{aligned}$$

Often, the response graph is not available and lamp operations must be inferred from data tables. Table 4-10 is an example of a flashlamp data sheet with photometric specifications. A complete data sheet will give the white light output, the time to the peak value, the time the bulb is at a level equal to or greater than one-half its peak value, the peak value, voltage range for operation, color temperature, and mechanical specifications. Since the typical data sheet gives light output as a specification, calculation is not usually necessary. If light output is not specified, a fair estimate of its value can be made by multiplying the specified half-peak time by the specified peak lumens. As is the case with all estimates, sometimes this will be very accurate, but more often it will be a little high or low.

The lamps listed in Table 4-10 with the same numbers are identical except for the color of their envelope. The clear bulb versions and the blue bulb versions have significantly different levels of light output and color temperature. A blue bulb

TABLE 4-10 TYPICAL FLASHLAMP DATA

Lamp Type[a]	Time to Peak (ms)	Half-Peak Duration (ms)	Peak Lumens (lm)	Light Output (lm · s)	Voltage (V)	Color Temperature (K)
1	17	15	1×10^6	16,000	45	3800
1B	17	15	550×10^3	10,000	45	5500
2	20	18	4.2×10^6	70,000	125	3800
2B	20	18	2×10^6	35,500	125	5500

[a] B lamp types are blue coated; all others have clear bulbs.

reduces the peak lm output as much as 50%. Apparent color temperature can be shifted up from 3800 K for the clear bulb to 5500 K by using a blue bulb. The color temperature for a lamp with a clear bulb is essentially the color temperature of the filament as it melts, which in turn is close to the actual filament temperature, as indicated in Figure 4-14.

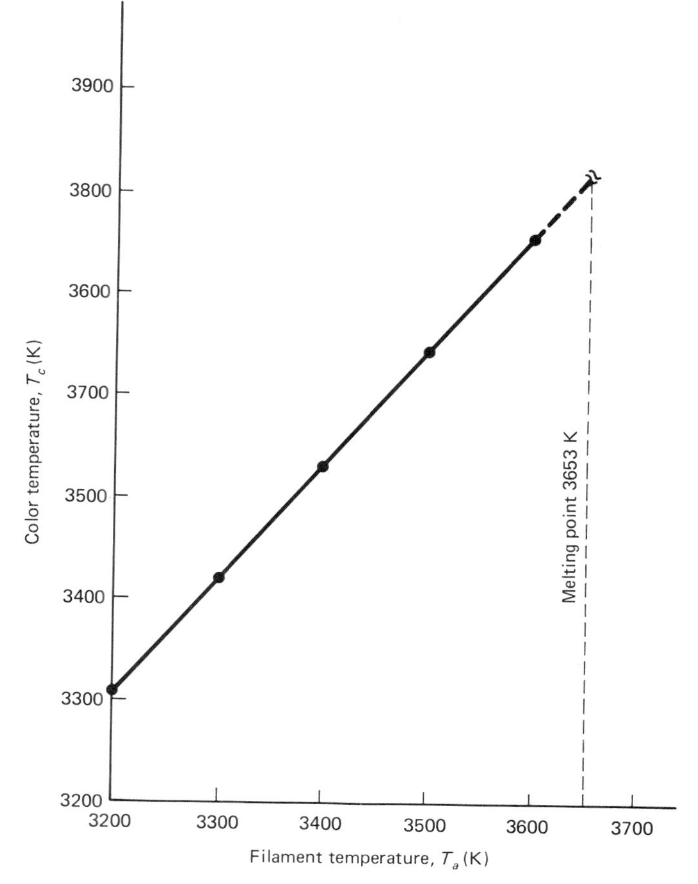

Figure 4-14 Color temperature versus actual temperature for a tungsten filament. Approximate equation for this temperature range: $T_c = (1.11)T_a - 242\text{K}$.

When a blue envelope is used, a chromatic balance is obtained which attenuates the longer wavelengths (reds) and passes relatively more of the shorter wavelengths (blues). A blackbody with a chromatic balance that has relatively less red and relatively more blue would be a blackbody at a higher temperature. It is for this reason that the color temperature specification is larger for the lamps with blue bulbs.

4-6 COLOR TEMPERATURE ESTIMATION

Many optoelectronic devices are characterized photometrically using incandescent lamps. If a technician finds it necessary to test these devices, it is important that the test conditions used by the manufacturer be duplicated. The specification that is often overlooked in this kind of setup is the color temperature of the source. A new clear glass tungsten lamp will have a color temperature of about 2850 K. Most testing of photometric devices is done with lamps operated very close to this color temperature.

Figure 4-15 illustrates the change of spectral emittance of a blackbody at two different wavelengths as a function of color temperature. The wavelengths selected are 656.3 nm and 435.8 nm. These wavelengths are both common test wavelengths. As the figure illustrates, the magnitude of the spectral emittance of the blue line approaches the magnitude of the spectral emittance of the red line as color temperature increases. We could find a temperature where the red and blue spectral emittances are the same. The graph of Figure 4-15 is restricted to color temperature values in the normal application range. The major point being made here is that the *apparent* color temperature of a source can be altered by changing the balance of the red and blue ends of the visible range of wavelengths.

The color temperature of an incandescent lamp can be varied in one of two ways. The lamp voltage can be changed, which will alter the lamp's filament temperature and therefore the lamp's color temperature. Filtering devices or system optics can alter the relative amplitudes of the red and blue ends of the visible spectrum and thus change the apparent color temperature. Figure 4-16 is a graph of a ratio calculated by dividing the spectral emittance at the red (656.3 nm) wavelength by the spectral emittance at the blue (435.8 nm) wavelength versus color temperature. This graph can be used to estimate the apparent color temperature caused by a blue filter.

EXAMPLE 4-7

Estimation of the Apparent Color Temperature of a Source Caused by Filtering

Given: **(1)** Clear lamp with a color temperature of 2800 K.
(2) Filter transmission at 435.8 nm is 0.90.
(e) Filter transmission at 656.3 nm is 0.55.

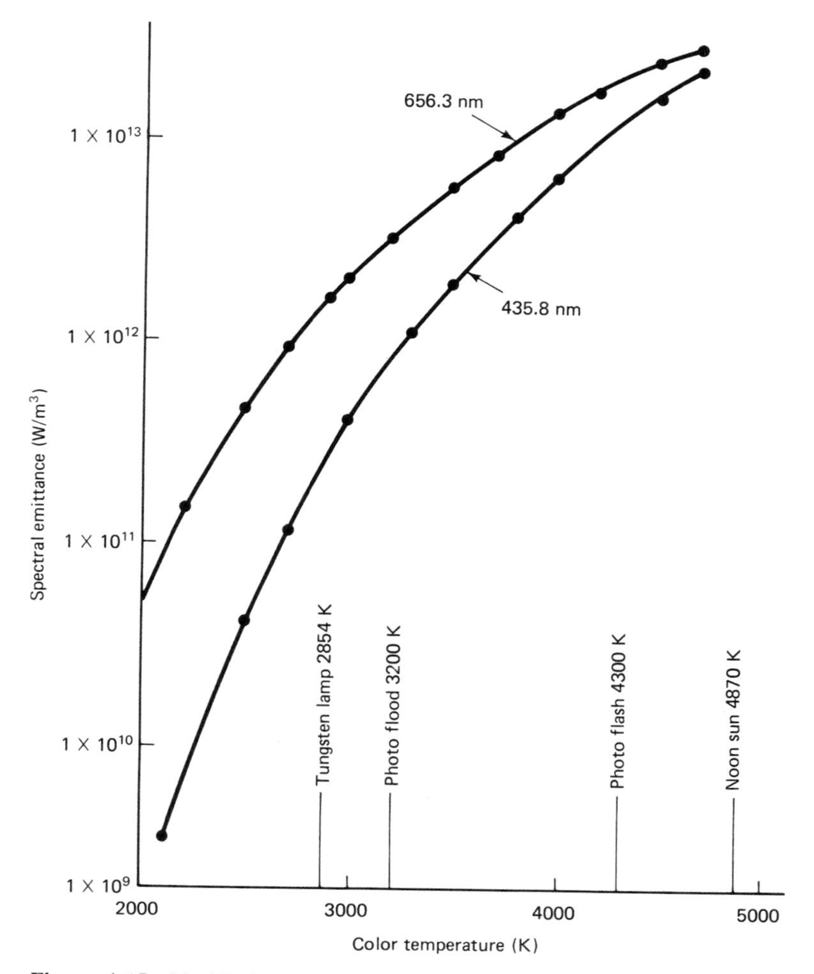

Figure 4-15 Blackbody spectral emittance at two wavelengths versus color temperature.

Find: The apparent color temperature of the source caused by the filter.

Solution: From Figure 4-16 the red-to-blue ratio at 2800 K is 6.70. Calculation of the red-to-blue ratio after the light passes through the filter:

$$\frac{6.70}{1} \times \frac{0.55}{0.90} = 4.09$$

To determine the color temperature, read the temperature corresponding to the calculated red-to-blue ratio of 4.09 from the graph of Figure 4-16. The answer is: T $\simeq$ 3250 K.

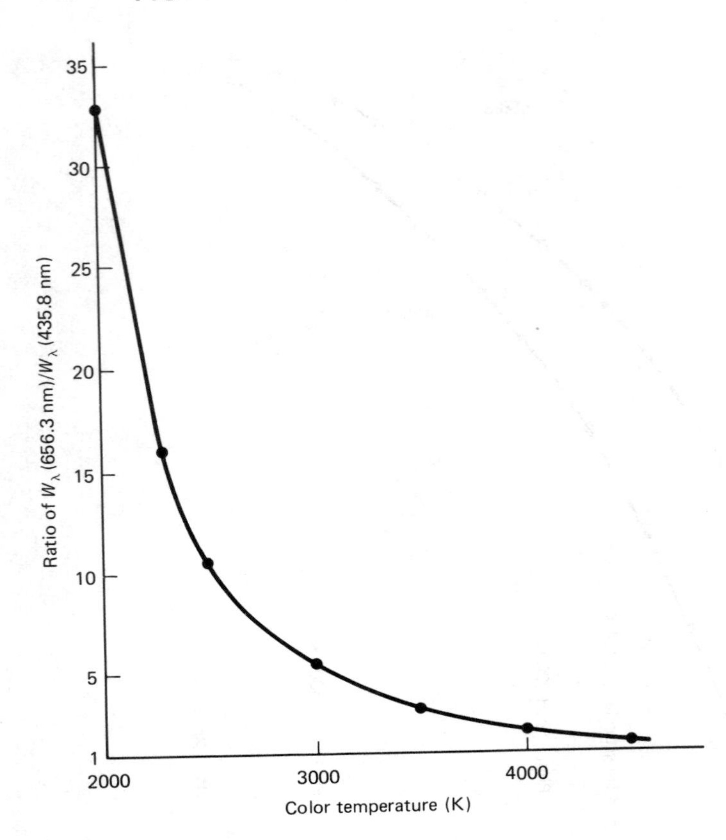

Figure 4-16 Ratio of the spectral emittance of a blackbody at two different wavelengths versus color temperature. $\lambda 1$, 656.3 nm; $\lambda 2$, 435.8 nm. ratio $= W_{\lambda 1}/W_{\lambda 2}$.

Example 4-7 is an illustration and review of the meaning of color temperature. The color balance between the red and blue ends of the visible spectrum determines a source's color temperature. The numerical value assigned to the color temperature is that temperature that a blackbody would be at with the same red–blue balance. Apparent color temperatures of a source can be changed by filtering to affect the source's red–blue balance. A blue bulb alters the color balance and the apparent color temperature is increased. High-color-temperature bulbs simulate daylight better do than lower-color temperature bulbs. To control color temperature, both source voltage and optical filtering must be controlled.

4-7 SUMMARY

In this chapter photometric units were introduced and the relationships between photometric and radiometric measurements at 555 nm were defined. The color temperature specification was discussed and its dependence on the color balance of the source output explained. Various lamps, lamp specifications, and formulas for predicting lamp performance caused by changes in lamp voltage were presented. Many of the ideas and concepts introduced in this and the preceding chapters will be used in subsequent chapters and in actual laboratory work. Especially important are the differences and similarities that exist between radiometric and photometric measurements, the fundamental electrical concepts of voltage, current, and power, and the effect of narrow-wavelength-range instruments on radiometric data.

PROBLEMS

1. A 555-nm source is delivering 17 mW to a detector, What is the value of the luminous flux intercepted by the detector measured in lumens?
2. How many lumens would a 555-nm source have to put out to develop 50 mW of radiant flux?
3. How many lumens does a 10-mW 470-nm source develop?
4. How much more power does a 450-nm source have to put out to develop the same brightness as that of a 560-nm source?
5. A 555-nm source is delivering 10 mW/cm^2 to a photometric detector that reads 6.83 lm/cm^2. How much flux density would a 680-nm source have to deliver to the same detector to obtain the same reading?
6. A point source develops 2 cd. What is its illuminance at 1 m? Compute the units of lm/cm^2.
7. The irradiance of a source measured at 1.5 m is 10 $\mu W/cm^2$. What is the source radiant intensity?
8. Figure 4-3 illustrates the luminous efficiency in lm/W versus temperature for a blackbody. The peak value of the curve is 93 lm/W. Practical incandescent lamps have color temperatures in the range 2000 to 3000 K. The efficiency of incandescent lamps will always be less than that of an ideal radiator. What would the absolute maximum values of luminous efficiency be in the practical temperature range?
9. What is the luminous efficiency of the four lamps described in Table 4-9? How do these luminous efficiencies compare with that of a blackbody?
10. The solid angle in steradians formed by a detector can be calculated from the Cartesian angle $\propto$ of the cone, formed by the detector as illustrated in Figure 4-7. A flat circular detector with a 1-cm^2 area is used at 1 m from a point source.
    (a) What solid angle does it intercept?
    (b) What Cartesian angle does it intercept?
11. The luminous intensity of a source changes significantly with each $1°$ change of Cartesian angle off-axis. What is the closest useful distance for measurement of luminous intensity with a 1-cm^2 detector?
12. Would the values of luminous intensity calculated based on the flux measurements of a detector placed very close to a source that creates an expanding beam be higher or lower than the correct value?
13. Derive the equation relating color temperature to voltage in a form such as that used for current and illumination. (*Hint:* Refer to the derivation of the power formula (4-19) in Section 4-4; assume that e is constant.)
14. Determine lamp current, MSCP, luminous flux, and color temperature for lamp D in Table 4-9 by calculation at 90% of rated voltage.
15. Determine lamp current, luminous flux, color temperature, and power for lamp B in Table 4-9 by calculation at a line voltage of 110% of rated.

16. If the voltage supplied to lamp A in Table 4-9 is decreased to 80% of its original value, find the new current power, resistance, color temperature, and actual filament temperature.

17. When operated at rated voltage and current, lamp C of Table 4-9 gives a measured luminous intensity of 2.41 lm/sr measured at $\frac{1}{2}$ m and a radiant intensity of 200 mW/sr on the same axis at the same distance. Determine the flux densities at 10-cm intervals out to 1 m. Start calculations at a distance of $\frac{1}{2}$ m. Calculate the ratio of E_v/H_e at each interval.

18. Use the same initial setup as in Problem 17. Imagine that a detector is fixed at the initial location and that the lamp voltage is reduced in $\frac{1}{2}$-V steps to a level of 2V. Calculate lamp current, lamp power, luminous intensity, and radiant intensity for each value of voltage.

19. Refer to the flashlamp specifications in Table 4-10. Calculate the luminous energy in lm · s for lamp type 2B. Use the time interval between the 50% intensity points in the calculation. Compare to the data sheet value.

20. Refer to the flashlamp specifications in Table 4-10. Calculate the luminous energy in lm · s, using the time interval between the 50% intensity points for the clear bulb, type 2. Compare to the specified value.

5

Gas Lamps

5-1 INTRODUCTION

In this chapter we describe briefly the physical processes of radiation and absorption of electromagnetic waves by gases and the interaction of radiation with solids. The physical principles used in these descriptions are drawn from the branch of physics called atomic or modern physics. Most of these principles were developed over a short period of time, from 1880 to 1910, by a small handful of scientists seeking to find ways to relate and describe various laboratory observations of interactions between radiation and gases, and radiation and solids. As is the case with all physics, the theories postulated and equations developed by these scientists provide tools that will predict the behavior of these systems.

In the years since these basic principles were developed, they have been used as the basis for new technological areas, such as atomic energy, semiconductors, and lasers. Most subsequent physics research has been involved in refining the original theories and equations and in developing clearer descriptions of many second- and third-order effects. Most of the images used in describing these phenomena are at best just convenient mental images that may or may not correspond to what actually occurs. Fundamental to these discussions is the relationship between matter and radiant energy, and there are contained in these discussions apparent contradictions to which no one at this time has the answers.

This chapter does not pretend to be in any way an exhaustive treatment of modern physics. The purpose of the chapter is to develop a set of images and concepts that will permit a functional understanding of gas lamps, gas lasers, and photovacuum photodetectors. Gas lasers and vacuum photodetectors are discussed in subsequent chapters.

5-2 SOME PHYSICAL OBSERVATIONS

In this section we describe two of these physical phenomena, the observation of which led to the development of the basic theories of atomic physics. All of these observations concern the interaction of radiant energy and the atomic particle called the *electron.* The two effects that we discuss are the photoelectric effect and the radiation absorption characteristics of low-pressure gases.

The *photoelectric effect* is a process in which electrons are emitted from a solid, usually a metal or an oxide, when the material is illuminated by radiation. This phenomenon was first observed in 1887 by Heinrich Hertz. The photoelectric process exhibits three basic characteristics:

1. The number of electrons emitted, which determines the amount of electron current flowing, is proportional to the intensity of radiation with a fixed wavelength.
2. Each material has a threshold wavelength of radiation. If the radiation has a wavelength longer than the threshold wavelength, no electrons are emitted.
3. The maximum velocity of the emitted electrons is independent of the radiation intensity, rather, the maximum velocity of the emitted electrons is inversely proportional to the wavelength of radiation. The shorter the wavelength of the radiation, the greater the velocity of the emitted particle.

Observations 2 and 3 cause some conceptual problems. The kinetic energy of a moving particle is directly proportional to the square of its velocity. Observations 2 and 3 indicate that the energies of the emitted electrons are not related to the total radiant energy but rather, to the radiant energy's wavelength. The increase in current noted in observation 1 is not due to the emission of particles with higher kinetic energy but rather, just to the presence of a greater number of emitted particles. Einstein, Millikan, and Planck all struggled with these observations and their analysis resulted in a new conceptual particle called the *photon* or *light quantum.*

Two mathematical equations were developed to describe these effects. These scientists conceived of radiation as consisting of small discrete packets of energy (photons) moving at the speed of light and having energies directly related to their frequency. The amount of energy contained in a photon is related to its frequency and therefore its wavelength by equations (5-1) and (5-2).

$$E = hf \tag{5-1}$$

where E is the photon energy, in joules
h is Planck's constant, 6.624×10^{-34} J · s
f is the frequency, in hertz

Equation (5-1) can also be expressed in terms of wavelength, where $f = \frac{c}{\lambda}$. Here c is the velocity of light, 3×10^8 m/s, and λ is the wavelength in free space, in meters.

Therefore,

$$E = \frac{hc}{\lambda} \tag{5-2}$$

$$= \frac{19.87 \times 10^{-26}}{\lambda} \text{ J}$$

When these photons of radiant energy hit the surface of the illuminated material, they transfer either all or none of their energy to the material's electrons. After the energy transfer the photons cease to exist. If an energized electron moves with enough energy in the correct direction, it can escape from the material. This process is described by

$$\tfrac{1}{2}mV^2 = hf - W \tag{5-3}$$

where $\frac{1}{2}mV^2$ is the kinetic energy of the energized electron
h is Planck's constant
f is the frequency of the photon, in hertz
W is the energy required to escape from the material surface, called the work function

Equation (5-3) can be rewritten in terms of wavelength using the same substitutions that were used to derive equation (5-2).

$$\tfrac{1}{2}mV^2 = \frac{hc}{\lambda} - W \tag{5-4}$$

The idea of a quantum is dramatically different from the usual mental images of radiation as a continuous wave or beam. How can radiation behave like a continuous electromagnetic wave when tested in lens and slit systems, and then behave like a particle when tested in electro-optical systems? However it is possible, this model is the key to visualizing the operation of most electro-optical devices.

The second set of phenomena that were examined by these scientists were those observed when low-pressure gases were subjected to radiation or when they were made to radiate. When a beam of white light is passed through a container holding a gas at low pressure and the resulting beam is observed with a wavelength-measuring instrument called a *spectrometer,* it is found that some wavelengths are significantly attenuated. This observed phenomenon is called *selective absorption.* If a low-pressure gas is made to radiate and the resulting light is analyzed with a spectrometer, it is found to consist of energy concentrations at very specific wavelengths and with little or no detectable energy at intervening wavelengths.

The wavelengths at which radiation and absorption take place are essentially the same for any given gas at a given pressure. Each specific gas has a characteristic array or pattern of wavelengths. If gas pressure is increased, radiated or absorbed energy will tend to cover a broader range of wavelengths and the value of the peak or principal wavelengths will change slightly.

It is apparent that there is something unique about the structure of each gas that results in its particular spectral pattern. The explanation proposed involves using the Bohr model of the atom and the equation for radiation quantum in modified form. The *Bohr model* of the atom envisions each discrete atom existing in isolation, with an orbital system of electrons surrounding a central nucleus consisting of protons and neutrons. In a low-pressure gas the atoms will tend to behave in the manner predicted by the Bohr model. Figure 5-1 represents the Bohr model of an isolated hydrogen atom.

A hydrogen atom consists of a single proton and a single orbital electron. When the atom is unexcited by any external energy, the electron will orbit about the proton in level $N = 1$ at radius r_0. This unexcited condition is called the *ground state* of the atom. For any orbital position of the electron the atom has an ionization energy given by

$$E_I = \frac{E_0}{N^2} \tag{5-5}$$

where E_I is the ionization energy
E_0 is a reference energy constant
N is the orbital number

When $N = 1$

$$E_I = E_0$$

If an electron absorbs an amount of energy equal to or greater than the ionization energy E_I, the electron will escape from the atom and the atom will become a positive ion. Energy increases having values less than E_I may result in the electron assuming a new orbital position at a greater distance from the nucleus. The electron cannot assume just any orbiting radius as it could in a gravitational system. In a Bohr atomic system there are only specific discrete permitted energy levels and radii. Figure 5-1 illustrates four of these permitted energy levels and radii for hydrogen, and equation (5-5) illustrates the relationship between the ionization energy of an electron and any permitted orbital level. The energy required for ionization when an electron is at a higher-number energy level will be less than the energy required when the atom is in its ground state. For this reason atoms with electrons in higher numerical orbital positions than those of the ground state are said to be in higher-energy states or to be excited. If it takes less energy to ionize the atom, the electron must have a higher original energy of its own. Each permitted energy level has its own specific ionization energy, and the difference between the individual ionization energy levels is the amount of energy the electron must absorb to make the transition from one permitted level to another. This is illustrated by

$$\Delta E = E_a - E_b$$

$$\Delta E = \frac{E_0}{N_a^{\,2}} - \frac{E_0}{N_b^{\,2}} \tag{5-6}$$

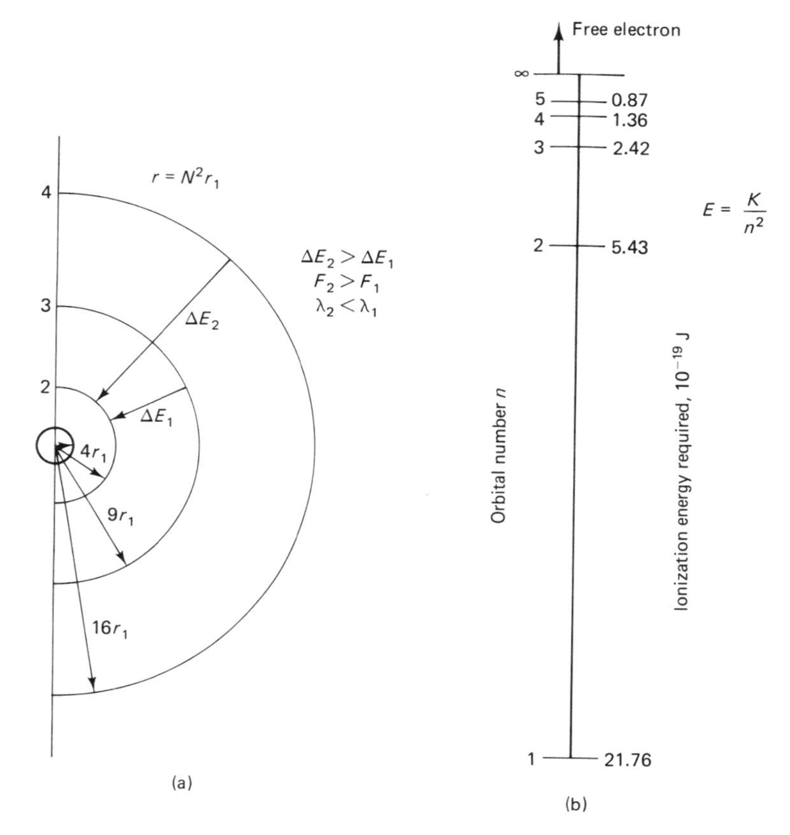

Figure 5-1 Permitted orbits and ionization energy of electrons in the Bohr model of a hydrogen atom: (a) electron orbits; (b) ionization energy required. Two possible energy changes, $\triangle E_1$, and $\triangle E_2$, shown.

where ΔE is the change in energy between levels a and b
E_a is the ionization energy of level a
E_b is the ionization energy of level b
N_a is the orbital number of level a
N_b is the orbital number of level b

Short of ionization the orbital electron of a Bohr atom can absorb only those quantum of energy that will cause it to assume specific permitted energy levels.

If an electron is subjected to no additional energy inputs after it has been raised to an excited state by absorption of energy, it will of its own accord fall back to the ground-state orbit. The amount of time an excited electron will stay in its excited state is a function of the material and the energy level. As an electron falls back to the ground-state energy level it will give off energy. An electron may fall directly back from the elevated level to level 1 or it may fall back by moving between individual intervening levels. For example, an electron that is raised to level 4 may follow any of the paths listed in Table 5-1. These four possible paths give rise to six different possible values of ΔE, as shown in Table 5-2.

TABLE 5-1 POSSIBLE TRANSITION PATHS FOR AN ELECTRON IN LEVEL 4

Path	States
1	4 to 1
2	4 to 3 to 1
3	4 to 3 to 2 to 1
4	4 to 2 to 1

TABLE 5-2 SIX POSSIBLE VALUES OF ΔE FOR AN ELECTRON RETURNING TO THE GROUND STATE FROM LEVEL 4[a]

Levels Changed	Quanta Energy (ΔE)
4 to 1	E_0 (135/144)
3 to 1	E_0 (128/144)
2 to 1	E_0 (108/144)
4 to 2	E_0 (27/144)
3 to 2	E_0 (20/144)
4 to 3	E_0 (7/144)

[a] Arranged in descending order of magnitude.

An electron returning to the ground state from level 4 can do so only by four possible paths, and in doing so it will emit one or more of the six possible quantum of energy. Similarly, the electron can only have been raised from level 1 to level 4 by one of the four possible paths, and in doing so it can only have absorbed one or more of the six possible quantum of energy.

The energy path processes illustrated by Tables 5-1 and 5-2 provide a schematic way of thinking about or explaining the interaction of low-pressure gases and radiation. A low-pressure gas will absorb from the white light only those quantum of energy that correspond to those values of ΔE permitted by Bohr atom theory. This idea is summarized and related to wavelength in equation (5-7).

$$E = \frac{hc}{\lambda} \tag{5-2}$$

$$\Delta E = E \text{ absorption or radiation}$$

Therefore

$$\Delta E = \frac{hc}{\lambda}$$

and

$$\lambda = \frac{hc}{\Delta E} \tag{5-7}$$

As equation (5-7) illustrates, each permitted electron-level transition has a single specific wavelength that will cause it to occur. When the low-pressure gas is subjected to the white light, it will absorb those wavelengths that correspond to permitted energy transitions. The emission properties follow the same general line of reasoning. If, for example, an electric current is created within the gas-filled enclosure, the electrical kinetic action within the current path will cause electrons to be raised to higher energy levels. As these elevated electrons return to the ground state, the excited gas will radiate energy at the permitted wavelengths. While all of the permitted transitions can and do take place, there are a few that are more likely to occur, and these tend to give the excited gas a characteristic color corresponding to the most likely or dominant wavelengths.

This discussion has proceeded from the assumption that the atoms being irradiated exist in isolation. This is a reasonable and workable model for low-pressure gases containing a single element. If gas pressure is increased or if the gas consists of molecular combinations of atoms, the result will be slightly different. When pressure is increased or when molecules are present, the atomic structure does not consist of discrete levels of energy but rather of bands of energy, each band having tightly grouped levels. For modeling purposes you can think of modifying the Bohr model of Figure 5-1 so that each level is replaced by a band (that is, a wider region) of energy. When the energy radiated or absorbed by this type of structure is analyzed with a spectrometer it will be found to have energy concentrations that cover a range of wavelengths rather than just discrete single wavelengths.

5-3 GAS LAMP OPERATION

Gas lamps are used both for illumination and as components in various laboratory test situations. A common laboratory application of a gas lamp is as a source for the optical surface testing process called *test plating*. In this process a source is selected that is monochromatic, or which has one visible wavelength that is so dominant that it appears to be monochromatic. This monochromatic light is used to illuminate the interface between a reference surface and the surface being tested. A visible interference pattern of dark and light lines is established which allows the operator to determine how closely the tested surface conforms to the reference surface. This technique can be used to evaluate both flat and curved polished surfaces. When using this method to evaluate a flat surface, the test plate would be a plane-parallel plate with one or both sides ground and polished to a very high degree of flatness. After carefully cleaning the reference surface of the test plate and the surface of the workpiece, the test plate reference surface is placed in contact with the polished surface of the workpiece. When evaluating curved surfaces the test plate used will be ground and polished to the desired radius of curvature. A concave test plate will be used to evaluate convex surfaces, and a convex test plate will be used to evaluate concave surfaces. For this technique to work effectively, the surface being tested must be highly polished and it must have dimensions that deviate from the test plate surface by only a few wavelengths. The spectral "lines" from a gas lamp can also be used to calibrate and test spectral measurement systems.

All gas lamps are electrical discharge devices in the sense that current flow is established through the gas. These lamps all have a very similar set of components. Gas lamps differ from one another in shape, size, power output, and the nature of the gas and therefore in the output spectrum. Gas lamps can be powered by ac or dc supplies. Direct-current supplies tend to be more expensive and are used only when strict stability of output is required. Some ac supplies are said to be "regulated," which usually means that the magnetic and resistive components of the supply have been fabricated in such a way that the output current of the power supply will remain essentially constant with changes of ac power line voltage in the normal range.

The basic components of the gas lamp are (1) the envelope; (2) the anode, which is the positive electrode; (3) the cathode, which is the reference electrode; and (4) the gas. The sequence of events in gas lamp operation is essentially the same regardless of lamp type:

1. A high-voltage potential is applied across the lamp.
2. The high initial voltage causes the ions and electrons in the gas to accelerate and collide at high kinetic energies.
3. The collisions cause the creation of additional ions resulting in the formation of a radiating discharge.
4. The discharge current reaches a specified value, and the potential across the lamp is reduced to a rated voltage.

There is usually a noticeable time period between the application of the high voltage and the establishment of a stable discharge between the lamp's electrodes. If stable illumination out of the lamp is required, it is good practice to wait 30 minutes or so after the discharge has been created before making measurements.

In many cases the gas in the lamp envelope is a low-pressure elemental gas. In other lamps, however, the gas is created in the lamp by heating a liquid or a solid. When a liquid or a solid is placed in an evacuated container, some molecules or atoms will naturally "boil" off the surface, creating a very low vapor pressure. The vapor pressure can be increased by heating the liquid or the solid and therefore increasing the number of atoms that "boil off." Lamps of this type have a reservoir area containing the material to be heated and some sort of heating element. The turn-on time for this type of lamp is usually longer than for plain gas lamps because of the heating portion of the operating cycle. These lamps are often sensitive to the physical position of operation because of the position of a heater and the material. Some sodium and mercury lamps employ a heating cycle. Lamps of this type also may employ a second gas, which is driven into a discharge stage and used to heat the solid or liquid. The transmission properties of an arc lamp envelope will affect the lamp's spectral output, and some lamps are available with a choice of envelope materials and filters that will provide various desired spectral characteristics.

There are four stages of discharge between the electrodes in a low-pressure gas lamp: (1) leakage, (2) Townsend, (3) glow, and (4) arc. The voltage–current characteristics of these four stages of discharge are illustrated in Figure 5-2.

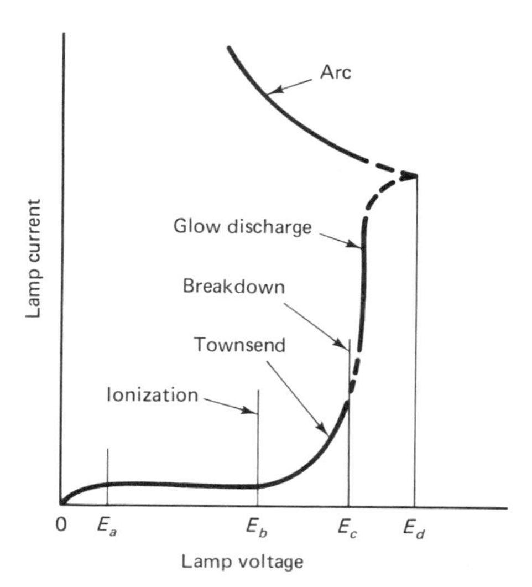

Figure 5-2 Stages of discharge in a low-pressure gas.

In the *leakage stage* the electronic particles contributing to conduction are free electrons and gas ions formed spontaneously in the gas through energy absorption from radiation outside the lamp. As voltage between the electrodes is increased from zero volts to voltage E_a, more and more of these free particles become involved in conduction. For voltages between E_a and E_b, essentially all the available free particles are involved, and so conduction remains constant at a very low value of constant current.

The voltage level E_b is the *ionizing voltage.* At this voltage level, ions and free electrons start being produced and the frequency of "collisions" between electrons and atoms increases. This process gives rise to the *avalanche effect,* where significant increases in current occur with small increases in voltage. In the voltage range between E_b and E_c the internal distribution of potential changes as terminal voltage is increased. When the lamp terminal voltage is equal to E_b (the ionization potential) the potential distribution is essentially linear, as illustrated in Figure 5-3. As the terminal voltage approaches voltage E_c, the greatest change in voltage occurs in the space close to the cathode. This very high voltage gradient imparts high acceleration to the positive ions toward the cathode and enhances emission of electrons from the cathode. When these processes become dominant the tube is said to be at the *breakdown* or *ignition potential.* The region of operation between terminal voltages E_b and E_c is called the *Townsend discharge region.*

In the terminal voltage range between E_c and E_d the lamp operates in the glow discharge stage. The *glow discharge* is a stage of conduction that follows the application of the breakdown voltage. The name of this stage of operation derives from the soft luminous glow of the gas in the lamp. Neon signs are examples of lamps operated in the glow discharge stage. This stage of operation is quite unstable and requires some sort of current regulation to be maintained. Figure 5-4 illustrates typical potential distributions and radiation distribution as a function of the distance between the cathode and the anode of the lamp. The radiation emitted from the gas is relatively uniform in intensity over the length of the lamp.

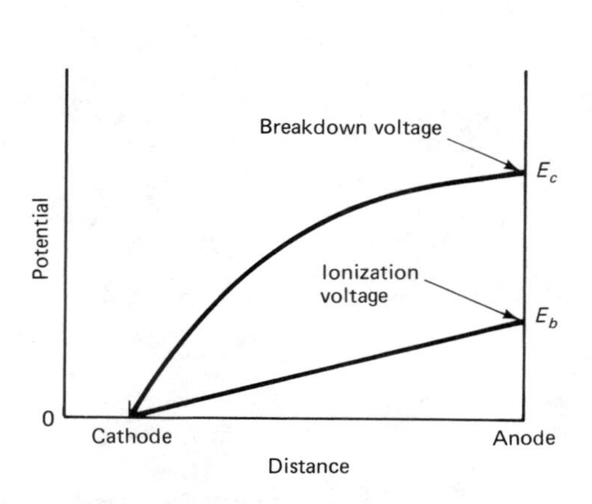

Figure 5-3 Potential distribution inside a gas lamp in the Townsend discharge stage of operation.

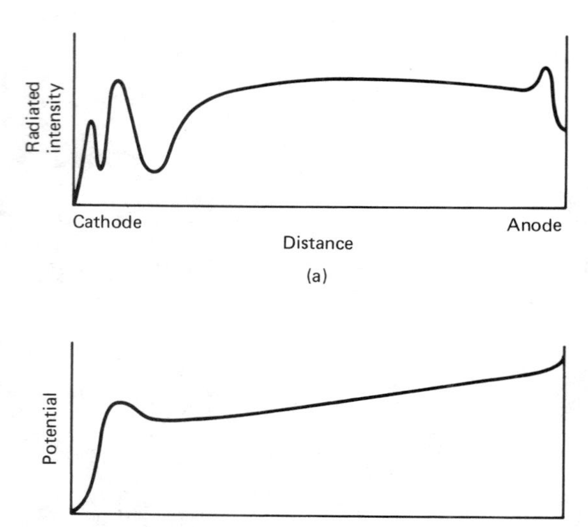

Figure 5-4 Glow discharge in a column of gas: (a) radiated intensity;
(b) potential.

The last stage of operation is the *arc*. An arc discharge has a high current density and a high temperature. It is a bright source of radiation with significant incandescence, and it has a negative resistance characteristic. A device is said to exhibit negative resistance when the terminal voltage decreases as the current increases.

The four stages of operation of a gas lamp can be characterized by their electrical terminal characteristics:

1. *Leakage:* increasing voltage, constant current
2. *Townsend:* increasing voltage and current, nonlinear
3. *Glow:* constant voltage, increasing current
4. *Arc:* decreasing voltage, increasing current

Practical lamps are operated in either the glow discharge or the arc region. The lamp manufacturer controls the lamp voltage and current characteristics by controlling lamp length, diameter, gas pressure, and gas type. The gas type used also determines the dominant spectral lines of the lamp.

For any selected gas pressure there is an electrode spacing that will yield a minimum breakdown voltage E_c. If the electrode spacing is made larger or smaller than this critical spacing, the magnitudes of the breakdown voltage will increase. Figure 5-5 illustrates the relationship that exists between breakdown voltage and electrode spacing for a fixed gas pressure. A graph with a similar shape would result if breakdown voltage were plotted against increasing pressure with electrode spacing fixed. The spacing that yields a minimum breakdown voltage is a function of the

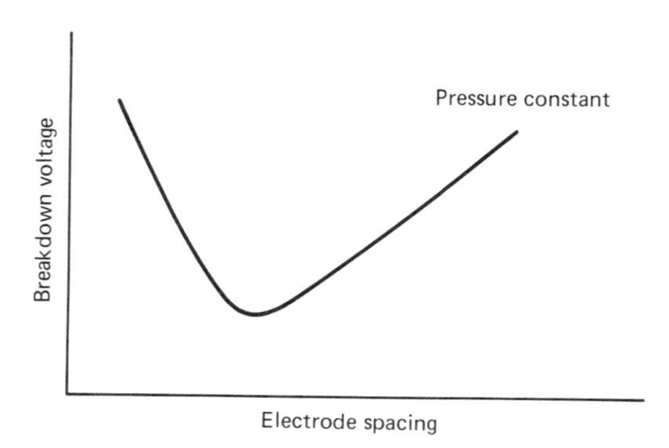

Figure 5-5 Breakdown voltage versus electrode spacing for a gas lamp.

average distance a molecule can travel in the gas before a collision occurs. This average distance is called the *mean free path.* Minimum breakdown voltage will occur at a physical distance between electrodes of about one mean free path. As gas pressure decreases, the mean free path will increase since fewer molecules are present. The mean free path for electrons is about six times as long as the mean free path for molecules.

In this section the general operating characteristics of gas lamps have been discussed. It has been shown that the electrical terminal characteristics and the radiation characteristics are established by the mechanical geometry of the lamp and the gas pressure. Lamps operate in one of two stages of lamp discharge: the glow stage, which exhibits an electrical characteristic of constant terminal voltage with increasing current, and the arc stage, which exhibits a negative resistance characteristic. The glow discharge stage of operation emits radiation with well-defined spectral lines and low incandescence. The arc discharge stage of operation emits radiation with a high level of incandescence on which are superimposed the spectral lines characteristic of the gas. A lamp operating in arc discharge will be much brighter and consume much more power than the same lamp operating in glow discharge.

5-4 PRACTICAL LAMPS

Most vendors differentiate between gas discharge sources and arc lamps. A *gas discharge source* is a lamp containing a low-pressure gas through which an electric current passes. The radiation produced is in the form of discrete lines of radiation; these lines are characteristic of the gas used. Figure 5-6(a) and (b) are the data sheets for two typical low-pressure gas discharge sources. As illustrated by the graph of typical irradiance in Figure 5-6(b), these lamps produce a well-defined spectral output. The 253.7-rm line is clearly dominant. This type of lamp is operating in the glow stage.

Arc lamps are different from gas discharge sources in three major ways. The pressure in the arc lamp is high, often reaching 60 atm or more in normal operation. The current density of the arc is very high, forming a hot plasma that yields a significant incandescent output in addition to the energy in spectral lines. The average luminous flux radiated by an arc lamp will be much higher than it will be for

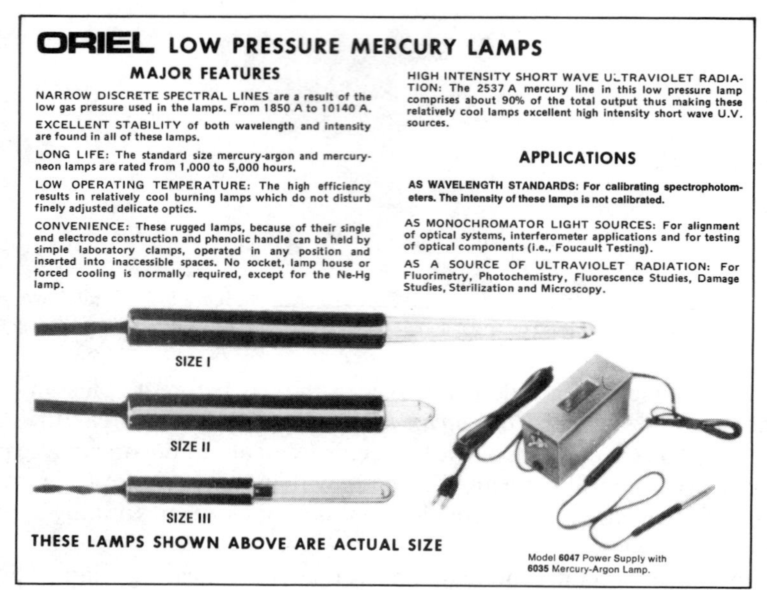

(a)

LOW PRESSURE MERCURY LAMP SPECIFICATIONS

MODEL NO.	SYMBOL	FILLING MATERIAL	PRICE	NOMINAL OPERATING CURRENT MA. A.C.	DISSIPATED WATTAGE APPROX.	ENVELOPE SIZE CLASS	ENVELOPE O.D. (MM.)	ENVELOPE LENGTH (IN.)	DIAMETER HANDLE (IN.)	OVERALL LENGTH (IN.)	RATED LIFE HOURS	REQUIRES POWER SUPPLY NO.
6035	*Hg(A)	Mercury(Argon)	**$79.00**	17	4.6	I	6.5	2⅛	⅜	4⅝	5000	**6047/6048**
6036	Hg(A)	Mercury(Argon)	**$83.00**	17	4.6	II	6.5	½	⅜	3	5000	**6047/6048**
6037	Hg(A)	Mercury(Argon)	**$150.00**	6	1.6	III	4	1	.235	2⅛	1000	**6043/6044**
6034	*Hg-Ne	Mercury(Neon)	**$124.00**	17	4.6	I	6.5	2⅛	⅜	4⅝	5000	**6047/6048**

Shipping weight on all lamps is 1 lb.

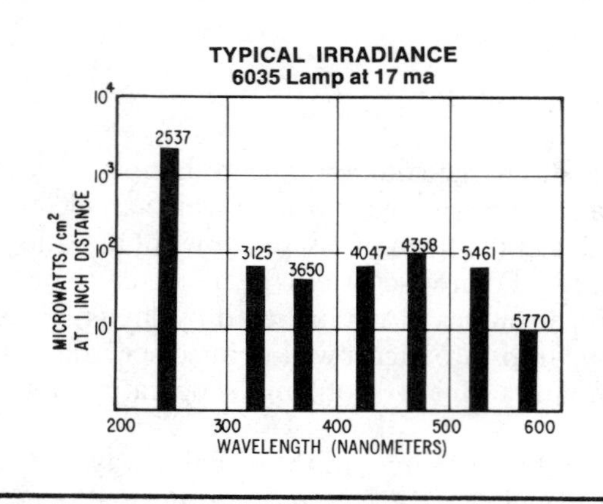

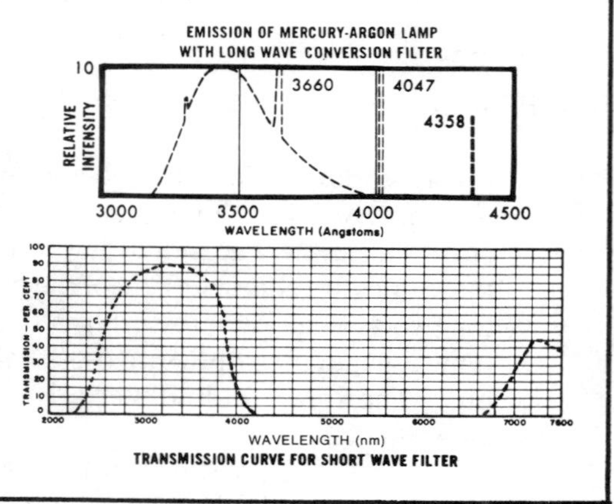

(b)

Figure 5-6 Low-pressure mercury lamps: (a) features and applications; (b) specifications. (Courtesy Oriel Corporation, Stratford, CT.)

a gas discharge source, and of course the electrical power consumed will also be much higher. Figure 5-7(a) and (b) present some typical data on arc lamps. Figure 5-7(a) gives a brief description of arc lamp operation and a tabulation of some typical specifications. Figure 5-7(b) illustrates spectral irradiance as a function of wavelength showing the pronounced spectral peaks superimposed on an incandescent irradiance level that is comparable in intensity to that from a quartz halogen incandescent lamp.

Arc lamps and gas discharge sources are examples of systems that provide a radiation output as the result of passing an electric current through a gas. Both types of sources have prominent lines or wavelengths of radiation. The arc lamp, in addition, supplies an appreciable amount of incandescent radiation. The gas in the lamp is the primary determinant of the lamp's spectral output. A Bohr atom–like process can be used to conceptualize the interaction of the gas with the electrical power and to justify the spectral output.

The last type of gas lamp of interest is the flashtube or flashlamp. Flashlamps are often used as components in laser systems. The major difference between flashtubes and the lamps just discussed is that flashtubes are turned on or off repeatedly. It is usually desired that the tube turn on rapidly, and a number of power supply and triggering schemes are used to achieve this end. A simple and common method of doing this is to keep a constant bias voltage applied across the tube near the breakdown voltage level. A pulse of voltage is then applied, which drives the tube into the radiating region.

Occasionally, a coil of wire is wrapped around the axis of the tube and subjected to a current pulse. The current pulse causes a magnetic pulse that accelerates some of the randomly moving particles in the tube in the axial direction. This increases the current flow and drives the tube into radiation.

Combinations of constant bias and magnetic pulse schemes are common. When a flashlamp is biased with a constant current and voltage it is said to be in "simmer" operation, rather like a pot that is going to boil over with just a little more heat. The lamp will flash with just a little increase of current caused by a magnetically generated force or an increase in terminal voltage.

5-5 SUMMARY

In this chapter radiation was presented as consisting of a stream of energy packets called photons. The amount of energy contained in a photon is a function of its wavelength. Atoms can absorb photons that will cause their electrons to change energy levels or to escape from an atom or a material. Electrons that remain within a given atom can absorb only photons, which have energies equal to the energy differences between the atom's permitted energy levels.

Atoms that have electrons at permitted energy levels above a reference level called the ground state are said to be excited. When the electrons of an excited atom return to the reference condition a photon will be emitted that has an amount of energy equal to the change in energy that the electron experiences. The emitted

ORIEL SHORT ARC LAMPS
XENON MERCURY and MERCURY (XENON)

These short arc lamps have two electrodes whose tips are closely spaced surrounded by Xenon and/or Mercury vapor at high operating pressures contained in a quartz envelope.

To start these lamps a very high voltage is required to ionize the gas and create a conduction pate. An arc is then established and maintained by a low voltage D.C. currrent flow.

The high pressures and corresponding high molecular densities allow the dissipation of large amounts of power in a small space. The light source thus created is extremely small and intense and is useful for any application requiring a high intensity on a small target or a highly collimated (parallel) beam.

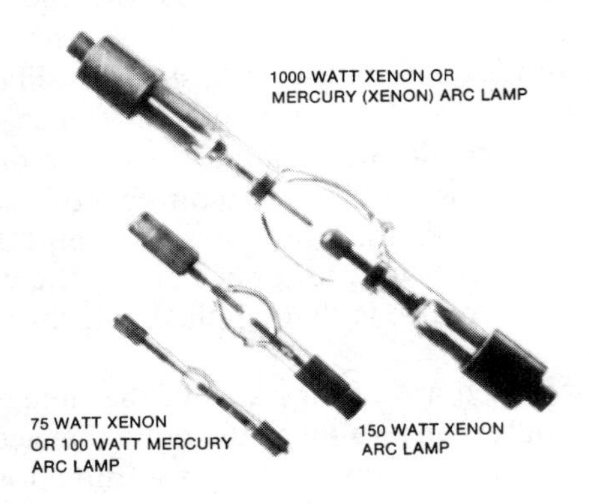

XENON ARC LAMPS

Xenon gas fills these lamps to a few atmospheres pressure when cold raising to 50-70 atmospheres when hot. The output spectrum combines thermal radiation from the glowing plasma with the Xenon line spectrum superimposed. The ultraviolet visible spectrum approximates a 6000°K color temperature continuum. The strongest Xenon lines lie between 0.75 and 1.0 micrometer followed by a declining continuum in the infrared.

The quartz envelope cuts off at about 0.19 micrometers in the U.V., absorbs between 2.5 and 2.8 and cuts off at about 4 micrometers in the infrared. The ozone free lamps begin to absorb at 0.3 and cut off at 0.25 micrometers in the ultraviolet.

See pages D-30, 31 for spectral irradiance curves.

MERCURY AND MERCURY (XENON) ARC LAMPS

When cold these lamps have a small pool of liquid mercury and a small amount of a rare gas, often argon. The rare gas is necessary to start the lamp and establish the arc. The heat generated then vaporizes the mercury until at operating temperature all of the mercury is vaporized. Starting from near atmospheric pressure when cold these lamps operate at 50-70 atmospheres when hot. Generally they take from ten to fifteen minutes to warm up and reach their final operating pressure.

The output spectrum is a series of mercury lines in the U.V. and visible from 0.24 to 0.7 micrometers followed by a continuum in the infrared to 2.5 micrometers.

The *mercury (Xenon) lamps* have a small amount of Xenon as a starting gas. The Xenon also speeds warm up, increases stability and lengthens lamp life.

This design is generally preferred for Mercury lamps of 1000 watts or more.

The output spectrum is similar to a mercury arc in the UV and visible regions and shows some weakened Xenon lines in the IR.

See pages D-30, 31 for spectral irradiance curves.

Wattage and Type	Oriel Model	Equivalent To	Voltage DC	Current (Amps)	Horizontal Intensity (CD)	Flux (Lumens)	Average Brightness (CD/mm²)	Effective Arc Size (mm) W x H	Average Life (Hours)	Bulb Diameter (mm)	Price
XENON ARC LAMPS											
75 Xe	6251	Osram XBO 75W/2	14	5.4	100	1000	400	0.25 x 0.5	400	10	**$161.00**
150 Xe	6253	Osram XBO 150W/1	20	7.5	300	3000	150	0.5 x 2.2	1200	20	**$205.00**
UV150 Xe	6254	Osram XBO 150W/4	20	7.5	300	3000	150	0.5 x 2.2	1200	20	**$340.00**
150 Xe	6256	Hanovia 901-C1	17-23	8.5-6.5	300	2200	96	.75 x 1.5	1000	20	**$134.00**
450 Xe	6261	Osram XBO 450W	18	25	1300	13000	350	.9 x 2.7	2000	29	**$462.00**
UV 450 Xe	6262	Osram XBO 450W/4	18	25	1300	13000	350	.9 x 2.7	2000	29	**$674.00**
*500 Xe	6265	Hanovia 959-C98	14-20	25-35	1500	9000	3500	0.3 x 0.3*	200	25	**$505.00**
1000 Xe	6269	Hanovia 976-C1	23	43.5	3000	30000	350	1.5 x 3.0	1000	38	**$417.00**
OF 1000 Xe	6271	Hanovia L5179	23	43.5	3000	30000	350	1.5 x 3.0	1000	38	**$429.00**
2200 Xe	6277	Hanovia 491C-1390	20-24	100	8000	81000	600	1.5 x 4.0	1000	57	**$962.00**
OF 2200 Xe	6279	Hanovia L5289-000	20-24	100	8000	81000	600	1.5 x 4.0	1000	57	**$997.00**
MERCURY AND MERCURY (XENON) ARC LAMPS											
100 Hg	6281	Osram HBO 100W/2	20	5	260	2200	1700	0.25 x 0.25	200	10	**$134.00**
200 Hg	6283	Osram HBO 200W/2	47-65	3.1-4.2	1000	10000	300	0.6 x 2.2	400	17	**$138.00**
200 Hg (Xe)	6291	Hanovia 901-B1	20-25	8-9.5	600	4500	190	0.75 x 1.5	1000	20	**$134.00**
350 Hg	6286	Osram HBO 350W	60-75	4.7-5.9	2000	19500	400	0.6 x 2.2	400	20	**$227.00**
500 Hg	6285	Osram HBO 500W/2	67-85	5.9-7.4	2850	30000	300	1.1 x 4.1	400	26.5	**$282.00**
1000 Hg (Xe)	6293	Hanovia 977-B1	30-38	26-33	5000	40000	320	1.5 x 3.0	1000	38.5	**$426.00**
OF 1000 Hg (Xe)	6295	Hanovia L5173	30-38	26-33	5000	40000	320	1.5 x 3.0	1000	38.5	**$490.00**
2500 Hg (Xe)	6278	Hanovia 929B-9U	45-55	45-55	12500	120000	660	2.0 x 5.0	1000	62	**$880.00**
OF 2500 Hg (Xe)	6297	Hanovia L5270	45-55	45-55	12500	122000	660	2.0 x 5.0	1000	62	**$968.00**

OF These lamps have an ozone free quartz envelope which absorbs below 300 nanometers.
UV These lamps have a special U.V. grade fused silica envelope with enhanced output between 185 and 250 nm.
* The 6265 500 watt Xenon lamp has an extremely small and bright arc but this lamp exhibits significant movement of the arc during operation.

(a)

Figure 5-7 Arc lamps: (a) features and specifications; (b) power output spectra. (Courtesy Oriel Corporation, Stratford, CT.)

ORIEL ARC LAMP SOURCE, POWER OUTPUT DATA

The absolute output spectra reproduced here were run in the Oriel Calibration Laboratory using spectral irradiance and thermopile standards traceable to N.B.S. Standards. This data should be regarded as typical and subject to variation from lamp to lamp. The instrument bandwidth is 2 Nanometers.

TO ESTIMATE IRRADIANCE FROM OTHER LAMPS

6261 450 watt Xenon: multiply 1000 watt Xenon by 0.29

6276 2500 watt Xenon: multiply 1000 watt Xenon by 2.2

6278 2500 watt Mercury (Xenon): multiply 1000 watt Mercury (Xenon) by 2.4

POWER OUTPUT IN COLLIMATED BEAM

The units shown on the vertical axis are microwatts per cm²-nm at a distance of 50 cm. from the bare lamp. To convert this to power in the collimated beam from the source multiply by the factors listed below:

Source Model No.	Lens f/no.	To obtain approximate Collimated Beam Power in milliwatts per Nanometer multiply units shown by:
6102	f/1.0	1.15 (F)
6104	f/0.7	2.0 (F)

F gives effect to the lens focal length shortening at shorter wavelengths. This can be taken as 1.0 above 500 nm Below 500 nm it varies as follows:

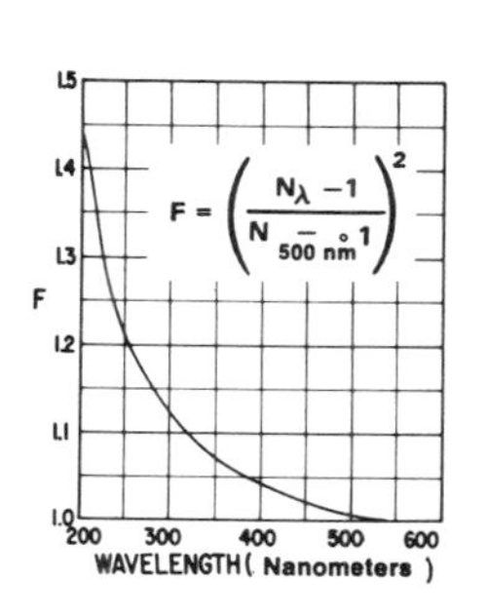

This data does not include power recovered by the spherical reflector which will add 25 to 30%

TYPICAL OUTPUT SPECTRA

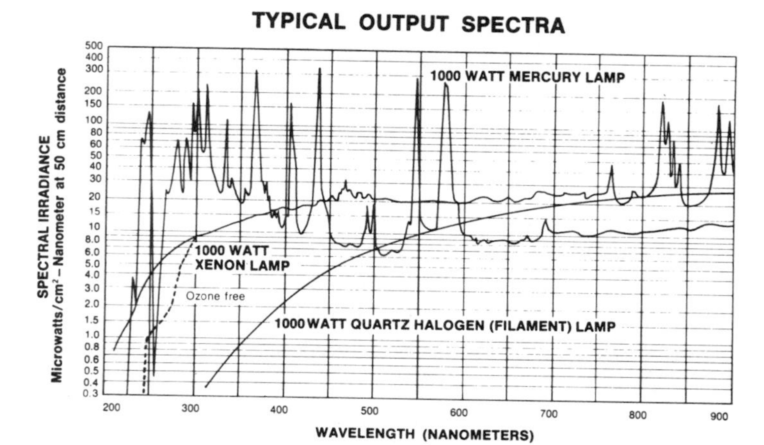

NOTE: The 1000 watt Xenon Lamp has about 15 times the output of the 150 watt Xenon Lamp in the visible and U.V.

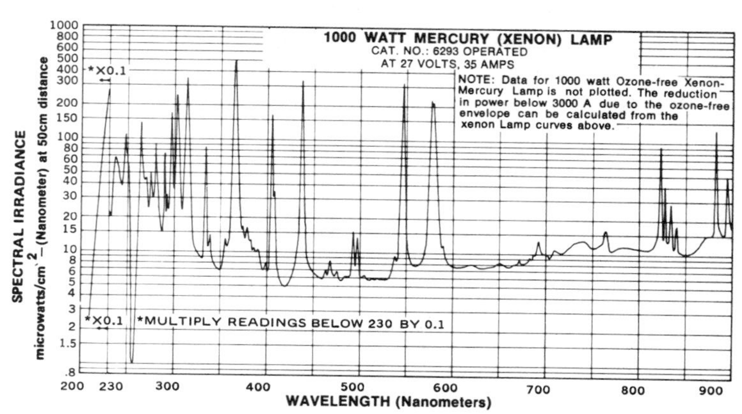

NOTE: The 1000 watt Mercury (Xenon) Lamp and the 1000 watt Mercury Lamp have similar U.V. and visible spectra.

ORIEL CORPORATION
15 MARKET ST., STAMFORD, CONN. 06902
(203) 357-1600 TWX 710-474-3563

(b)

Figure 5-7 *(continued)*

photons will be at a specific wavelength that is a function of the energy change that takes place. Gas lamps exploit this characteristic of atoms to provide radiation.

PROBLEMS

1. If an atom had five permitted energy levels, how many spectral lines could it produce?
2. What change in energy level would be required to emit a photon of 555 nm?
3. What range of energies do the visible wavelengths 400 to 700 nm represent?
4. Given: The mercury argon lamp of Figure 5-6 operated at 17 mA.
   (a) What will the wavelength of the prominent line be, in nanometers?
   (b) If stronger visible wavelength lines are desired, what must be done?
   (c) What limits are imposed on lamp current?
5. Refer to Figure 5-7.
   (a) List all the bulbs with a luminous flux of greater than 30,000 lm and a bulb life greater than 1000 h.
   (b) Which bulb on your list has the greatest luminous efficiency?
6. How does an arc lamp differ from a gas discharge lamp?
7. Why do mercury and sodium lamps have a heating cycle?
8. How is the mercury in the arc lamps of Figure 5-7(a) heated?
9. What is simmer operation?
10. Why do flashlamps sometimes have a coil of wire wrapped around the tube?
11. You have three mercury arc lamps of the same external physical size and with essentially the same spectrum. The lamps exhibit three different breakdown and operating voltages. What is probably different about the lamp structure?
12. Rank the following three lamps in terms of energy in the visible range of wavelengths: 1000-W quartz halogen, 1000-W xenon, and 1000-W mercury. Refer to Figure 5-7(b).
13. Name some applications of low-pressure mercury lamps.

6

Gas, Solid-State, and Liquid Lasers

6-1 INTRODUCTION

Lasers as a family of devices have found widespread use throughout the industrial and scientific environments. Laser devices are used in the operating room to repair eye tissue and in the machine shop to drill, cut, and weld. The information carried by optical fiber communication systems is usually put into the cable by a "semiconductor" laser. Continuous-wave gas lasers are used in the optical laboratory to align, test, and characterize optical elements and devices.

We confine this chapter to a discussion of the operation of gas, liquid, glass, and other crystal material lasers, such as those used in an optics laboratory or industrial environment. Glass- and crystal-based lasers are called *solid-state lasers.* Solid-state lasers should not be confused with semiconductor lasers, which are discussed in Chapter 13.

The word *laser* is an acronym for "light amplification by stimulated emission of radiation." Even though the output radiation of many laser devices does not fall in the visible range of wavelengths, the word *laser* is used to describe them. It is not uncommon to hear references to an infrared or ultraviolet "laser."

6-2 COMPONENTS OF A GAS LASER

The central element of a *gas laser* is a gas-filled tube much like a gas lamp. It is within this tube that the stimulated emission and light amplification takes place. The low-pressure gas-filled tube is excited into emission by a high-value dc voltage or by a radio-frequency (RF) voltage source. Figure 6-1 illustrates some typical gas-filled tubes and the dc and RF connections. The dc excitation of the gas takes

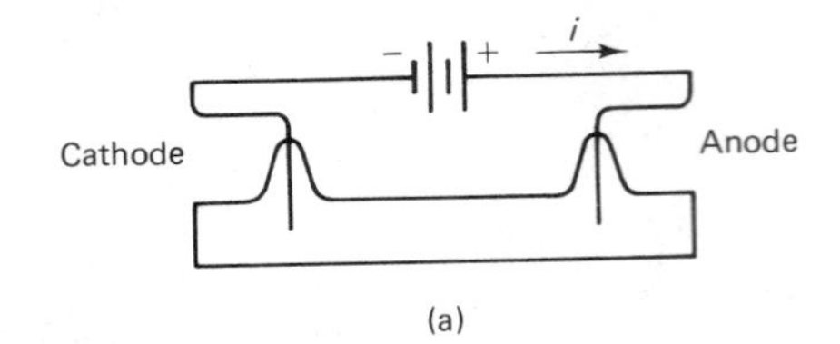

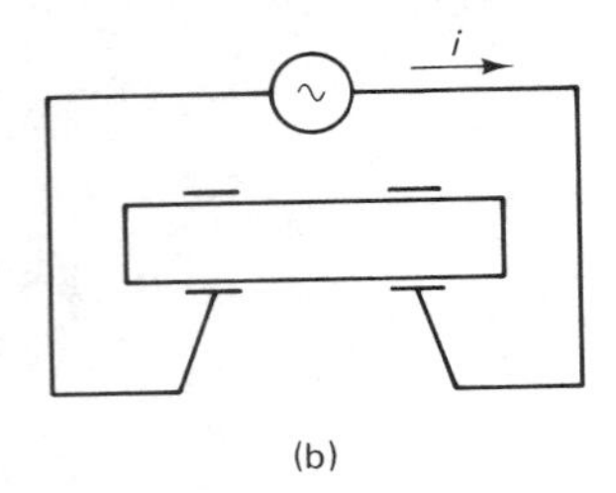

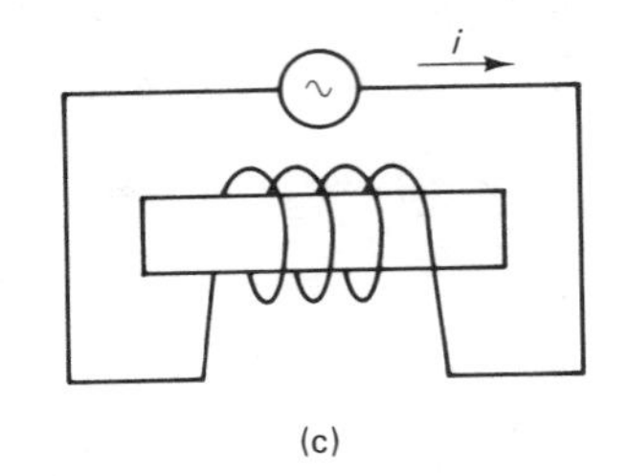

Figure 6-1 Electrical excitation of a gas laser medium: (a) dc current flow; (b) RF capacitive coupled current flow; (c) RF inductive coupling.

place through electrodes extending into the gas just like the electrodes of a gas lamp. A typical laboratory low-power helium–neon laser will use a dc voltage of 7000 V to start the arc in the gas and have an operating voltage of 1800 V. Current flow will be about 5 mA. This type of laser will deliver 3 mW of radiated power for an electrical power input of 9 W. Radiation will be emitted from the gas when electrons make energy transitions. This process was discussed in Chapter 5.

RF voltage sources cause excitation by capacitive coupling or induction. In these laser types the atoms are excited by the rapidly changing RF electric or magnetic fields. RF-excited lasers do not have electrodes embedded in the glass; rather, energy is transferred through the glass by capacitive coupling or by magnetic induction. Coils of wire about the glass tube or capacitive plates on the tube provide the means of energy transfer. RF-excited lasers typically use a power-supply frequency in the range of 20 to 30 kHz. In general, the RF voltages used are much lower than the dc voltages used, but the RF current will be much higher than the typical dc current.

The next key elements in a laser are the end mirrors. Reflecting mirrors are placed at each end of the excited gas tube. Figure 6-2 illustrates some typical mirror configurations. The radiation emitted from the excited gas tube is reflected back and forth between the end mirrors. One or both of the mirrors is designed to allow some transmission, and this transmitted radiation through the mirrors becomes the laser's output. The radiated output may be passed through a lens system to obtain a particular beam configuration. Figure 6-3 illustrates a lens system used to achieve a collimated beam.

Gas lasers can contain two or more gases. The applied electric field causes the atoms of one of these gases to become excited. Collisions of the excited gas with the other gas force it into excitation and emission. The emitted photons reflecting back and forth between the mirrors cause the amplification process of stimulated emission to take place. Stimulated emission takes place when a photon falls on an atom that is in an excited state. When this occurs, the atom immediately makes an

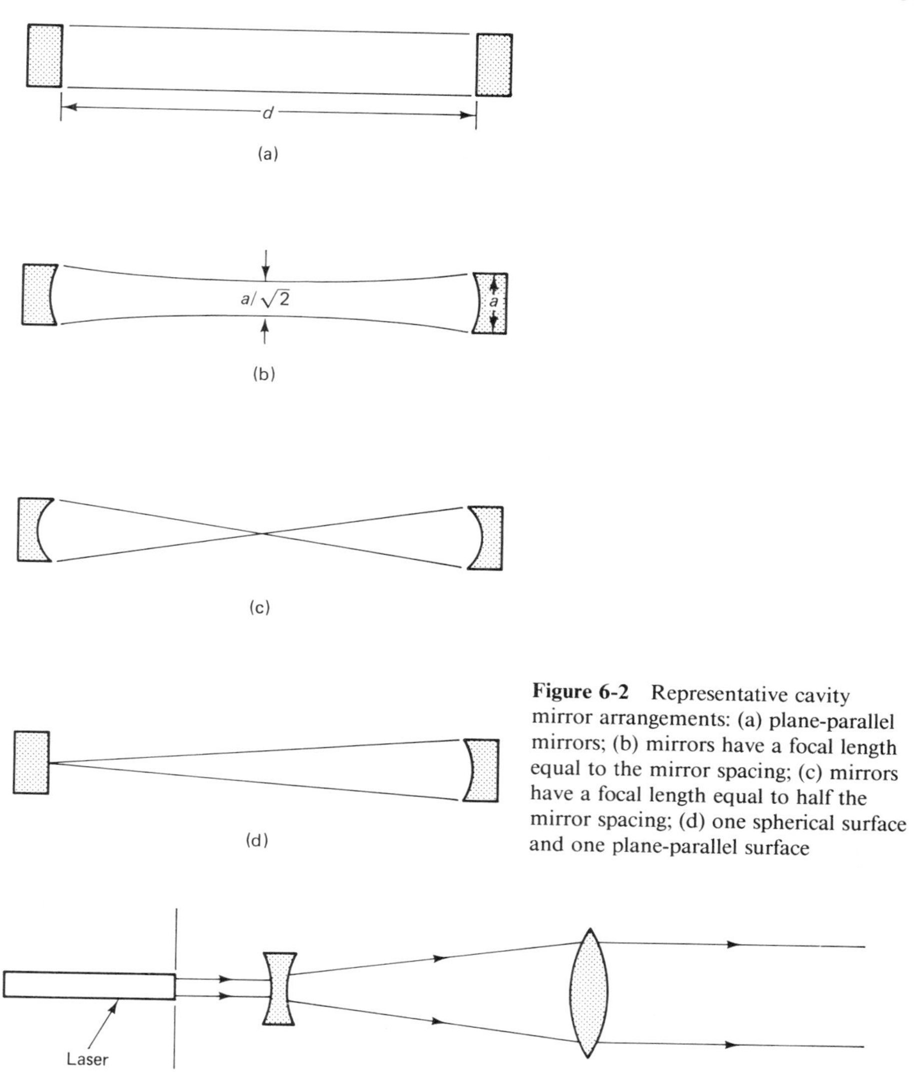

Figure 6-2 Representative cavity mirror arrangements: (a) plane-parallel mirrors; (b) mirrors have a focal length equal to the mirror spacing; (c) mirrors have a focal length equal to half the mirror spacing; (d) one spherical surface and one plane-parallel surface

Figure 6-3 Laser optics expand and collimate the laser beam.

energy-level transition. The transition causes the atom to emit a photon of the same wavelength and phase as the incident photon.

The *helium–neon laser* is very common and it will be used as an example of this process. Figure 6-4 is an energy-level diagram of the gases in a helium–neon laser. The helium–neon laser operates with an electrical discharge through the gas-filled tube. This discharge energy causes the electrons of the helium atoms to be raised to a higher energy level called the *metastable state*. Then, via collision, these excited helium atoms cause neon atoms to become excited. When conduction is well established, the majority of the atoms in the gas will be in the metastable state.

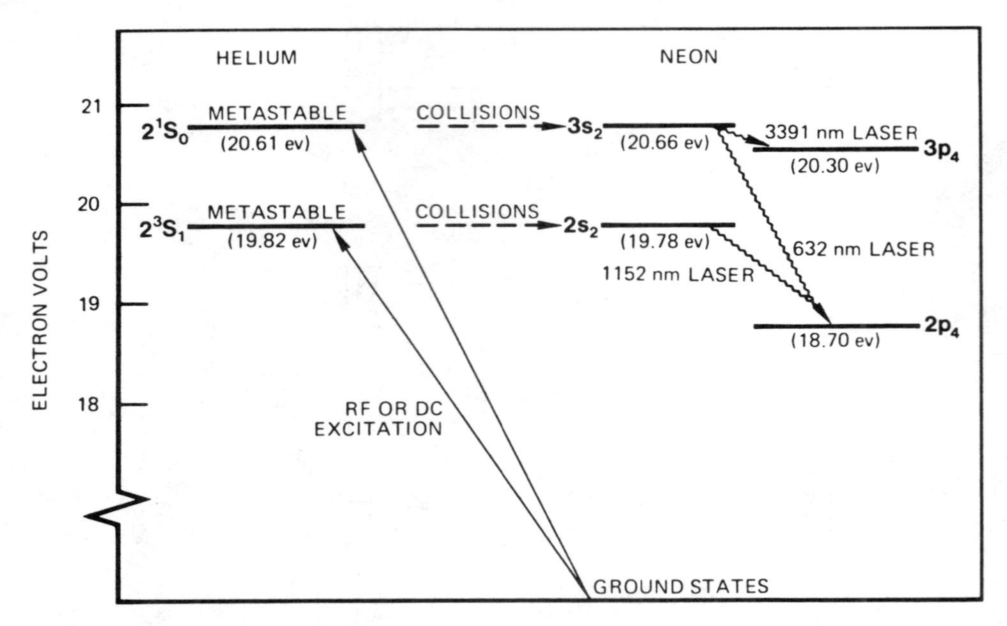

Figure 6-4 Helium–neon laser energy diagram illustrating three possible wavelengths of radiation. (Courtesy Spectra-Physics, Inc., Mountain View, CA.)

This condition is called a *population inversion* because a normal unexcited gas would have a majority of atoms in the ground state.

The electrons of the excited neon atoms can go through a number of different energy transitions emitting energy at different wavelengths. Three transitions are illustrated in Figure 6-4. It is these neon atom transitions that cause the radiation output. As was the case with gas lamps, several different transitions are possible and there will be some dominant wavelengths. Radiation from the neon atoms at one particular wavelength is optically enhanced by the reflection characteristics of the end mirrors. Through the process of stimulated emissions this enhanced wavelength becomes the dominant radiation output. Continuous-wave operation is achieved when the reflecting mirrors at each end of the gas tube form a cavity resonator that stores most of the photons and reflects them back and forth through the tube. The probability that any given photon will collide with an excited neon atom on one pass through the tube, causing stimulated emission, is low. Therefore, photons must pass through the tube many times to cause a stable level of stimulated emission. For this reason only a small amount of transmission can be allowed through the end mirrors.

The most likely transition in the helium–neon system is the one corresponding to 3391 μm. To obtain the more commonly desired 632.8-nm red output, the end mirrors have to be designed to reflect this desired wavelength. The reflected photons will cause the stimulated emission of photons that have the same wavelength and phase. This results in an output beam that is essentially monochromatic and spatially coherent.

The gases used in the laser tube, the excitation level achieved, and the transmission and reflection characteristics of the end mirrors all affect the laser's output. Table 6-1 lists some common lasers and their wavelength and power ranges.

TABLE 6-1 TYPICAL LASERS

Usual Name or Active Ion	Wavelength (nm)		Power Range Average Power
	Range	Common	
	Gas and Vapor Medium		
Argon	350–528	488, 514.5	2 mW–20 W
Argon–fluoride	—	193	1 W–50 W
Argon–krypton	450–670	—	500 mW–6 W
Carbon dioxide	9000–11,000	9600, 10,600	20 W–15 kW
Carbon monoxide	5000–7000	—	—
Copper vapor	510–578	—	—
Deuterium fluoride	3600–4000	—	10 mW–100 W
Gold vapor	—	628	—
Helium–cadmium vapor	—	325, 442	2 mW–50 mW
Helium–neon	543, 594, 604, 633, 1152, 1523, 3391		100 μW–50 mW
Hydrogen fluoride	2600–3000	—	10 mW–150 W
Krypton	350–800	647.1	5 mW–6 W
Krypton fluoride	—	248	1 W–100 W
Nitrogen	—	337	1 mW–330 mW
Xenon	—	540	—
Xenon chloride	—	308	1 W–150 W
Xenon fluoride	—	351	1 W–30 W
	Impurity-Doped Crystal		
Alkali halide (F-center)	1430, 1580, 2300, 3500		1 mW–100 mW
Holmium	—	2060	—
YAG–neodymium[a]	—	1064, 1320	40 mW–600 W
Ruby (chromium)	—	694	Pulsed
	Impurity-Doped Glass		
Erbium	—	1540	—
Neodymium	—	1060	Pulsed
	Dye in Liquid		
	Dye used determines wavelength: 300–1000		50 mW–50 W

[a] YAG, yttrium–aluminum–garnet

6-3 GENERAL LASER PRINCIPLES

As pointed out in Section 6-2, lasing action occurs through the interaction of two systems. The atomic system in which energy transitions give rise to photons and the optical cavity created by the mirrors. In this section we look at each of these phenomena in more depth and try to develop a better sense of the problems and pitfalls inherent in laser design and operation.

First we look at the interaction of radiation and the atomic system a little more closely. We have already established that a photon will be emitted when an electron undergoes the transition from a higher energy level to a lower energy level. This process is summarized by the equation

$$hf = E_2 - E_1 \tag{6-1}$$

where h is Planck's constant
f is the photon frequency
E_2 is the energy of level 2
E_1 is the energy of level 1

In a working system some sort of excitation will have raised the electrons from level E_1 to E_2 prior to time $t = 0$. An average number of these particles per unit time will then begin to make the transition from E_2 to E_1. For any given atomic system there will be a characteristic transition rate.

When we examine the spectral lines radiated by an excited gas we find that the radiated lines are not truly monochromatic but that they cover a narrow range of frequencies. This deviation from a single line is called *lineshape broadening.* Two processes cause lineshape broadening. One process is called *homogeneous broadening* because it is characteristic of all the atoms in the system. It is a function of the atomic systems finite lifetime in the emitting state, which is given the symbol τ. If this were the only process at work, the frequency distribution of the emitted line could be described by

$$A(\omega) = \frac{K}{(\omega - \omega_0)^2 + (1/\tau)^2} \tag{6-2}$$

where $A(\omega)$ is the amplitude of the radiation at each value of ω
K is a constant of proportionality
$\omega = 2\pi f$
$\omega_0 = \dfrac{E_2 - E_1}{h} 2\pi$ (h is Planck's constant)
τ is the finite lifetime of the emitting state

Readers familiar with electronics will recognize equation (6-2) as the same as that for the response of a resonant *RLC* circuit. A graph of this function is shown in Figure 6-5.

The other broadening process is called *inhomogeneous broadening,* in which the spectral line is broadened by individual atomic effects. In crystals, different atoms may have slightly different transition energies due to their surroundings. In gases the various atoms and molecules are moving in a variety of directions at different velocities. This results in Doppler shifts in frequency. The Doppler shift process is described by

$$f = f_0 + \frac{vf_0}{c} \tag{6-3}$$

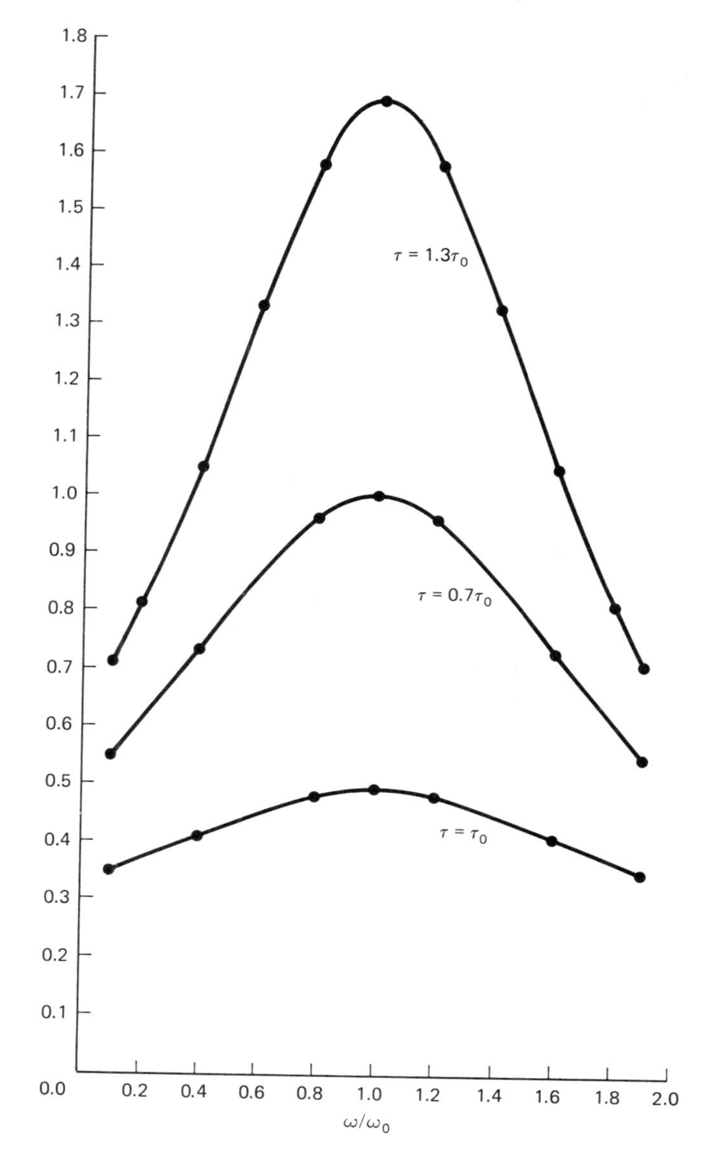

Figure 6-5 Homogeneous line shape broadening caused by different values of atomic lifetime τ.

where f is the resultant frequency

f_0 is the frequency of a stationary particle

$$f_0 = \frac{E_2 - E_1}{h} = \frac{c}{\lambda_0}$$

v is the component of the particle velocity along a line to the observer and can be positive or negative

c is the speed of light

The half-peak width of the spectral line caused by the Doppler effect alone can be predicted by calculation using

$$\Delta f = 2f_0 \left(\frac{KT}{M}\right)^{1/2} \tag{6-4}$$

where $f_0 = c/\lambda_0$
K is a constant, 165.8×10^{-15} amu/K
T is the cavity temperature
M is the atom's atomic mass (20 for neon)

For gas lasers this temperature-dependent process is the dominant effect causing line broadening. This line broadening is a function of temperature, which controls the degree of electron motion. Helium–neon lasers will have line half-power widths of about 1.1×10^9 Hz to 1.4×10^9 Hz. The broadened spectral line can be thought of as a gain curve under which it is possible to obtain a laser output. Each permitted line of the gas will have its own line width. In a practical laser, there will be losses due to the mirrors and to scattering in the gas. This will affect the amplitude of the various gain curves.

The second important subsystem of the laser is the cavity created by the mirrors. For an oscillation to exist, the round-trip cavity length must be equal to an integral number of wavelengths. As a practical matter, the integer number will be very large. The following equations illustrate the relationship between frequency and cavity length.

$$2L = M\lambda \tag{6-5}$$

where L is the cavity length
M is any integer
λ is the wavelength

$$2L = \frac{Mc}{f}$$

Therefore,

$$f = \frac{Mc}{2L} \tag{6-6}$$

where f is a possible lasing frequency.

There are as many possible resonating frequencies as there are integers, with each frequency separated by one integer value. These individual resonances represent possible lasing frequencies permitted by the cavity characteristics. The following equation allows us to predict the frequency spacing of these cavity resonances.

$$\Delta f = \frac{c}{2L} \tag{6-7}$$

Laser energy can be radiated at those cavity resonant frequencies that also lie inside the gain curve of one of the permitted lines. But which cavity resonances or modes and which lines will cause radiation? Figure 6-6 illustrates two of the gain curves that could exist for a helium–neon laser. One curve is centered at 1.969×10^{14} Hz and the other at 4.738×10^{14} Hz. On the same graph, cavity resonances spaced every 500 MHz are shown. Three cavity resonance peaks fall inside the 13.5% width of the 1.969×10^{14} Hz gain curve. Five cavity resonance peaks fall inside the 4.738×10^{14} Hz gain curve. Clearly, we do not want all of these output wavelengths. The laser designer optimizes a design for the desired wavelength by controlling the gas mixture, the excitation, the reflection characteristics of the cavity mirrors, and possibly by putting a filter in the path. For example, if the 4.738×10^{14} Hz line were the desired output frequency, the gain of the system would be maximized at 4.738×10^{14} Hz and the potential line at 1.969×10^{14} Hz, and all other lines would be rejected. The remaining 4.738×10^{14} Hz line, however, would really consist of a number of closely spaced lines caused by the cavity resonances falling under the gain curve. If fewer lines are desired for a given gain curve, the cavity length will have to be made smaller, which will space the cavity resonances farther apart.

In Chapter 5 it was pointed out that there are many possible energy transitions that a gas atom can go through. An obvious question is: Can every possible transition be made a lasing wavelength? The practical answer to this question is "no." Only those transitions with relatively long lifetimes will yield spectral lines that can be converted to a useful output. The population of possible energy transitions can

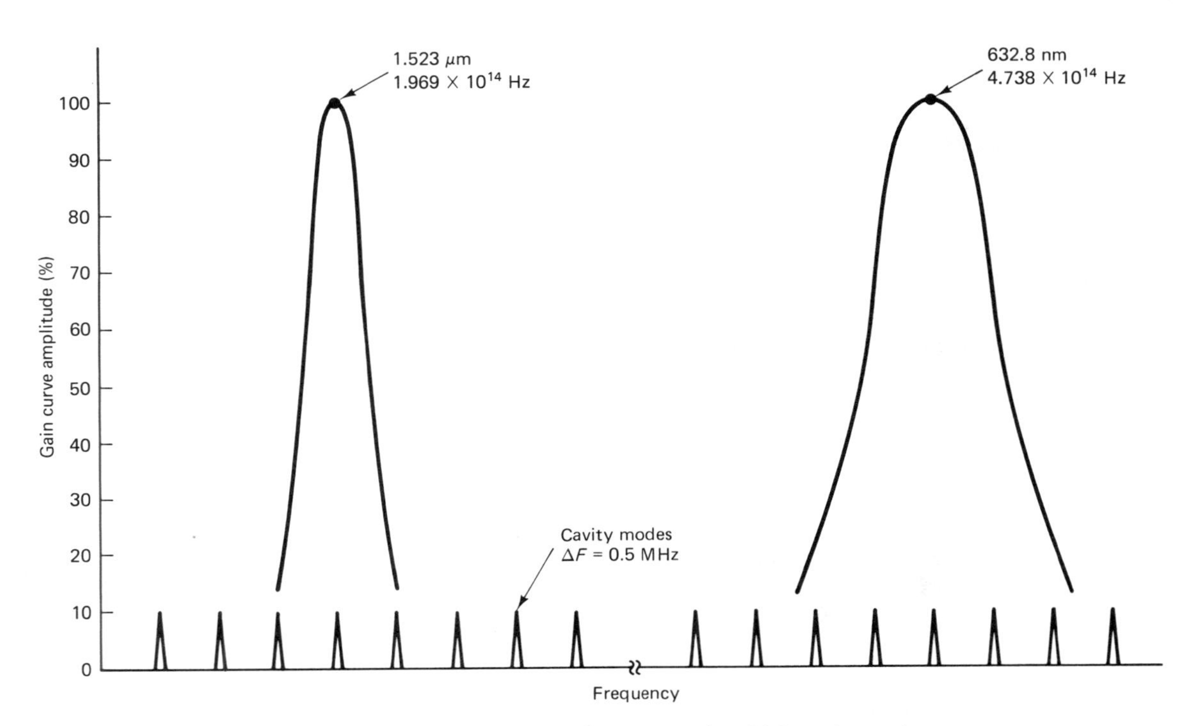

Figure 6-6 Two line-width gain curves and multiple cavity modes.

be thought of as consisting of fast transitions that cannot be used and slow transitions that can be used for light amplification by simulated emission.

Useful laser energy is obtained when the gain of the cavity is adjusted to select one of the possible lasing lines. In a gas laser the selected line will have a Doppler-broadened line width or gain curve. The length of the cavity will determine how many cavity resonances will fall inside the gain curve of the line. The resulting laser output will consist of the product of the gain curve and the cavity modes. The resultant simultaneous emissions are called the *longitudinal modes* of the laser.

In addition to the longitudinal modes, the laser cavity may cause a number of spatial modes. These modes are called the *transverse electric and magnetic* (TEM) *modes.* The various possible TEM modes are identified by subscript number, and they describe the energy distribution across the front of the beam. Figure 6-7 illustrates various modes. In practice the desired mode is usually TEM_{00}, which is a single beam with a Gaussian energy distribution. The subscripts on the other modes indicate the number and location of the energy nulls on the wavefront. The TEM modes are subscripted TEM_{HV}, where the number in the H position indicates the number of nulls on the horizontal axis and the number in the V position indicates the number of nulls on the vertical axis. For example, the TEM_{12} mode has one horizontal null and two vertical nulls. A mode not illustrated that often occurs is the *donut mode,* which forms a full circle of radiation with a hole in the middle; by

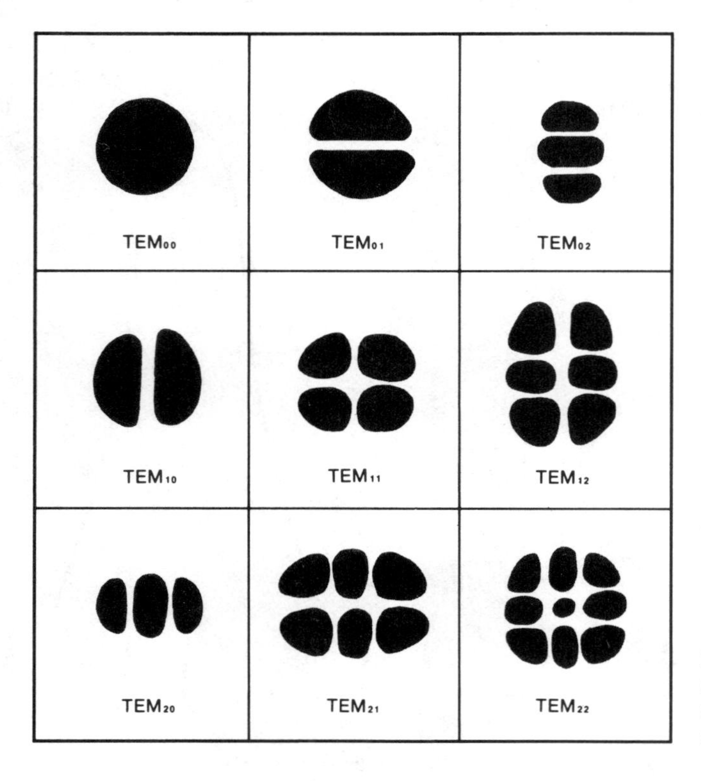

Figure 6-7 Transverse electrical and magnetic modes of a laser cavity. (*Melles Griot Optics Guide 4,* 1988; courtesy Melles Griot, Irvine, CA.)

convention this mode is designated TEM_{10}. To reiterate, however, in general, the TEM_{00} mode is the normal and desired spatial mode because this mode will exhibit minimum diffraction loss, has minimum divergence, and can be focused to the smallest possible spot.

The frequency-selective control of cavity gain is a primary method used to manipulate laser output. A recent advertisement describes a helium–neon laser that may be adjusted to emit each of 14 different lines in the wavelength range 611.8 to 1206.6 nm. This tunable laser contains an adjustable frequency selective filter and two sets of mirrors. The adjustment of the filter and the substitution of the mirrors allow the operator to select the desired gain curve (line) of operation. These 14 lines will have different amplitudes as well as different frequencies. Just as the various gas lamps discussed in Chapter 5 have dominant high-intensity lines, the gas mixtures in the gas lasers will also have wavelengths of higher and lower intensity.

6-4 OTHER LASER TYPES

Thus far helium–neon lasers have been used as an example of gas lasers. In a helium–neon laser the possible transitions are established by the neon gas. The other important gas laser types are the argon laser, the carbon dioxide laser, and the nitrogen laser.

The *argon laser* works with pure argon and its output wavelengths are characteristic of that gas. There are nine possible lines in the range 350 to 528 nm. The dominant transition is at 488 nm. Argon lasers are often used in medical settings as a surgical instrument.

The *carbon dioxide laser* and the *nitrogen laser* are molecular lasers rather than atomic lasers. In this type of structure the output is due to the relative motion of the atoms in the molecules. The electrons in the individual atoms remain in the ground state and do not contribute to the laser output. Carbon dioxide lasers can develop very high levels of radiation at very long wavelengths, 10.6 μm and 9.6 μm being common. These lasers can be excited by electrical discharge. Output power may be increased by adding other gases, such as helium or nitrogen, to the mixture. These additional molecules increase the number of collisions and raise the energy output. Power levels in the kilowatt range are possible with carbon dioxide lasers. These very powerful lasers are used in drilling and cutting operations, where the beam literally vaporizes the material.

Many industrial applications require high-powered pulsed lasers. A common technique used to cause the high-powered pulses is called *Q switching*. In this context Q can be thought of as being a synonym for cavity gain. A Q-switched laser is one in which the gain of the cavity is manipulated to cause the pulse. The sequence of events is as follows:

1. Do something to reduce cavity gain to zero. For example, cover up one of the mirrors.

2. Pump the laser medium to a much-higher-than-normal population inversion.
3. Restore cavity gain abruptly and obtain a high-energy pulse.

Figure 6-8 illustrates the *Q*-switching process. These very high powered pulses exist for only short intervals of time, such as 5 to 10 ns. *Q* switching can be achieved by physically rotating one of the cavity mirrors or by controlling attenuation in the optical path with an electronically controlled attenuating device.

Lasers can also be pulsed simply by pulsing the power supply current. The

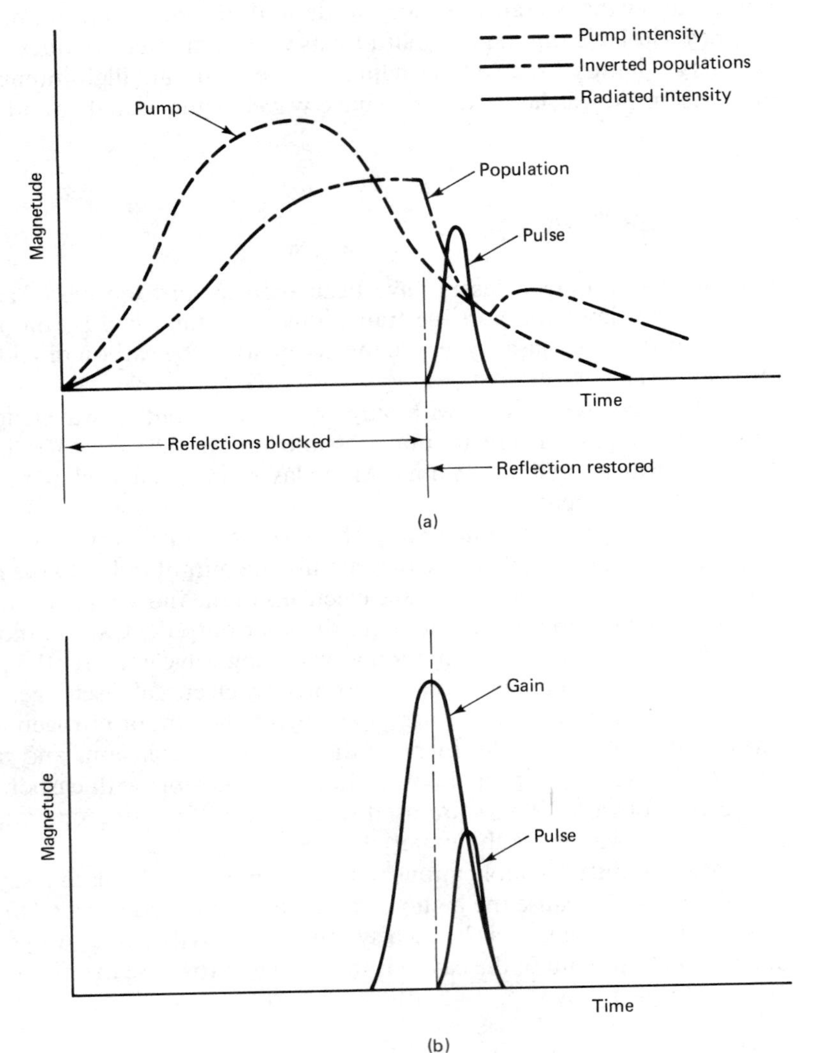

Figure 6-8 *Q*-switched laser operation: (a) pump, population, and radiation; (b) cavity gain and radiation.

power supply is often called the *pumping source.* Lower levels of peak radiated output will be obtained by this method of modulation than with Q switching.

Liquid lasers are lasers where the lasing elements are dye molecules suspended in a liquid. This liquid dye mixture is pumped through the laser cavity. Liquid lasers are excited or optically pumped by flashlamps or gas lasers. Argon and nitrogen lasers are also used to optically pump dye lasers.

The emitted wavelength from the cavity of the dye laser is selected by a *tunable filter.* A common filter would be a rotatable Brewster window constructed of several layers of birefringent crystal. The position of the filter is adjusted to maximize cavity gain at the wavelength of interest. The dye material used establishes the range of possible wavelengths. Dyes are available that provide outputs from 300 to 1000 nm. Changing dyes will alter both laser wavelength and output power.

Solid-state lasers are historically the oldest type. A ruby rod was used in the first successful laser. Crystal lasers and glass lasers are fabricated using ruby rods, garnet rods, and doped glass. The yttrium–aluminum–garnet (YAG) rod and glass rods doped with neodymium (Nd^{3+}) are important laser materials. Crystal lasers are pumped by flashlamps or other lasers. Lamps and laser pumping sources are selected which have dominant line outputs at those wavelengths that will be absorbed by the solid material, causing the desired excitation levels in the material. The wavelengths radiated by mercury discharge lamps are well matched to the absorption characteristics of the ruby rod used in the ruby laser. Xenon flashlamps are used with YAG lasers.

Ruby lasers typically radiate a 694.3-nm wavelength. YAG lasers operate well at 1.0641 μm. Nd^{3+} glass lasers have a typical output of 1.059 μm. The actual peak output of Nd^{3+} lasers will vary slightly with the type of glass used. Ruby lasers are often used in eye surgery to weld the tissue. YAG lasers are much more powerful and are used in the electronics industry to trim and shape material when fabricating devices with thin deposited films.

In summary: Lasing mediums can be solid, liquid, or gas. Within all these mediums photons are created by either the energy transitions of electrons in atoms or by the energy changes of atoms in molecules. For these transitions to take place, the lasing medium must be pumped or excited. The pumping process raises the electrons or atoms to higher-than-normal energy states. As these particles return to the ground state, photons are emitted which may be used for the laser output. Gas lasers are usually pumped by establishing an electrical discharge through the gas. Solid-state lasers are pumped by either flashlamps placed in close physical proximity to the solid rod or by other lasers. Dye lasers may be pumped by flashlamps or by gas lasers. The lasing medium is mounted in a laser cavity containing two end mirrors and a frequency-filtering device. The gain of the cavity as a function of frequency is controlled by the characteristics of the lasing medium and pump, the reflection characteristics of the end mirrors, and the frequency response of the filtering device. These parameters are all adjusted to favor one of the possible transition frequencies over all the others. Photons of the selected frequency then reflect back and forth through the cavity replicating themselves. A small percentage of the reflected energy is allowed to exit the cavity, and this exiting energy becomes the laser's radiated output.

6-5 LASER SPOT SIZE AND SAFETY

The radiation out of the laser can occur continuously or in pulses. A laser that operates continuously is called a *continuous-wave laser* and usually has a low instantaneous and average output power. A *pulsed laser* is one where the lasing action is turned on and off, creating pulses of radiation. These radiated pulses can have large instantaneous values of power. The average power, however, may still be quite low because the pulses occur infrequently.

The radiation output of a laser operating at a single wavelength can be focused down to very small spot sizes, in this way very high flux densities are achieved. The spot size to which a collimated laser beam can be focused is a function of the beam's diameter, wavelength, and the focusing optics used. Figure 6-9 illustrates a focused laser beam.

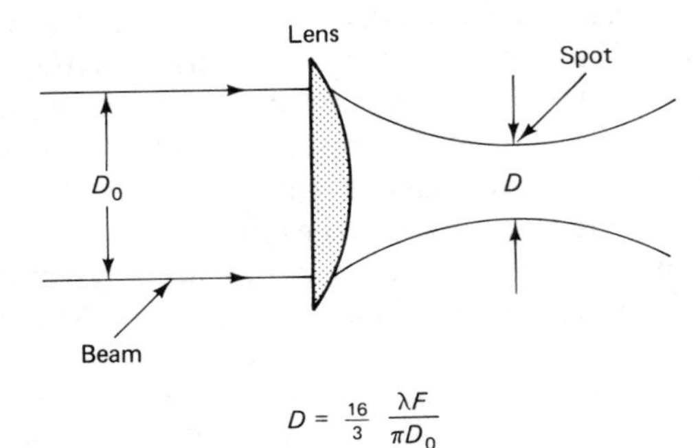

Figure 6-9 Laser beam and minimum spot size.

The smallest possible spot size may be estimated as follows:

$$D = \frac{16}{3} \frac{\lambda F}{\pi w} \tag{6-8}$$

where D is the focused spot diameter
λ is the laser wavelength
F is the lens focal length
w is the laser beam width at the point where the intensity is at 13.5% of maximum intensity

---

EXAMPLE 6-1

Calculation of Minimum Focused Laser Spot Diameter and Flux Density

Given: **(1)** Power out = 0.5 mW
(2) Laser beam diameter = 0.44 mm
(3) Wavelength = 543.5 nm
(4) Lens focal length = 17 mm

Find: **(a)** The focused spot diameter.
(b) The flux density.

Solution: **(a)** $D \simeq \frac{16}{3} \times \frac{\lambda F}{\pi w}$

$$\simeq \frac{16}{3} \times \frac{(543.5 \times 10^{-9}\text{ m})(17 \times 10^{-3}\text{ m})}{\pi\,(0.44 \times 10^{-3}\text{ m})}$$

$$\simeq 35.6 \times 10^{-6}\text{ m}$$

$$\simeq 35.6\ \mu\text{m}$$

(b) $H = \frac{p}{(\pi/4)D}$

$$= \frac{0.5\text{ mW}}{(\pi/4)(35.6\ \mu\text{m})^2}$$

$$= 502\text{ kW/m}^2 = 50.2\text{ W/cm}^2$$

The power in the laser beam can be reduced with a circular aperture that has a diameter smaller than the beam diameter. The ratio of the transmitted flux to the incident flux is given by

$$\frac{\phi_e}{\phi_i} = 1 - \exp\left(\frac{-2D^2}{w^2}\right) \tag{6-9}$$

where ϕ_e is the transmitted beam flux
ϕ_i is the original beam flux
D is the aperture diameter
w is the beam diameter

Like other forms of radiation, the laser is capable of doing damage to the human body without causing immediate pain. A sunburn is a familiar example of radiation damage. Your eyes are most susceptible both because of their structure and because of the environment in which the laser is used. Long-wavelength infrared lasers will have most of their energy absorbed by the fluid in the eye and are quite capable of doing serious damage to the cornea. Shorter-wavelength lasers such as the "red" helium–neon laser can penetrate the eye and do damage to the retina. Only very small amounts of energy are required to do damage to the eye. The key number is energy density, and experiments are still being conducted at this time to determine safe and unsafe levels.

The threshold dose for long wavelengths appears to be a flux density of about 100 mW/cm^2, a number easily reached with a small beam size with even a very low power laser. Short-wavelength lasers operating in the visible region, such as helium–neon lasers, may cause trouble with flux density levels as low as 10^{-16} W/cm^2. A 5-mW bench laser will have a spot area of about 0.03 cm^2, yielding a flux density of about 0.2 W/cm^2. Exposure to this level of flux density for 5 μs would yield an energy density of 10^{-6}/cm^2.

When working with lasers, don't look into the beam! Be sure you know where the beam reflections are going. This is a particularly dangerous problem when working with invisible wavelengths. If any chance of eye injury exists, wear safety glasses.

6-6 SUMMARY

A laser is a device consisting of two key elements: a lasing medium and a cavity with end mirrors. The stimulated emission of the lasing medium at a single wavelength is caused by reflecting photons of that wavelength between the mirrors. This is what causes the laser to radiate monochromatic radiation. The "light fantastic" of the laser, like the old magician's trick, is done with mirrors.

PROBLEMS

1. What does the acronym *laser* mean?
2. Describe what occurs when an atom has electrons in excited levels and a photon of the appropriate energy is incident on it.
3. What are the key elements of a laser system?
4. Why is only a small portion of the cavity radiation emitted?
5. What is line broadening?
6. State two causes of line broadening and name the cause that is dominant in gas lasers.
7. Line width is related to the center frequency and two physical parameters of the system. What are the two physical parameters?
8. **(a)** What is the Doppler effect?
   (b) Describe a common Doppler experience.
9. If the temperature of a cavity tuned to frequency f_0 is raised, what will happen to the line width?
10. In electronic systems Q is defined as center frequency divided by bandwidth.
    (a) How is line width Q affected by increasing mass?
    (b) How is line width Q affected by increasing temperature?
11. What happens to cavity resonant frequency spacing as cavity length is increased?
12. Estimate cavity resonant frequency spacing for cavity lengths of 30 cm, 20 cm, 10 cm, 5 cm, and 1 cm.
13. What is a longitudinal mode?
14. What is a TEM mode?
15. What characteristic of the atom determines which lines can be used in a laser?
16. How is the lasing frequency selected from all the possible lasing frequencies?
17. Will the laser output always consist of a single line?
18. What energy transitions cause laser operation in a carbon dioxide laser?
19. How can the frequency of a dye laser be changed?
20. How is a crystal laser pumped?
21. Calculate the smallest spot that can be obtained with a red helium–neon laser. Assume a beam diameter of 0.5 nm and a lens focal length of 15 mm.
22. Why is Q switching used?
23. What sequence of events takes place in a Q-switched laser?

7

Vacuum Photodetectors

7-1 INTRODUCTION

Vacuum photodetectors are electronic devices that use the photoelectric effect to cause voltages and currents that are proportional to input flux density. These devices are very sensitive, with fast response times. Although the vacuum photodetector has been superseded by various semiconductor devices in routine applications, it can still be found in a variety of laboratory apparatus. In this chapter the theory of operation of the vacuum photodetector and photomultiplier is discussed. In Section 7-4 we explore the operational characteristics of these devices in a simple series circuit.

7-2 PHOTOTUBE PROCESSES

The interaction of radiation and solid surfaces discussed at the beginning of Chapter 6 has been exploited to develop a family of photodetector devices. The simplest of these devices is the vacuum photodiode pictured in Figure 7-1. A vacuum photodiode is constructed with two electrical elements, called the *cathode* and the *anode.* Both of the electrical elements are enclosed in an evacuated glass tube. The cathode is constructed of a metal surface curved in shape and coated with an oxide. The anode of a vacuum photodiode is fabricated as a thin post or rod positioned at the electronic focal point of the cathode. An electrical voltage is applied across the vacuum photodiode with the most positive voltage connected to the anode and the least positive voltage connected to the cathode.

The vacuum photodiode is positioned in the optical system in such a way that the radiation to be measured will fall on the cathode. The radiation falling on the

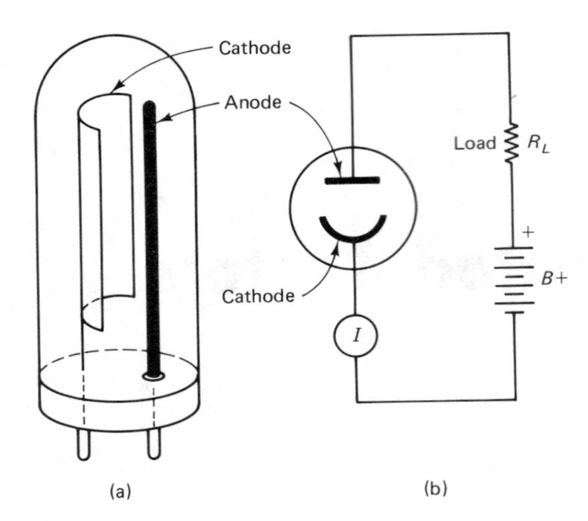

Figure 7-1 (a) Construction and (b) schematic diagram of a phototube circuit.

cathode causes electrons to be emitted from the cathode surface. The positive-to-negative electric field that exists between the cathode and the anode causes the emitted electrons to be attracted to the anode. This cathode-to-anode current flow is proportional to the incident radiation.

The emission of electrons from the cathode requires that the incident photon contain sufficient energy to overcome all the forces that bind the electrons to the atoms of the cathode material and the cathode surface. The surface binding force is the result of the positive charge on the surface caused by the escaping electrons, this positive charge attracts the escaping electrons back to the surface. The total energy required to overcome the atomic and the surface binding forces is called the *work function, W*. An electron that is successful in escaping the surface will be moving at a velocity greater than zero, so it will have a kinetic energy, E_K. The total energy of the photon absorbed by an escaping electron provides the energy needed to overcome the work function, and the excess energy above this amount provides the kinetic energy to the electron. This idea is summarized as follows:

$$hf = W + E_K \tag{7-1}$$

where hf is the energy of the incident photon absorbed
W is the material's work function
E_K is the electron's kinetic energy after escape

Clearly, for any fixed material and incident photon there will be some maximum kinetic energy, as is illustrated by the equation

$$E_{K\max} = hf - W \tag{7-2}$$

Normally, these energies are expressed in units of electron volts. The electron volt is the amount of energy equal to the change in energy of one electron when it moves through a potential of 1 V. There are 1.6×10^{-19} J per electron volt. If we were to apply a voltage across the tube so that it opposed the movement of electrons (anode negative, cathode positive), we could offset the electron's kinetic energy.

If the applied voltage were selected so that it just exactly offsets the maximum possible kinetic energy, conduction could be reduced to zero. This voltage value is called the *stopping voltage.* Example 7-1 below illustrates these ideas.

EXAMPLE 7-1

Solving for the Stopping Voltage of a Vacuum Photodiode

Given: An incident wavelength of 550 nm. The material is potassium, which has a work function of 1.55 eV.

Find: The stopping voltage.

Solution:

$$E_{K\max} = hf - W$$

$$f = \frac{c}{\lambda} = \frac{3 \times 10^{8}\ \text{m/s}}{550 \times 10^{-9}\ \text{m}}$$

$$= 545.5 \times 10^{12}\ \text{Hz}$$

$$h = 6.63 \times 10^{-34}\ \text{J} \cdot \text{s} \times \frac{1\ \text{eV}}{1.6 \times 10^{-19}\ \text{J}}$$

$$= 4.144 \times 10^{-15}\ \text{eV} \cdot \text{s}$$

$$E_{K\max} = (4.144 \times 10^{-15}\ \text{eV} \cdot \text{s})\,(545.5 \times 10^{12}\ \text{Hz}) - (1.55\ \text{eV})$$

$$= 2.26\ \text{eV} - 1.55\ \text{eV}$$

$$= 0.710\ \text{eV}$$

In other words, an electron trying to move against a negative 0.710 V loses all of its kinetic energy. Therefore, the stopping potential for this tube is a 0.710 V.

Another interesting characteristic of the vacuum photodiode illustrated by equation (7-1) is the concept of the threshold frequency. The *threshold frequency* is the frequency of an incident photon below which there is no emission from the cathode. If the incident photon has an energy just equal to the work function energy, the cathode's electron will have enough energy just to rise to the surface of the material. The electron will not have any kinetic energy to make its escape. In other words, if the incident photon has energy just equal to the work function, no emission will occur. Example 7-2 illustrates this idea.

EXAMPLE 7-2

Solving for the Threshold Frequency and Wavelength for Cesium

Given: The work function of cesium is 1.36 eV.

Find: **(a)** f_0, the minimum (threshold) frequency.
(b) λ, the maximum (threshold) wavelength.

Solution:

$$hf = W + E_K$$

$$E_K = 0 \text{ at } f_0$$

$$hf_0 = W$$

$$\textbf{(a)}\ f_0 = \frac{W}{h} = \frac{1.36 \text{ ev}}{4.144 \times 1^{-15} \text{ eV} \cdot \text{s}}$$

$$= 328.2 \times 10^{12} \text{ Hz}$$

$$\textbf{(b)}\ \lambda_0 = \frac{c}{f_0} = \frac{3 \times 10^8 \text{ m/s}}{328.2 \times 10^{12} \text{ Hz}}$$

$$= 914 \text{ nm}$$

The graphs in Figure 7-2 illustrate the fundamental terminal characteristics of a vacuum photodiode. Figure 7-2(a) illustrates the effect of photon frequency on device current. Below the threshold frequency f_0, the current is zero. Figure 7-2(b) illustrates the effect of photon frequency on device stopping voltage. As photon frequency increases, the kinetic energy of the emitted electrons also increase and therefore the magnitude of the voltage required to stop conduction increases. Figure 7-2(c) illustrates the fact that the number of emitted electrons and therefore the magnitude of the device current is directly related to the intensity of the radiation. Figure 7-2(d) is the forward transfer characteristic of the vacuum photodiode. This figure illustrates the relationships that exist between terminal voltage, current, and flux density at a fixed photon frequency. First, observe that the current characteristics are essentially flat above the knee voltage. At voltages above the knee voltage all of the electrons being emitted are involved in conduction and the photodiode is said to be *saturated.* Circuits that use vacuum photodiodes are designed so that the photodiode will operate in the saturated region. Second, notice that the current caused by flux density H_2 in the saturated region is essentially twice as large as the current produced by flux density H_1. This is the same idea as that conveyed by Figure 7-2(c). Third, notice that the stopping voltage is the same for both flux densities. Stopping voltage is a function of photon frequency and it is not affected by the quantity of photons present.

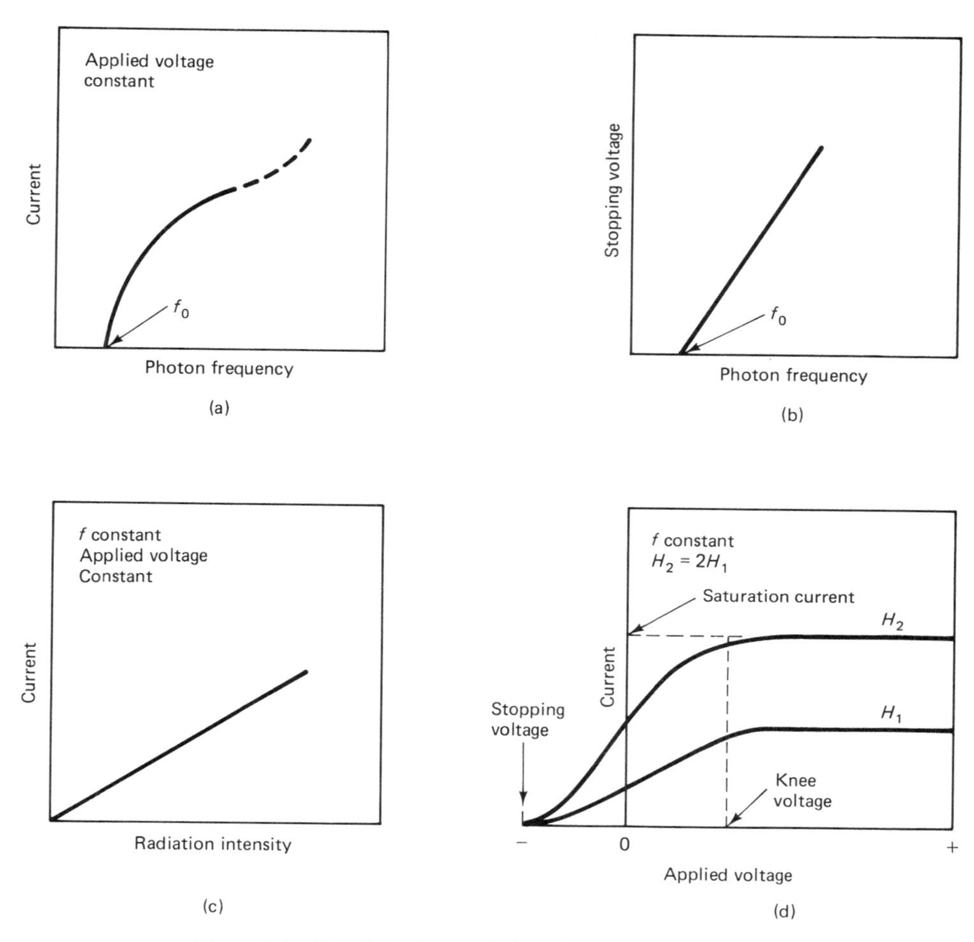

Figure 7-2 Terminal characteristics of a vacuum photodiode.

Figure 7-3 illustrates the forward transfer characteristic curves that would be found on a manufacturer's data sheet. The load resistances plotted on the curves represent the series resistance placed in the circuit with the tube [refer to Figure 7-1(b)]. The light flux on this particular curve is specified in lumens. The flux measurement would not be made at a fixed wavelength but rather, at a fixed color temperature. A typical source color temperature used for testing photodiodes would be 2854 K. A photodiode will act as a linear producer of current in response to increasing flux from a source with a fixed color temperature. Also illustrated is the low current sensitivity of these devices. For example, with this tube, if a 25-MΩ resistor and a 250-V power supply are used, 0.1 lm of flux will cause a current flow of only 4.4 μA. This is a sensitivity of 44 μA/lm.

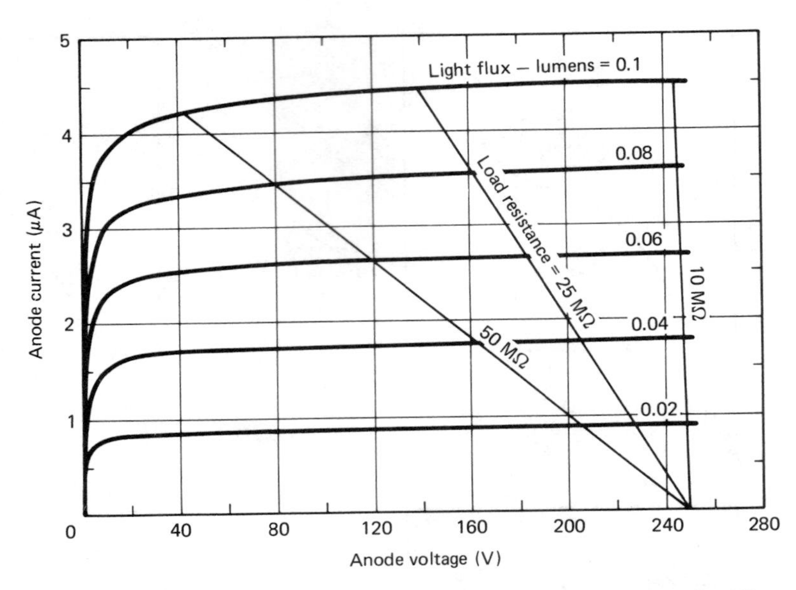

Figure 7-3 Transfer characteristic of a vacuum photodiode with three load lines illustrated.

7-3 PHOTOMULTIPLIERS

The relatively low current sensitivity of the photodiode has led to a number of modifications of the tube, all of which attempt to provide more tube current per unit of flux. While the actual implementation can become very sophisticated, almost all of these solutions revolve around a phenomenon called *secondary emission.* This is a process that occurs when high-velocity electrons collide with a material that has a relatively low work function. When this occurs, the kinetic energy of the moving electron is transferred to electrons in the material, causing two or more electrons to be dislodged with each collision.

Figure 7-4 schematically illustrates this process. Here surface C_0 is a photocathode, surfaces C_1, C_2, and C_3 are secondary emission cathodes, and the final surface, A_0, is the anode. The power supply consists of a string of dc voltages. Photocathode C_0 is at the lowest potential and supply E_1 causes cathode C_1 to be at a positive potential with respect to C_0. Supplies E_2 and E_3 cause cathodes C_2 and C_3 to be at successively higher potentials, and finally, supply E_4 causes the anode to be at the highest positive potential.

The operation of the device is as follows. Flux enters the tube and causes photoemission from cathode C_0; these emitted electrons are accelerated by the voltage difference between C_0 and C_1, causing them to collide with C_1 at a high velocity. The collision causes the secondary emission from C_1 of more electrons than arrived from C_0. These secondary electrons are accelerated between C_1 and C_2, and additional secondary emission takes place at C_2. Finally, the electrons from C_2

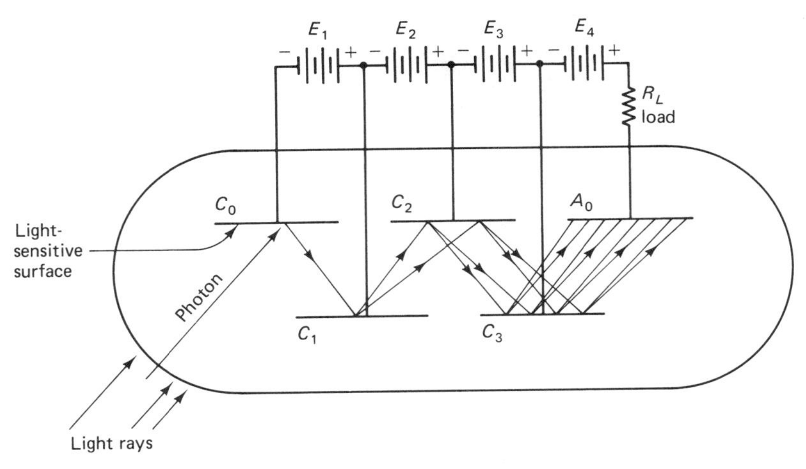

Figure 7-4 Drawing of a photomultiplier illustrating the increase in the number of electrons at the dynodes C_1, C_2, and C_3.

cause secondary emission from C_3, all of whose electrons are then collected by the anode. Essentially, secondary emission is used as a process to multiply the number of electrons available for current flow.

Figure 7-5(a) and (b) are portions of a data sheet for a type 4818 photomultiplier tube that employs the process just described. The characteristic range of values for equipment design listed in 7-5(a) illustrates just how effective this process is. The cathode sensitivity is listed as 6×10^{-5} A/lm, which is 60 μA/lm and is typical of a photocathode. The anode sensitivity is listed, however, as 300 A/lm, an increase of 5 million times. There is no question that secondary emission is a powerful way to increase sensitivity. The increased sensitivity is not free. Figure 7-5(b) shows a picture of the internal structure of the tube. In addition to the photocathode and the anode, this tube has nine secondary emission cathodes, called *dynodes.* This means that both the tube and its power supply are much more complex than they are for a simple diode. A schematic of the photomultiplier tube and its associated power supply are also illustrated in Figure 7-5(b). An additional complication with photomultiplier tubes is the very high power supply voltage used, in this case 1250 V.

The tube shown in Figure 7-5(b) has a circular cage focus structure. A tube structure with a potentially higher gain and faster response time would be the linear focused structure. A linear focused tube also has a different cathode structure. In the linear focused tube the cathode material is deposited in a very thin layer at one end of the tube (Figure 7-6). Electrons are accelerated away from the opposite side of the cathode through tightly coupled dynodes located so that the reflected electron beam is focused onto the next dynode. Other structures using wire grids between the dynodes are also used. These grid structures tend to be slower and less efficient than the focused structures.

In summary: A photomultiplier is an electronic device that uses the photoelec-

RCA

Electronic Components

Photodetector

4818

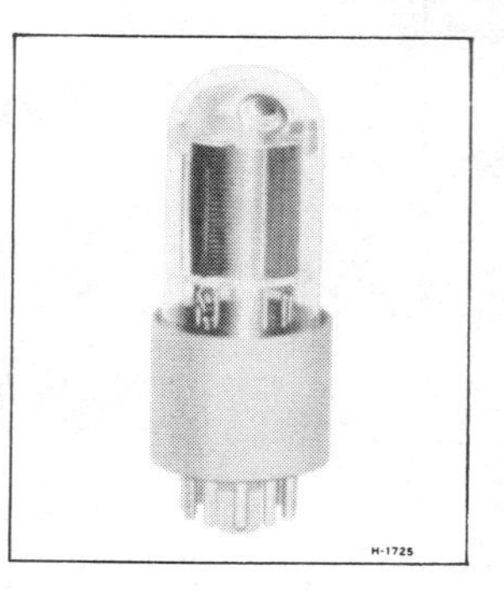

Photomultiplier

Variant of 1P28 Having a Bialkali Photocathode

- Spectral Response Range — 200 to 650 nm
- Anode Current Drift — ± 1.5% maximum for an initial anode current of 3 μA
- High Current Amplification — 5×10^6 at 1000 volts
- Fast Time Resolution Characteristics — Anode Pulse Rise Time, 1.6×10^{-9} s at 1250 volts

  Electron Transit Time, 1.6×10^{-8} s at 1250 volts

RCA 4818* is a 1-1/8" diameter, 9-stage, side-on photomultiplier designed for spectroscopic and other low light level detection systems.

This tube employs a bialkali photocathode, an ultraviolet-transmitting glass envelope, and cesium-antimony dynodes.

General Data

Spectral Response	See Figure 1
Wavelength of Maximum Response	400 ± 50 nm
Cathode, Opaque	Potassium-Cesium-Antimony (Bialkali)
Minimum projected length	0.94 in (2.4 cm)
Minimum projected width	0.31 in (0.8 cm)
Window	Ultraviolet-Transmitting Glass (Corning[a] No.9741), or equivalent
Index of refraction at 589.3 nanometers	1.47
Dynodes:	
Substrate	Nickel
Secondary-emitting surface	Cesium-Antimony
Structure	Circular-Cage, Electrostatic-Focus Type
Direct Interelectrode Capacitances (Approx.):	
Anode to dynode No.9	4.4 pF
Anode to all other electrodes	6.0 pF
Maximum Overall Length	3.68 in (9.3 cm)
Maximum Seated Length	3.12 in (7.9 cm)
Maximum Diameter	1.31 in (3.3 cm)
Base	Small-Shell Submagnal 11-Pin, (JEDEC Group 2, No.B11-88) DAP (Di-Allyl Phthalate) Non-Hygroscopic Material
Socket	Amphenol[b] No.78S11T, or equivalent
Magnetic Shield	Millen[c] No.80801B, or equivalent
Operating Position	Any
Weight (Approx.)	1.6 oz

*Formerly Dev. Type C7045G.

Maximum Ratings, Absolute-Maximum Values[d]

DC Supply Voltage:			
Between anode and cathode	1250	max.	V
Between dynode No.9 and anode	250	max.	V
Between consecutive dynodes	250	max.	V
Between dynode No.1 and cathode	250	max.	V
Average Anode Current[e]	0.5	max.	mA
Ambient Temperature	85		°C

Characteristics Range Values for Equipment Design

Under conditions with dc supply voltage (E) across a voltage divider providing 1/10 of E between cathode and dynode No.1; 1/10 of E for each succeeding dynode stage; and 1/10 of E between dynode No.9 and anode, and at a temperature of 22° C.

With E = 1000 volts (Except as noted)

	Min.	Typ.	Max.	
Anode Sensitivity:				
Radiant[f] at 400 nm	–	2.7×10^5	–	A/W
Luminous[g] (2854° K)	100	300	1200	A/lm
Cathode Sensitivity:				
Radiant[h] at 400 nm	–	5.4×10^{-2}	–	A/W
Luminous[j] (2854° K)	2.5×10^{-5}	6×10^{-5}	–	A/lm
Quantum efficiency at 400 nm	–	16.5	–	%
Anode-Current Drift:[k]				
For an initial anode current (I_b) of 3 μA	–	–	±1.5	%
Current Amplification	–	5×10^6	–	
Anode Dark Current at 1000 Volts	–	2×10^{-9}	1.5×10^{-8}	A
Equivalent Anode Dark Current Input[m] at 1000 Volts	–	6.6×10^{-12}	–	lm
Anode Pulse Rise Time[n], at 1250 Volts	–	1.6×10^{-9}	–	s
Electron Transit Time[p], at 1250 Volts	–	1.6×10^{-8}	–	s

This product data sheet is to be used in conjunction with the RCA publication "Tips on the Use of Photomultipliers" – PIT-711.

Printed in U.S.A./6-72
4818

(a)

Figure 7-5 Photomultiplier data sheet: (a) specifications; (b) schematic and mechanical drawing. (Courtesy Burle Industries, Inc., Lancaster, PA.)

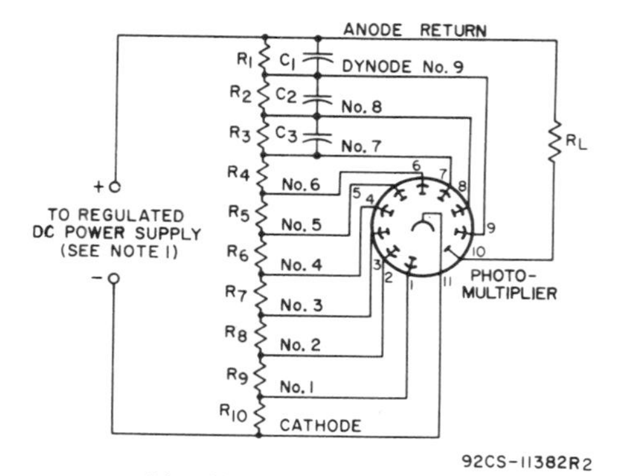

C1: 0.05 μF, 500 volts (DC working)
C2: 0.02 μF, 500 volts (DC working)
C3: 0.01 μF, 500 volts (DC working)
R1 through R10: 20,000 to 1,000,000 ohms

Note 1 – Adjustable between approximately 500 and 1250 volts.

Note 2 – Capacitors C1 through C3 should be connected at tube socket for optimum high-frequency performance.

Figure 8 – Typical Voltage-Divider Arrangement

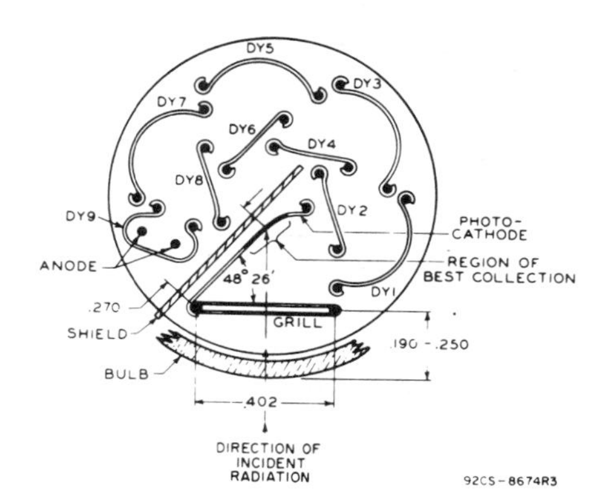

Figure 10 – Detail A – Top View

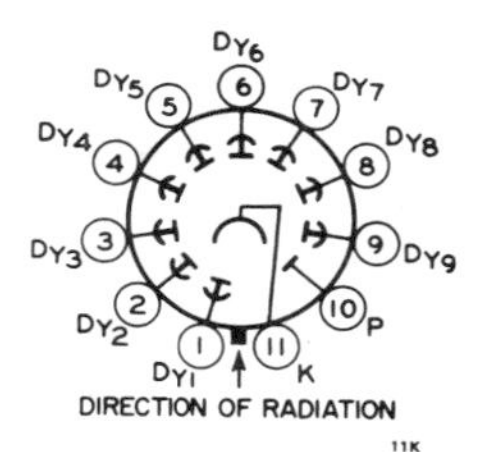

Pin 1: Dynode No.1
Pin 2: Dynode No.2
Pin 3: Dynode No.3
Pin 4: Dynode No.4
Pin 5: Dynode No.5
Pin 6: Dynode No.6
Pin 7: Dynode No.7
Pin 8: Dynode No.8
Pin 9: Dynode No.9
Pin 10: Anode
Pin 11: Photocathode

Figure 11 – Basing Diagram – Bottom View

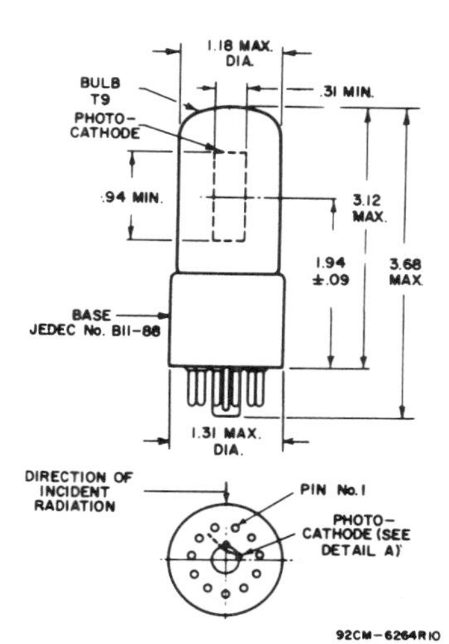

℄ of bulb will not deviate more than 2° in any direction from the perpendicular erected at center of bottom of base.

Figure 9 – Dimensional Outline

Dimensions are in inches unless otherwise stated. Dimensions tabulated below are in millimeters and are derived from the basic inch dimensions (1 inch = 25.4 mm).

Inch Dimension Equivalents in Millimeters					
Inch	**mm**	**Inch**	**mm**	**Inch**	**mm**
.09	2.3	.31	7.9	1.31	33.2
.190	4.8	.402	10.2	1.94	49.2
.250	6.3	.94	23.8	3.12	79.2
.270	6.8	1.18	29.9	3.68	93.4

(b)

Figure 7-5 *(continued)*

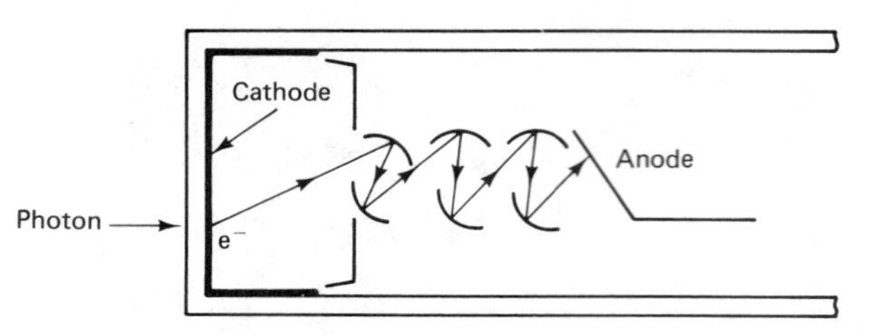

Figure 7-6 Photomultiplier with end feed cathode and a linear focused structure.

tric effect to generate a cathode current proportional to the incident flux. This initial cathode current is multiplied by use of the secondary emission process to achieve a greatly increased sensitivity.

7-4 PHOTOTUBE CIRCUITS AND DATA SHEETS

Photodiodes come in a wide variety of physical sizes and operating voltages. A photodiode may be as small as 33 mm in height with an operating voltage of 15 V, or it may be several inches high, with an operating voltage of 250 V or more. Figure 7-7(a) and (b) are pages from a data sheet describing a physically small low-voltage phototube. As the typical spectral response curve of Figure 7-7(b) illustrates, photodiodes have a response that varies with input wavelength.

The spectral response of a photodiode is governed by the emission characteristics of the oxide used to coat the cathode and the transmission characteristics of the transparent envelope enclosing the cathode and the anode. On the data sheet of Figure 7-7(a) the luminous sensitivity of the R1842 is listed as 60 μA/lm. This is a fairly typical value for any photodiode. A photodiode with meaningful spectral response in the visible range of wavelengths will usually have a sensitivity in the range 30 to 60 μA/lm. Great caution should be exercised in trying to apply this sensitivity specification. Using the R1842 data as an example, the absolute maximum allowable cathode current is 0.75 μA. Combining this limit with the sensitivity specification of 60 μA/lm yields a maximum permissible illumination level of 12.5 mlm. The surface area of the cathode is specified as 16.5 mm^2, so a flux of 12.5 mlm would correspond to a flux density of 758 lm/m^2. Radiometric sensitivities are also listed for the specific wavelengths of 254 nm and 340 nm. Radiometric sensitivities for other wavelengths could also be obtained from the spectral response graph of Figure 7-7(b).

A specification called the *dark current* is also listed. Dark current is the current that will flow when the tube is shielded from radiation, that is, when the tube is in the dark. The magnitude of the dark current will vary with temperature and terminal voltage. The 10-pA value shown here was obtained when terminal voltage was 15 V dc and the temperature was 25°C. The magnitude of the dark current essentially establishes the minimum level of radiation that can be detected. The desired photon-generated current must be larger than the dark current before it can be detected reliably. The dark current is always flowing. The trick is to arrange

SPECIFICATIONS

GENERAL:	R1826	R1842	Units
Spectral Response	185～350	185～650(S-5)	nm
Wavelength of Maximum Response	240	340	nm
Photocathode Material	Cs-Te	Sb-Cs	—
Minimum Useful Size	3 x 5.5	3 x 5.5	mm
Window Material	UV glass	UV glass	—
Anode Shape	Plate	Plate	—
Recommended Operating Voltage	15	15	Vdc
MAXIMUM RATINGS (Absolute Maximum Values):			
Anode Supply Voltage	20	20	Vdc
Peak Cathode Current	0.75	0.75	μA
Peak Cathode Current Density	0.05	0.05	μA/mm²
Average Cathode Current	0.25	0.25	μA
Ambient Temperature	50	50	°C
CHARACTERISTICS (at 25°C, 15 Vdc):			
Luminous Sensitivity (2856K)	—	60	μA/lm
Radiant Sensitivity at 254 nm	20	30	mA/W
at 340 nm	—	40	mA/W
Dark Current	10	10	pA Max.

FIGURE 1
Dimensional Outline (Unit: mm)

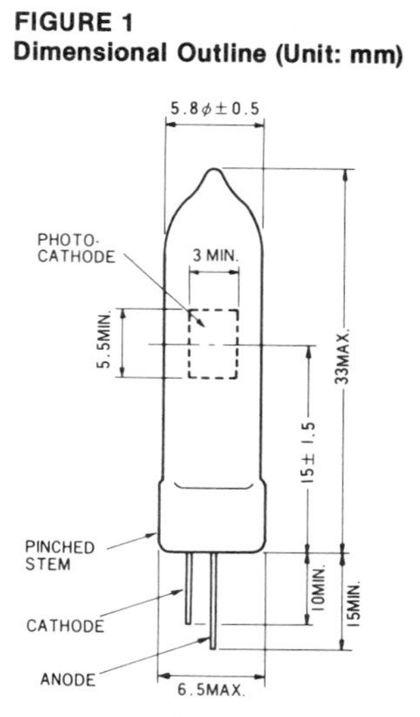

(a)

FIGURE 2
Typical Spectral Response

FIGURE 3
Typical V-I Characteristic

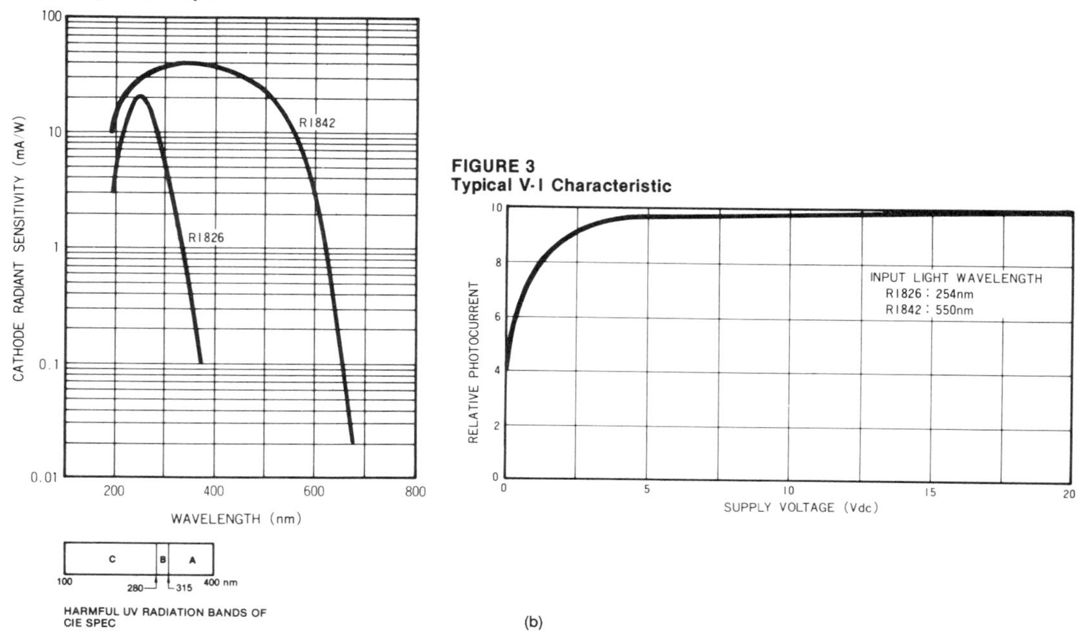

(b)

Figure 7-7 Low-voltage phototube: (a) specifications; (b) response curves. (Courtesy Hamamatsu Corporation, Bridgewater, NJ.)

things so that the dark current is only a small portion of the total current flow. Dark current can vary significantly from tube to tube of the same part number. Dark current is also temperature dependent, and to a certain extent, voltage dependent. Dark current for a particular tube will change as the tube ages and as it heats up in normal operation. Most measurement systems using phototubes will provide some method of zeroing out the system's indicated response to the dark current prior to making a measurement. When using these systems it is usually necessary to adjust the zeroing correction repeatedly for dark current.

As the typical *V–I* characteristics of Figure 7-7(b) illustrate, the current through this photodiode is relatively independent of the voltage across the tube as long as the voltage across the tube exceeds the 5-V dc knee voltage. Figure 7-3, discussed earlier, shows a typical *V–I* characteristic for a high-voltage photodiode. The rated anode voltage for this photodiode is 250 V. *V–I* response curves are shown for various levels of incident flux in lumens, and each of the response curves has the same classic shape as that illustrated in Figure 7-7(b). Also drawn on this set of curves are three sloped lines representing values of external circuit resistances of 50 MΩ, 25 MΩ, and 10 MΩ, respectively.

The circuit diagram of Figure 7-1 illustrates how the power supply and external resistors would be connected to the photodiode. The photodiode and the resistor form a simple series circuit across the dc power supply represented by the battery symbol. The sum of the voltage drops across the photodiode and the load resistor must be equal to the supply voltage; it is assumed that there is no drop across the current meter. The current flow in the circuit is due to the photocurrent characteristics of the photodiode. The voltage drop across the load resistor will be equal to the tube current multiplied by the resistor value. The *Kirchhoff voltage law* (KVL) equation summarizes the voltage–current relationships for this circuit:

$$E = V_T + R_L I_T \tag{7-3}$$

where E is the dc supply voltage, called "B+" on the diagram
V_T is the tube voltage
R_L is the load resistance
I_T is the tube current

The load line is plotted on the *V–I* characteristic curve in such a way that all of these values can be determined from the graph. All of the load lines intersect the horizontal zero current axis at the supply voltage value of 250 V. The slope of the individual lines is a function of the resistance of the individual load resistors. Example 7-3 illustrates how this kind of plot can be used to determine the voltage across and the current through the photodiode. A common procedure in algebra is to find the solution to two equations by plotting the equations on a common set of axes and finding the intersection of the lines. The process followed in Example 7-3 is the same. The curved lines represent the *V–I* equation of the tube and the straight lines represent the *V–I* equation of the load resistors. Where these lines intersect is the solution that will satisfy KVL equation (7-3).

EXAMPLE 7-3

Voltage and Current Values in a Phototube Circuit

Given: The characteristic curves of Figure 7-3, flux levels of 0.04 lm and 0.08 lm, and a 50-MΩ resistor.

Find: The tube current I_T, tube voltage V_T, and resistor voltage V_R at:
(a) 0.04 lm
(b) 0.08 lm

Solution: **(a)** Data from the intersection of the 50-MΩ resistor curve and the 0.04-lm curve—approximate values:

Current coordinate = 1.75 μA
Voltage coordinate = 165 V

$$I_T = 1.75\ \mu\text{A}$$
$$V_R = I_T R_L \qquad \text{Ohm's law}$$
$$= 1.75\ \mu\text{A} \times 50\ \text{M}\Omega$$
$$= 87.5\ \text{V}$$

$$V_T = E - V_R \qquad \text{KVL}$$
$$= 250\ \text{V} - 87.5\ \text{V}$$
$$= 162.5\ \text{V}$$

Note: The calculated value of V_T is in good agreement with the value of V_T (165 V) read from the graph. The current coordinate will be the current through the tube and the resistor. The voltage coordinate is the voltage across the tube.

(b)

$$I_T = 3.5\ \mu\text{A}$$
$$V_T = 80\ \text{V}$$
$$V_R = I_T R_L$$
$$= 3.5\ \mu\text{A} \times 50\ \text{M}\Omega$$
$$= 175\ \text{V}$$
$$V_T = E - V_R$$
$$= 250\ \text{V} - 175\ \text{V}$$
$$= 75\ \text{V}$$

Note: Conclusion V_T will be about 80 V and the current will be about 3.5 μA.

Like most forms of engineering analysis, load-line analysis does not yield exact answers but rather, practical answers that will be close to those actually measured. When you stop to consider the tolerances on the tube, resistor, and power supply, you will realize that there is not much point in trying to obtain answers any more exact than those obtained by just reading the graph. The calculation of V_T is an insurance step used to protect against gross errors when reading the graphed values.

The circuit analyzed in Example 7-3 yields a change in the value of the voltage across the tube as a function of a change in input flux. The magnitudes of the voltage and flux changes can be used to calculate a circuit specification called *voltage sensitivity,* which is defined by

$$S_v = \frac{\Delta V}{\Delta F\text{v}} \tag{7-4}$$

where S_v is the voltage sensitivity
ΔV is the change in tube voltage
$\Delta F\text{v}$ is the change in luminous flux

EXAMPLE 7-4

Calculating the Sensitivity for Example 9-1

$$S_v = \frac{165\text{ V} - 80\text{ V}}{0.04\text{ lm} - 0.08\text{ lm}}$$

$$= 2125\text{ V/lm}$$

An examination of the other two load resistor lines shows that they will yield much smaller changes in voltage for the same change in flux; in fact, the 10-MΩ line is so close to vertical that it really is not possible to determine a ΔV value by graphic analysis. These lower values of resistance, however, do have the advantage of yielding a more linear change in current with respect to illumination because they intersect portions of the V–I curve where tube current is less dependent on tube voltage. Examination of the 50-MΩ resistor V–I curve intersections shows that as luminous flux increases, the portion of the V–I curve intersected becomes progressively less linear. This discussion illustrates an unfortunate but generally common problem concerning detector sensitivity and linearity. In general, the greater the sensitivity, the lower the linearity.

There are two major sensitivity specifications associated with photodiode circuits: current sensitivity and voltage sensitivity. *Current sensitivity* is specified as a characteristic of the tube alone. Current sensitivity represents the expected change in diode current as a function of a change in input illumination with a load resistance of zero ohms. The current sensitivity of a tube is specified assuming that the tube terminal voltage does not change and that the tube is being operated above the knee voltage. You could estimate the current sensitivity of the high-voltage diode by drawing a vertical line on Figure 7-3 at a desired fixed terminal voltage and reading off the corresponding changes in voltage and current. Quite often a typical value of current sensitivity is listed on the data sheet. The tube described by the data sheet of Figure 7-7(a) lists a luminous current sensitivity of 60 μA/lm when tested using a 2856 K lamp as a source.

Voltage sensitivity is a characteristic of the tube and its associated circuitry. Voltage sensitivity represents the expected change in diode terminal voltage as a function of a change in input illumination. A rough estimate of voltage sensitivity can be made by multiplying the current sensitivity by the load resistance. A better estimate of the voltage sensitivity can be made by drawing the load line on the characteristic curves and determining the corresponding terminal voltages as a function of changes in input illumination.

Both sensitivity specifications assume that the tube exhibits a linear change in current in response to changes in input illumination. This is a valid assumption for small changes in illumination and "small" values of load resistance. Figures 7-8 through 7-11 illustrate the typical response characteristics of a high-voltage vacuum diode with characteristics similar to those of Figure 7-3.

Figure 7-8 illustrates the manner in which tube current can be expected to change in response to changes in input flux density. Note that for flux densities below 0.08 lm, the current flow caused by any given level of illumination is almost independent of the load resistance. At high levels of illumination, however, the circuit with the smaller load resistor yields a higher current. If you refer back to Figure 7-3, you will see why this occurs. At high levels of illumination the load line of the larger load resistor is closer to the knee voltage, so the circuit yields a lower current.

An alternative way to look at the current response of a tube is by examining the manner in which current sensitivity varies as a function of tube current. Figure 7-9 is a graph of current sensitivity versus tube current. An ideal detector would have a constant fixed value of current sensitivity. The circuit using a 25-MΩ load resistance yields a current response most like that of an ideal detector, and a sensitivity slightly greater in magnitude than that which occurs when using a 50-MΩ load resistance.

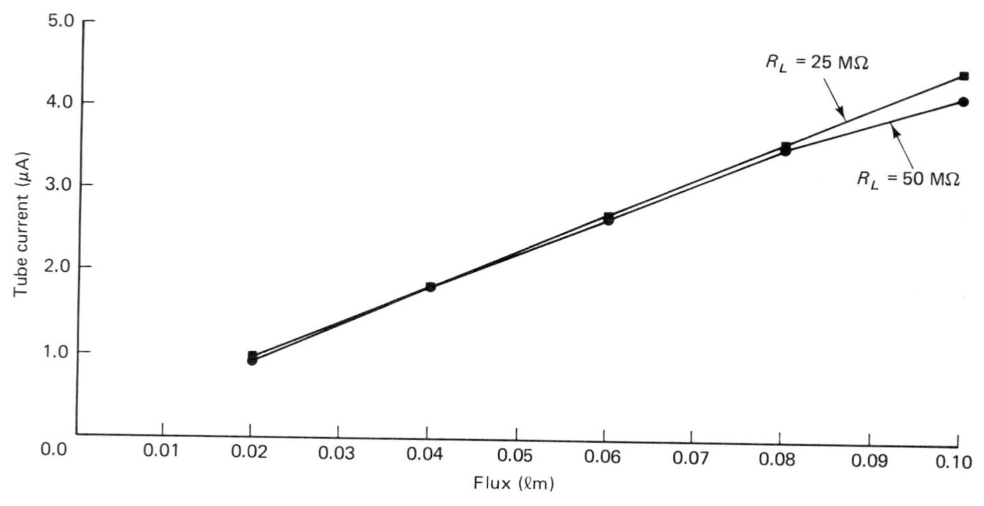

Figure 7-8 Vacuum photodiode current versus flux for two different values of load resistance.

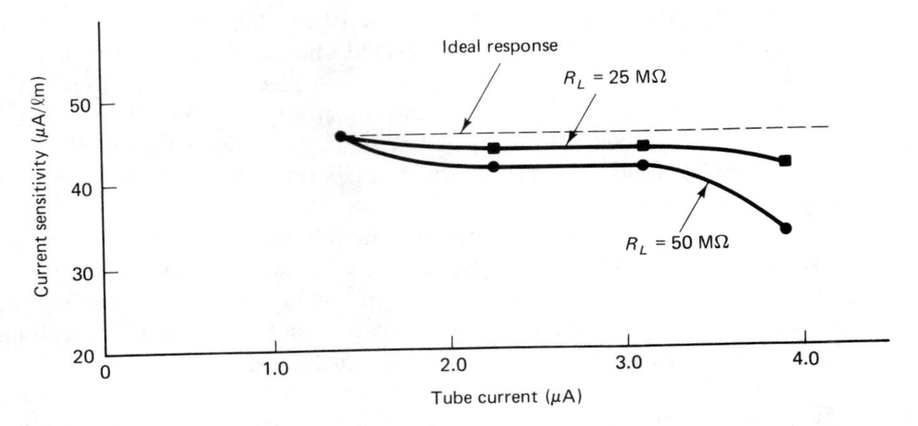

Figure 7-9 Vacuum photodiode current sensitivity versus photodiode current for two different values of load resistance.

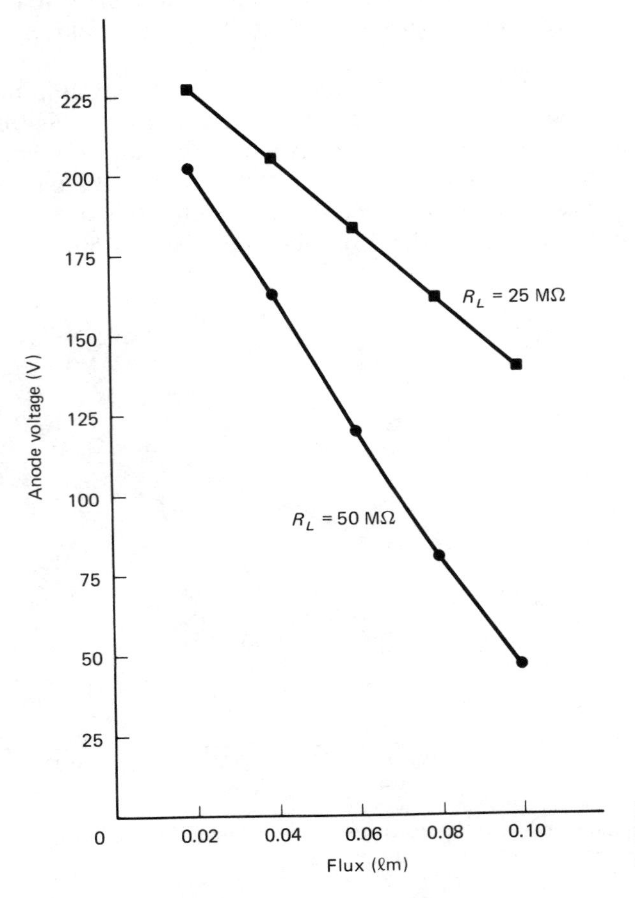

Figure 7-10 Vacuum photodiode anode voltage versus flux for two different values of load resistance.

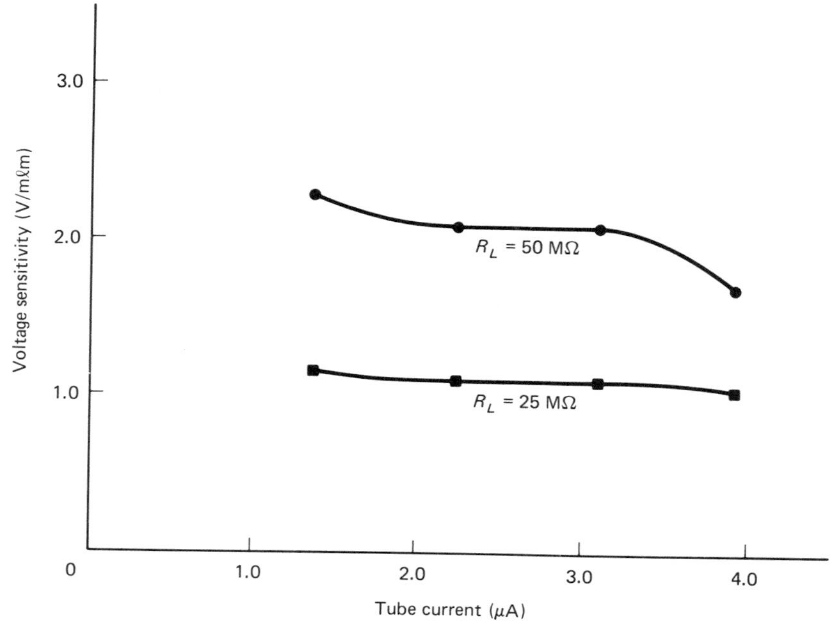

Figure 7-11 Vacuum photodiode voltage sensitivity versus photodiode current for two different values of load resistance.

Figure 7-10 illustrates the manner in which anode voltage changes in response to changes in flux density. The anode voltage decreases as the flux density increases. The larger load resistance yields smaller values of anode voltage for any given flux density. But the larger load resistance also yields a larger change in anode voltage per unit change in flux density. Figure 7-11 illustrates the variation in voltage sensitivity for the two different values of load resistance. As was the case with the current sensitivity, the smaller load resistor yields a more constant sensitivity factor. Unlike current sensitivity, there is a large difference in the magnitude of the voltage sensitivities. The 50-MΩ resistor is twice as large as the 25-MΩ resistor and the voltage sensitivities caused by the two resistors also exhibit a 2:1 ratio.

As a summary of the foregoing discussion, we can conclude:

1. Photodiode current will increase linearly with increasing flux density if the load resistance is small.
2. Ideal, specified, current sensitivity has a constant value that is independent of load resistance.
3. Smaller values of load resistance will yield current sensitivities most like the ideal.
4. Practical circuits depart from the ideal most noticeably at large and small values of tube current.
5. As a practical matter there is little change in the value of current sensitivity as a function of load resistance at intermediate current values.

6. Anode voltage will decrease with increasing flux density.
7. Voltage sensitivity is directly proportional to load resistance.
8. Small values of load resistance will yield the most constant value of voltage sensitivity and the most linear change in voltage and current as a function of changing flux density.

Because the current characteristics of these circuits are relatively independent of component values, instrumentation circuits that monitor photodiode current are easy to design and maintain. This type of instrumentation is most suitable for applications where the flux being monitored has a constant value for relatively long periods of time, and when changes in flux density occur, they are relatively large. The biggest disadvantage associated with using photodiode current as the detected quantity is its small magnitude, which can be difficult to measure. In addition, the photodiode current will change as the diode changes temperature and as it ages. These changes in current are not a result of the incident flux and must be compensated for in some way.

Circuit voltage characteristics are much more sensitive to component values than are current characteristics. To obtain linear voltage operation, the load resistance needs to be selected with some care. Instrumentation that monitors changes in voltage is most suitable for applications where the flux being monitored can be expected to undergo frequent rapid changes in value. An oscilloscope or chart recorder would be a typical readout instrument. The readout instrument would be insulated from the high average anode voltage with a coupling capacitor and possibly an amplifier. When using anode voltage changes as the circuit output, a well-regulated photodiode power supply is required because changes in supply voltage will be felt directly at the circuit output. The biggest advantage in using anode voltage change as the circuit output is the fact that small changes in flux density will cause easy-to-measure large changes in voltage. For example, the photodiode of Figure 7-3 will experience an 11-V anode voltage change in response to a 0.01-lm change in flux when a 25-MΩ resistor is used.

7-5 SUMMARY

Vacuum photodetectors are electronic devices that can cause a voltage or current output which is linearly related to the incident flux. Photomultipliers are vacuum photodetectors that use the process of secondary emission to increase the current sensitivity of the detector. The response of these detectors is wavelength sensitive. Luminous characteristics are specified using sources operated at a specific color temperature, whereas monochromatic sensitivities are specified with graphs of response versus wavelength. The current sensitivity of these devices is relatively independent of circuit component values, with only a slight decrease in sensitivity occurring with increasing load resistance. Voltage sensitivity is a circuit characteristic and is directly related to the size of the load resistance. As a rough estimate, the operating voltage sensitivity will be equal to the product of the load resistance and the specified current sensitivity. Greater linearity will be obtained with the smaller values of load resistance.

Current operation is most suitable for metered steady-state measurements. Voltage operation is most suitable for pulsed operation using an oscilloscope or chart recorder as a readout device.

PROBLEMS

1. Refer to Figure 7-3. On a single piece of graph paper, plot phototube voltage versus light flux for each of the three plotted values of load resistance and the five values of light flux. You will have to calculate tube voltage for the 10-MΩ load.
   (a) Which plot is most linear?
   (b) Which plot is most sensitive?
2. Describe the process of secondary emission.
3. Refer to the schematic in Figure 7-5(b) labeled "typical voltage divider arrangements." Assume that the designer has selected resistors R_1 to R_{10} and R_L to be of such a size that each voltage increase between any two elements is the same [see the note on Figure 7-5(a)]. Assume that the supply voltage is 1250 V. What will the voltage across every resistor be?
4. To obtain the absolute maximum rated average anode current for a type 4818 photodiode, what is the maximum luminous flux in lumens that can fall on the cathode? What must the light-source color temperature be?
5. **(a)** For the Hamamatsu R1842, what is the minimum detectable luminous flux from a source with a color temperature of 2856 K?
   (b) For the type 4818 photomultiplier, what is the minimum-detectable luminous flux from a source with a color temperature of 2854 K?

   Assume as a rule of thumb that the photodiode current caused by illumination must be at least equal in magnitude to the device's dark current.

8

Thermal Detectors

8-1 INTRODUCTION

Thermal detectors are a class of devices that operate by converting incident radiant energy to heat and then to a measurable electrical quantity. These devices use a surface coating that mimics the behavior of the ideal blackbody. Thermal detectors are used as radiometric detectors in laboratory and calibration equipment. In this chapter we discuss four common thermal detectors and their specifications: (1) bolometers, (2) thermistors, (3) thermopiles, and (4) pyroelectric detectors.

Bolometers and thermistors exhibit a change in resistance when subjected to radiation. Thermopiles exhibit terminal voltages that are proportional to the incident radiation's intensity. Thermopiles can monitor constant radiation levels, but they exhibit very low cutoff frequencies and are not suitable for monitoring rapid changes in radiation. Pyroelectric detectors exhibit a change in terminal voltage when subjected to a change in radiation. This change in voltage is not fixed and will decay to zero over time. Thus pyroelectric detectors can be used to monitor changes in radiation levels but are not suitable for monitoring fixed levels of radiation. At the radiation levels normally used in a laboratory, all four devices will be coupled to some sort of electronic amplifying circuits.

Thermistors and thermopiles of a more robust construction are used as heat monitors in industrial and process control tests. The physical temperatures of furnaces and ovens are often monitored with these detectors. The thermistors and thermopiles used in optical setups are fragile devices constructed using thin-film deposition techniques. Properly coated thermopiles and pyroelectric detectors are essentially ideal radiometric detectors. The major limitation on the range of wavelengths to which they respond is the net optical transmission characteristic established by windows and lenses used in the detector package and the laboratory setup.

To interpret a detector's data sheet, an understanding of detector specifications is required. In the next three sections we define these specifications, with particular attention being paid to the optical specifications of the detector head in Section 8-3 and noise specifications in Section 8-4.

8-2 GENERAL DETECTOR SPECIFICATIONS

In an optical detector the incident radiation causes a change in the detector element. This change manifests itself as a change in conductivity, a change in voltage, a change in current, or a change in stored electrical charge. The change experienced by the detector is usually converted to an electrical signal that has an amplitude related to the amplitude of the radiation. This electrical signal can be quite small, so it is normally amplified. The amplified signal is then used to drive a meter, a chart recorder, or a similar indicating element, which is calibrated in radiation units.

Many optical detector systems actually use two identical detectors mounted in close physical proximity so that both detectors are at the same ambient temperature. One of these detectors will be exposed to the radiation to be measured and the other will be shielded from the radiation. The electrical signal developed is obtained by subtracting the response on the reference detector from the response of the irradiated detector. Practical systems will have a zeroing procedure during which both detectors are shielded (darkened) and any signal indication is zeroed out.

The most important optical detector specifications are responsivity (sensitivity), spectral response, time constant, noise equivalent power (NEP), detectivity, normalized detectivity D^*, and detector field of view. These specifications are defined below.

> *Responsivity:* The ratio of an optical detector's electrical output to its optical input, the precise definition depending on the detector type; generally expressed in amperes per watt or volts per watt of incident radiant power. *Sensitivity* is often used incorrectly as a synonym. (*Source:* ANSI/IEEE Standard 100-1984. Reprinted by permission of the IEEE Standards Department.)

The responsivity R is defined quantitatively as

$$R = \frac{dr}{d\phi_e} \tag{8-1}$$

where R is responsivity
dr is the change in detector voltage or current
$d\phi_e$ is the change in flux falling on the detector

> *Spectral Response* (Spectral Sensitivity Characteristic) The relation between the radiant sensitivity and the wavelength of the incident radiation, under specified conditions of irradiation. *Note:* Spectral sensitivity characteristic is usually measured with a collimated beam at normal incidence. (*Source:* ANSI/IEEE Standard 100-1984. Reprinted by permission of the IEEE Standards Department.)

Thermal detectors have a wide flat spectral response which is limited primarily by the transmission characteristics of the "windows" used in the package. Semiconductor detectors and vacuum photodiodes have spectral responses that are generally characteristic of the materials used to fabricate the detector.

The detector time constant represents the step response of the detector to a step change in radiation level, and may be specified indirectly in terms of frequency. Many detectors are illuminated with a chopped beam. A *chopped beam* is broken into pulses (modulated) either mechanically with a rotating shutter or by modulating the source power. The detector will behave like a low-pass filter circuit and it will develop different outputs for different chopping or modulation rates. Detector-time response may also be specified in terms of rise times. These time-related specifications are described below.

> *Time Constant* The value T is an exponential response term, $A\ \exp(-t/T)$. *Note:* For the output of a first-order system forced by a step or an impulse, T is the time required to complete 63.2% of the total rise or decay. (*Source:* ANSI/IEEE Standard 100-1984. Reprinted by permission of the IEEE Standards Department.)

The time constant time T is sometimes called the $1/e$ time or the $(1 - 1/e)$ time. On a Bode frequency diagram for a first-order low-pass filter, the −3-dB frequency, f_b, is related mathematically to the time constant.

Bandwidth, Low-Pass, −3-dB Frequency:

$$f_b = \frac{1}{2\pi T} \tag{8-2}$$

where f_b is the −3dB frequency of a first-order low-pass system, the system bandwidth.

> *Pulse Rise Time* (Semiconductor Radiation Detectors) (t_r) The interval between the instants at which the instantaneous value first reaches specified lower and upper limits, namely, 10 and 90% of the peak pulse value unless otherwise specified. *Notes:* In the case of a step function applied to an RC low-pass filter, the rise time is given by $t_r = 2.2RC$. (*Source:* ANSI/IEEE Standard 100-1984. Reprinted by permission of the IEEE Standards Department.)

$$t_r = 2.2T \tag{8-3}$$

These time and frequency specifications are illustrated in Figure 8-1. The data sheet value of the time constant is dependent on the equivalent resistance and capacitance of the detector under specified test conditions. The useful time constant of the device will be affected by the circuitry to which the device is connected. The response of a complete measurement system will always be slower than the specified time constant of the detector.

> *Noise Equivalent Power* (NEP) At given modulation frequency, wavelength, and for a given effective noise bandwidth, the radiant power that produces a signal-to-noise ratio of 1 at the output of a given detector. *Notes:* (1) Some manufacturers and authors

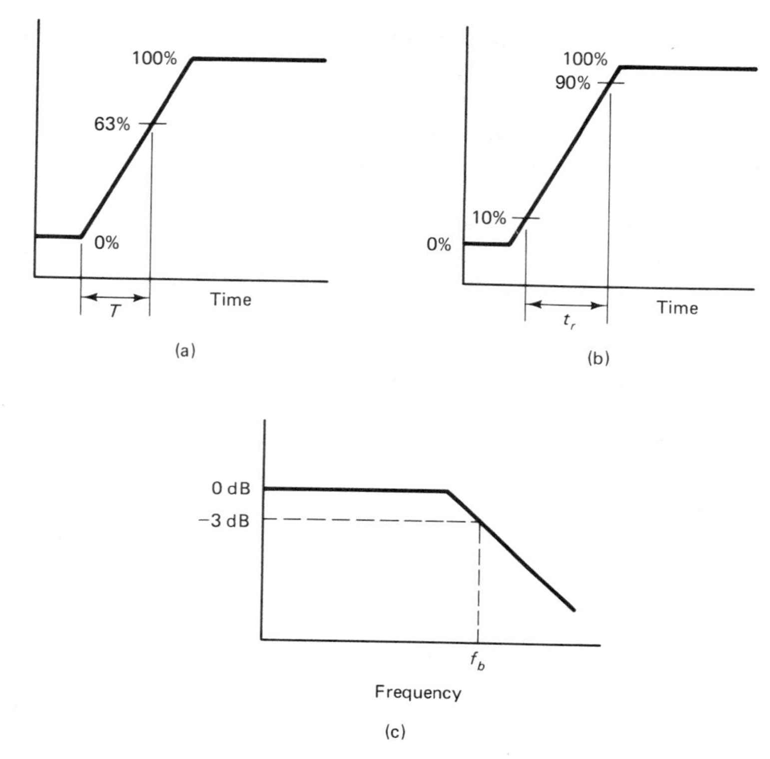

Figure 8-1 Pulse and frequency specifications of detectors: (a) response time; (b) rise time; (c) frequency response, bandwidth.

define NEP as the minimum detectable power per root unit bandwidth; when defined this way, NEP has units of watts/(hertz)$^{1/2}$. Therefore, the term is a misnomer, because the units of power are watts. (2) Some manufacturers define NEP as the radiant power that produces a signal-to-dark current noise ratio of unity. (*Source:* ANSI/IEEE Standard 100-1984. Reprinted by permission of the IEEE Standards Department.)

In addition to the definitions above, you may encounter a definition of NEP as the power that will produce a voltage signal-to-noise ratio of 1. When comparing NEP values, you must determine, first, if they relate to the same definition and second, if the measurement conditions are comparable.

Some detector–window combinations do not have flat spectral responses, so it is necessary to know the nature of the incident radiant energy used to establish these specifications. Sources used for this purpose emit either incandescent radiation of a fixed color temperature, or they have a monochromatic output.

Because the time constant characteristics will affect the magnitude of the device output, the modulation frequency and wave shape must also be specified. The magnitude of the detected signal and noise power, noise current, or noise voltage will depend on the electrical frequency bandwidth of the measurement system.

In the definition for NEP this is called the *effective noise bandwidth.* Only NEP values measured or adjusted for the same bandwidth can be directly compared.

NEP is a specification that attempts to describe the minimum useful signal level that a device can develop. The units of measurement are watts of incident radiant flux that give rise to a signal-to-noise ratio of 1:1. There is some ambiguity in practice as to which kind of signal-to-noise ratio will be used as a basis for measurement. Ratios of noise power to signal power, noise current to signal current, and noise voltage to signal voltage are all used. The NEP value represents a threshold level of performance, so the smaller the number, the better.

Detectivity The reciprocal of noise equivalent power. (*Source:* ANSI/IEEE Standard 100-1984. Reprinted by permission of the IEEE Standards Department.)

Detectivity is sometimes incorrectly used for D^*, defined below.

D^* (pronounced "D-star") A figure of merit used to characterize detector performance. Defined as the reciprocal of noise equivalent power (NEP) normalized to unit area and unit bandwidth.

$$D^* = \frac{[A(\Delta f)]^{1/2}}{\text{NEP}} \tag{8-4}$$

where: A is the area of the photosensitive region of the detector and Δf is the effective noise bandwidth. Synonym: specific detectivity. (*Source:* ANSI/IEEE Standard 100-1984. Reprinted by permission of the IEEE Standards Department.)

D^* is a figure of merit that increases as NEP decreases. Large values of D^* are better than small values. Because D^* is normalized to area, it allows the specification of a number of devices of similar construction, with one specification. As was the case with NEP, the conditions of measurement must be known if device comparisons are going to be made. D^* appears to be a more common specification than NEP.

All of the specifications above presume that the incident radiation from the test source is the only radiation causing the detector to respond. Many thermal detectors respond in a uniform manner to a very wide range of wavelengths. When testing these devices the ambient background energy can add to the source energy. The impact of the ambient effects on the measurement is dependent on such detector characteristics as the field of view and angle of view. Detectors with large angles of view will see more of the ambient background sources. Also, the magnitude of the radiation from background sources relative to the magnitude of the radiation from the source under test is an important consideration. It may be to your advantage to use optical filters when evaluating monochromatic sources to exclude the effects of background radiation.

The *field of view* of a detector is an area in space from which the detector receives power. The angle of view is an angle measured relative to the detector's surface that defines the boundaries of a volume in space from which energy can reach the detector. Ideally, the only radiating source in the field of view will be the test source. Background sources are sources other than the test source which cause power to fall on the detector. The temperatures of objects in the background should be too low to cause detectable radiation. The field of view and angle of view are discussed further in the next section. All of the specifications discussed in this section may be found on data sheets for optical detectors discussed in this and subsequent chapters.

8-3 DETECTOR OPTICS

In the discussion of incandescent lamps in Chapters 3 and 4, it was pointed out that most of their radiated energy falls in the infrared portion of the spectrum. Thus the detectors discussed in this chapter are used primarily as radiometric infrared detectors, and it is important that the optics used with these radiometric detectors be able to work effectively in the infrared portion of the spectrum. Note that the basic principles of optics reviewed in this section are applicable to all regions of the electromagnetic spectrum.

It is the job of the detector's optical system to concentrate the radiation on the detector in a predictable manner. Mirrors and lenses can be used to accomplish this task. As pointed out in Chapter 1, first-surface gold mirrors will yield the best infrared results, but aluminum coatings can also be used. Most optical glass has poor transmission qualities at the very long wavelengths of 1 to 10 μm, so various special materials are used for instrument optics in these ranges (Figure 8-2).

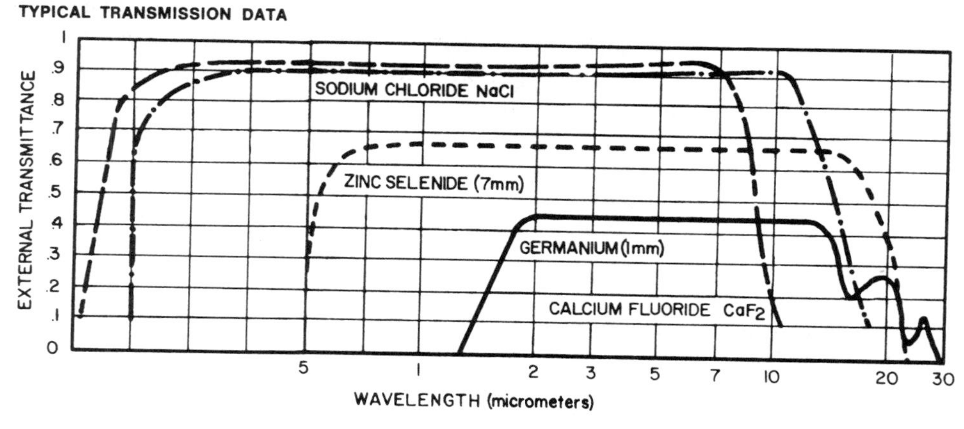

Figure 8-2 Transmission data for typical window materials. (*Oriel Optics and Filters,* Vol. III, 1984; courtesy Oriel Corporation, Stratford, CT.)

In addition to the focusing elements and windows in the detector head, filters may be used to limit detection to specific wavelengths. The major optical specifications applied to the detector head in addition to spectral transmission are (1) F number, (2) angle of view θ_v, and (3) field of view or spot size S. Figure 8-3(a) illustrates a detector head that uses a lens to focus the incident radiation on the

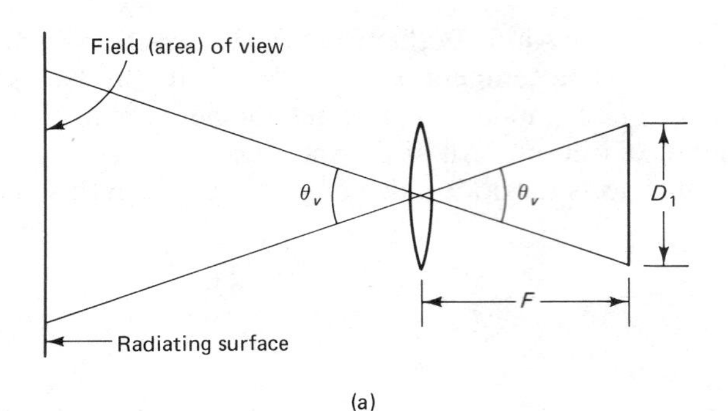

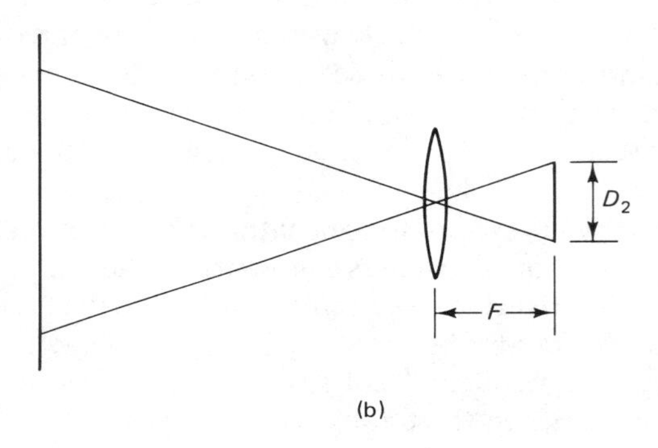

Figure 8-3 Relationship between angle of view, focal length, and detector diameter: (a) long focal length; (b) short focal length.

active surface of the detector. The detector is located at the focal plane of the lens so that the entire active surface is irradiated. The illustrated vertical dimension of the detector, D_1, and the focal length, F, of the lens form an angle θ_v called the *angle of view.* For small angles of less than 30°,

$$\theta_v = D/F \text{ radians} \tag{8-5}$$

For angles greater than 30°,

$$\theta_v = 2\tan^{-1}\frac{D/2}{F} \tag{8-6}$$

where θ_v is the angle of view
D is the detector's vertical dimension
F is the focal length of the lens

Equation (8-5) is justified for small angles because the arc length projected by the angle θ_v at the distance F is essentially the same length as the chord D.

Figure 8-3(b) illustrates a detector head with the same angle of view as that of Figure 8-3(a) but with a smaller detector and a shorter focal length. Both of the detectors of Figure 8-3 can be thought of as projecting a vertical dimension d onto the surface of the extended source as illustrated. The detector has a two-dimensional surface, so an area is actually projected onto the surface of the extended source. If the detector is circular, you can envision a cone extending out from the surface of the detector; this cone may be referred to as the *cone of acceptance*, the *solid angle of acceptance*, or the *solid angle of view*. Any radiation that occurs inside this cone can cause flux to fall on the surface of the detector. In this case the solid angle is envisioned as having the detector at the apex of the cone formed by the solid angle and the extended source at the base of the cone. This is the reverse of our original definition for a solid angle in Chapter 3, where the solid angle was defined with a point source at the apex and the detector surface at the base. In either case the boundaries of the solid angle are established by an internal Cartesian angle. For the cases illustrated in Figure 8-3, this Cartesian angle is called the *angle of view* of the detector. The area projected onto the surface of the source is called the *field of view* or *spot size*, and is assigned the symbol S. The perimeter of this area is established by the intersection of the solid angle of view with the surface.

The field of view is the surface from which energy is collected to excite the detector. In Figure 8-3, if both detectors have the same shape, they will both have the same size field of view. This means that both detectors will receive the same amount of flux and therefore develop the same signal voltage. The noise voltage developed by the detectors, however, is proportional to their area, so the smaller detector will develop the smaller noise voltage. This means that the smaller detector will exhibit a better signal-to-noise ratio. The F number of the optical detector system is the ratio of the system focal length to aperture diameter. Because short-focal-length systems yield superior signal-to-noise performance, it is desirable to have low-F-number optical detector systems.

The field of view projected by the detector will have the same shape as the detector element. If the detector is circular, the field of view will be circular. If the detector is square, the field of view will be square. Referring back to Figure 8-3, if the detector is circular, the dimension d will be the diameter of the field of view. Knowing that the field is circular, we can compute its area S in terms of the angle of view and the distance R.

$$d = \theta_v R \qquad \text{small-angle formula}$$

$$S = \frac{\pi d^2}{4} \qquad \text{area of circle}$$

$$S = \frac{\pi \theta_v^2 R^2}{4} \qquad \text{circular detector small angle} \tag{8-7}$$

Using the same process for large angles yields

$$S = \pi \left(\tan \frac{\theta_v}{2} \right)^2 R^2 \tag{8-8}$$

If the detector is sqaure, the dimension d would represent one side of a square field of view. The area of a square is simply the product of the lengths of the sides. So the spot size S for a square detector is calculated as follows. For small angles

$$S = \theta_v^2 R^2 \qquad \text{square detector} \tag{8-9}$$

For large angles

$$S = 4\left(\tan \frac{\theta_v}{2}\right)^2 R^2 \qquad (8\text{-}10)$$

Occasionally, a detector is rectangular. A rectangular detector will have different principal x and y dimensions, and it will have two different angles of view, corresponding to these two different dimensions. The x dimension will give rise to an angle of view θ_x and the y dimension will give rise to an angle of view θ_y. The field-of-view area will be computed in terms of both of these angles and the distance R. For small angles

$$S = \theta_x \theta_y R^2 \qquad (8\text{-}11)$$

For large angles

$$S = 4\left(\tan \frac{\theta_x}{2}\right)\left(\tan \frac{\theta_y}{2}\right) R^2 \qquad (8\text{-}12)$$

In each of the six equations for the field of view above, the spot size is directly related to the square of the distance from the detector to the surface of the extended source and the angle of view of the detector. Figure 8-4 illustrates two detector heads. Figure 8-4(a) illustrates a detector head with a lens such as we have just discussed. Figure 8-4(b) illustrates a simpler head design which contains a plane-parallel window and a detector. When the dimensions of the detector and the window and the spacing between them are known, the angle of view for this design can be determined. The equations for a circular design are presented below, where D_w represents the window diameter, D represents the detector diameter, and x represents the distance between the detector's active surface and the front surface of the window. For small angles

$$\theta_v = \frac{D_w + D}{x} \quad \text{rad} \qquad (8\text{-}13)$$

For large angles

$$\theta_v = 2 \tan^{-1} \frac{D_w + D}{2x} \qquad (8\text{-}14)$$

For the cases we have been discussing, the source area is larger than the field of view. This is sometimes called the *near-field condition.* For all distances that yield a field of view smaller than the source area and for sources that are diffuse, the power incident on the detector will be constant and independent of the distance between the source and the detector. For these cases the power incident on the detector can be calculated as

$$\phi_e = \tau L_e \omega S \qquad (8\text{-}15)$$

where ϕ_e is the incident power
τ is the transmission ratio of the optics
L_e is the source radiance

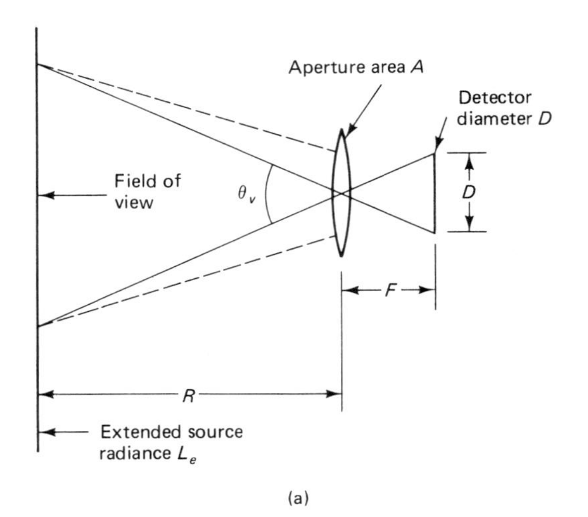

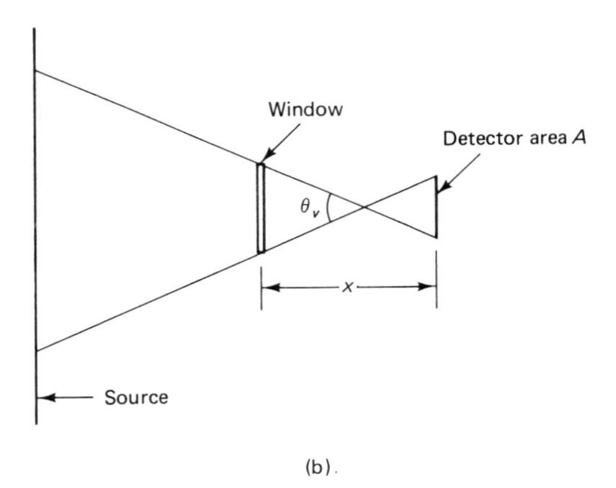

Figure 8-4 Detector field of view on a large radiating surface: (a) angle of view and field of view, focused system; (b) angle of view and field of view with a simple window system.

ω is the solid angle of view of the detector
S is the field of view

Recall the definition of the solid angle of view:

$$\omega = \frac{A}{R^2}$$

where A is the entrance aperture area of the detector head
R is the distance between the detector aperture and the source

The equation used for S is dependent on the detector shape and the magnitude of the angle of view. For purposes of illustration the equation for incident power on a circular detector with a large field of view is derived below. The other five possible equations for incident power can be derived in a similar manner by substituting the various equations for S derived previously.

$$\phi_e = \tau\, L_e\, \omega\, S$$

$$\phi_e = \tau \frac{L_e A_2}{R}\, \pi \left(\tan \frac{\theta_v}{2} \right)^2 R^2$$

$$\phi_e = \tau L_e A \pi \left(\tan \frac{\theta_v}{2} \right)^2 \tag{8-16}$$

As equation (8-16) illustrates, the incident power is independent of the distance R, the two R^2 terms divide out, and the incident power is directly proportional to the source radiance and a set of terms with values that are established when the detector head is fabricated. All other equations for near-field incident power will have the same form.

EXAMPLE 8-1

Computation of the Power Falling on a Square Detector Caused by an Extended Source

Given: **(1)** $L_e = 100\ \text{mW/sr} \cdot \text{cm}^2$
(2) $\theta_v = 10\ \text{mrad}$
(3) $A = 16\ \text{mm}^2$
(4) $\tau = 1.0$

Find: The power received.

Solution:

$$\phi_e = \mathrm{r} L_e A \theta_v^2$$

$$= 1.0 \times \frac{100\ \text{mW}}{\text{sr} \cdot \text{cm}^2} \times 16\ \text{mm}^2 \times (10 \times 10^{-3}\ \text{rad})^2$$

$$= 2\ \mu\text{W}$$

This example illustrates that the received power is a function of the detector's aperture, angle of view, and the source radiance. As a practical matter, you would be more likely to use the instrument power reading to determine surface radiance than to use radiance to determine power.

Measurement of the power received from a source with an area a_t that is smaller than the spot size S represents a separate problem. When the source area a_t is smaller than the spot size, the incident power on the detector will decrease as the distance between the detector and the source increases.

$$\phi_e = \mathrm{r} L_e \omega a_t$$

Substituting for ω gives

$$\phi_e = \mathrm{r} L_e \frac{A}{R^2}\, a_t \tag{8-17}$$

But $I_e = L_e a_t$; therefore,

$$\phi_e = I_e \frac{A}{R^2} \tag{8-18}$$

where ϕ_e is the received flux (power)
I_e is the intensity of the source
A is the aperture area
R is the distance to source
τ is the transmission factor
ω is the solid angle of view of the detector

Since the intensity of the source and the aperture area of the detector are fixed, equation (8-18) shows that the received power will decrease with the square of increasing distance. This is the inverse-square law as it applies when the spot size of the detector is larger than the source area.

Objects in the solid angle of view of the detector behind the small source can also cause power to fall on the detector. It is important when measuring a small source that all surfaces in the solid angle of view be at very low color temperatures compared to the source's color temperature, and that reflections be suppressed. It is often necessary to surround the source and the detector system with nonreflecting baffles to obtain accurate measurements.

The power in a collimated beam can be measured if the beam can be expanded or focused so that it just fills the aperture area of the detector. With collimated beams, it is important not to exceed the flux density ratings of the detector. Attenuation caused by additional optical elements in the beam path between the beam and the detector will have to be compensated for. The power measured for a collimated beam should be constant and independent of distance at lens bench distances for 1 or 2 m.

The boundary between the near-field region of measurement and the far-field region of measurement remains to be defined. This boundary can be determined by comparing detector spot size S at a given distance R to the source surface area a_t facing the detector. The critical distance R_c occurs when $S = a_t$.

$$S = a_t \qquad \text{when } R = R_c \tag{8-19}$$

Two equations are derived below, first for a small angle of view and a square detector, and second, for a large angle of view and a circular detector.

Derivation of critical distance R_c for a square detector and a small angle of view:

$$S = a_t$$

$$\theta_v^2 R_c^2 = a_t$$

$$R_c^2 = \frac{a_t}{\theta_v^2} \tag{8-20}$$

where S is the spot size of the detector
R_c is the critical distance
a_t is the source area
θ_v is the angle of view

Derivation of critical distance R_c for a circular detector and a large angle of view:

$$S = \pi\left(R \tan \frac{\theta_v}{2}\right)^2$$

$$S = a_t \qquad \text{when } R = R_c$$

Therefore,

$$R_c^2 = \frac{a_t}{\pi(\tan \theta_v/2)^2} \tag{8-21}$$

If R is equal to or less than R_c, the source should be considered an extended source and the measurements made are near-field measurements. If R is greater than R_c, the measurements made are far-field measurements and the small source can be considered a point source.

An ideal extended source will yield measured values of flux that are constant and independent of distance. An ideal point source will yield measured values of flux that decrease according to the inverse square law. In practical measurement situations the measured values vary from both of these idealized cases in the region around the critical distance. Measured values will show good agreement with the predictions of the extended source equations for distance values that are 75% or less of the critical distance. Measured values will show good agreement with the predictions of the inverse-square-law equations for distance values that are 130% or more of the critical distance.

EXAMPLE 8-2

Critical Distance and Source Area

Given: The angle of view is specified as 80°.

Find: The maximum distance that a circular 10-cm^2 source (a source of radius 1.78 cm) can be from a circular thermopile detector and still be considered an extended source.

Solution: Find the critical distance R_c.

Note: This is a large-viewing-angle problem:

$$R_c^2 = \frac{a_t}{\pi(\tan \theta_v^2)^2}$$

$$= \frac{10\ \text{cm}^2}{\pi(\tan 80°/2)^2}$$

$$= 2.13\ \text{cm}$$

Conclusion: The source must be 2.13 cm or closer to the detector before near-field measurements can be made. Invoking the rule of thumb just discussed, measurements inside the critical distance should be made at distances less than 75% of R_c.

$$R = 0.75 \times 2.13\ \text{cm}$$

$$= 1.6\ \text{cm or less}$$

This small distance clearly presents practical setup problems. For this source a detector with a much smaller angle of view should be used to determine radiance in the near field.

A similar approach could be taken to find how far away from the source the detector would have to be for the source to be considered a point source. One of the characteristics of a point source is that the flux density decreases as a function of 1 over the distance squared. This is called the *inverse-square law.* This effect can rapidly lead to power levels that are too low to measure reliably because the signal cannot be distinguished from the noise.

Figure 8-5 illustrates how calculated and measured values of detected power compared in an actual test setup. In this test setup the greatest deviations from the calculated values occurred at distances near 22.5 cm. This was the calculated critical distance. For distances of less than 17 cm and greater than 30 cm, there was no serious difference between measured and calculated values. For this test setup, the source appeared to be an extended source for distances that are 75% or less of the critical distance. The source appeared to be a point source for distances greater than or equal to 130% of the critical distance. The source used in this exercise was a diffuse circular source with a 12-cm radius. The rated aperture of the circular detector was a 1 cm^2 and its angle of view was specified as 56°.

A given source may appear as either a point source or an extended source depending on its size and location with respect to the detector. There is a critical distance near which the source appears to be neither a pure point source nor a pure extended source. Measurements should be made outside of this ambiguous region.

When making near-field measurements it is important not to get too far away, but it is also possible to get too close. An ideal diffuse extended source has the same flux crossing any unit of area. Practical sources unfortunately usually depart from this ideal. If sufficiently small subareas of the surface are examined, it will be found that there are subareas where the flux is more intense than the average and others where the flux is less intense than the average. As the detector comes very close to the surface, there is an increasing chance that the radiation intercepted by the detector's reduced spot size will be caused by one of these atypical subareas of the

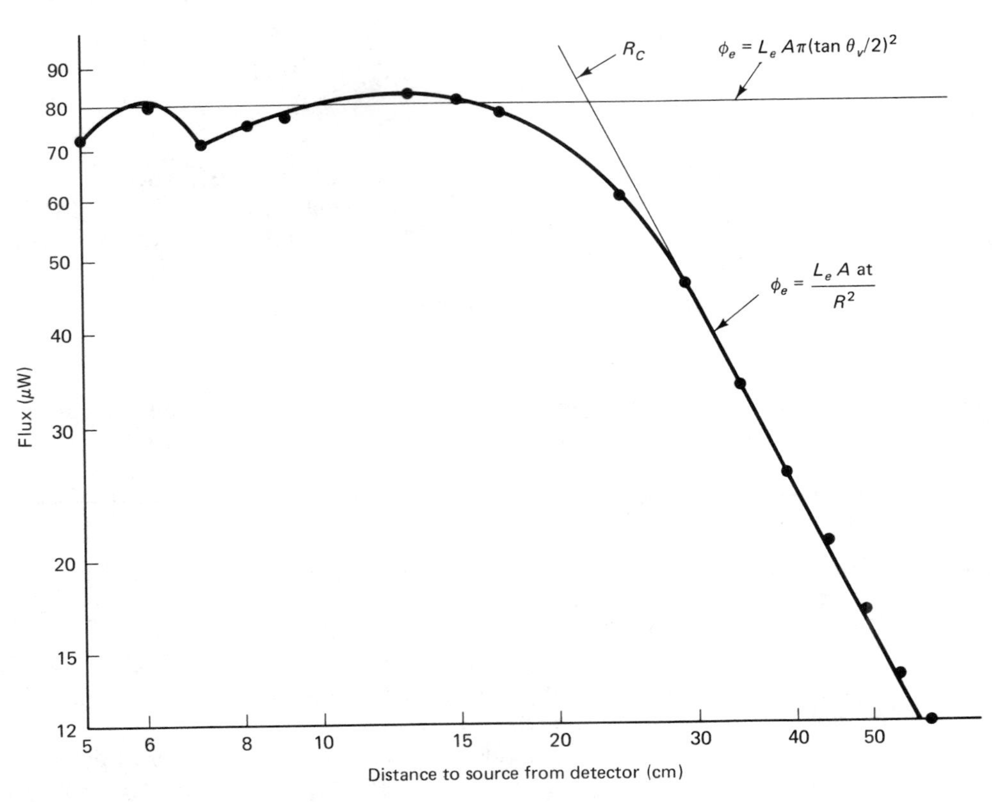

Figure 8-5 Comparison of measured and computed values of flux.

source. A set of measurements made at varying distances in the near field, which all yield approximately the same flux value, is your best indication that you are neither too close nor too far away.

In the inverse-square-law region it is also possible to be too close or too far away. A distance that is equal to 130% of the critical distance establishes the "too close" boundary. The "too far" boundary is established by a number of phenomena that often occur simultaneously. This region is characterized by a very small constant value of measured flux or a series of very small flux readings which tend to vary in an unpredictable manner. The noise characteristics of the detector and its associated instrumentation contribute to this problem. In addition, laboratory settings often have a variety of background radiating and reflecting surfaces that will cause radiation to fall on the detector as the distance between the detector and the source increases.

As a result of the discussion in this section you should be able to define reasonable flux measurement distances for a given source and detector pair. If you know the aperture and angle of view of the detector, it is possible to make a series of flux measurements from which the source radiance can be calculated. Given a source of known area, it is also possible to make a series of measurements from which the field of view and the angle of view of the detector can be determined.

Flux, measured in a collimated beam, should be constant and independent of the distance from the source. System optics are often adjusted so that the beam just fills the detector when the detector is axially centered on the beam. If measured values of flux vary significantly with distance in this kind of system, we must conclude either that the beam is not collimated or that the detector is not maintaining axial alignment with the beam.

8-4 NOISE CALCULATIONS

System noise levels establish the minimum level of measurement. Estimates of these noise thresholds and of signal-to-noise ratios can be made using data sheet specifications. Rigorous noise analysis is a complicated and time-consuming process best left to number-crunching computer programs. Our discussion will limit itself to some comparatively simple techniques (compared to rigorous techniques). These techniques will allow us to predict a reasonable value for signal-to-noise ratio if we know incident power or to estimate minimum measurable values of incident power. The NEP and D^* specifications will be used in these calculations.

Before noise calculations can be performed, an estimate of the noise bandwidth must be made. Noise bandwidth is seldom the same as the electrical bandwidth of the system. The noise bandwidth that needs to be determined is the bandwidth that results from the combined effects of the detector, its amplifier, and any associated amplifiers and instruments in the signal measurement path. If the net resultant electrical bandwidth has a response curve that falls off at a rate of -18 dB/octave or greater above the high-frequency cutoff point, the noise bandwidth and the electrical bandwidth are essentially the same. If the net resultant electrical bandwidth has a response curve that falls off at a rate of -6 or -12 dB/octave, a correction factor will have to be employed. Figure 8-6 illustrates these three conditions for the same cutoff frequency. Noise bandwidth is calculated by multiplying the electrical bandwidth f_b by a bandwidth correction factor. Table 8-1 lists the required correction factors.

Thermopiles have very small electrical cutoff frequencies, typically on the order of 5 Hz with slopes of -6 dB/octave. The instruments and amplifiers connected to the thermopile will have much higher cutoff frequencies. When making estimates of noise bandwidths for thermopiles, the product of the -6 dB/octave correction factor and the thermopile cutoff frequency will yield a realistic value of noise bandwidth.

The definition section of this chapter illustrated that there are several definitions for NEP and D^* which differ from the IEEE Standard. The data sheet of Figure 8-12 (pp. 198-99) utilizes one of these common alternative definitions for NEP. It is not immediately apparent how the definition for D^* used on this data sheet corresponds to the IEEE standard.

The derivations below illustrate how the data sheet definitions of D^* and NEP were arrived at. After the derivations have been completed, two numerical examples will be done using these data sheet values.

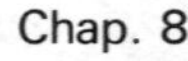

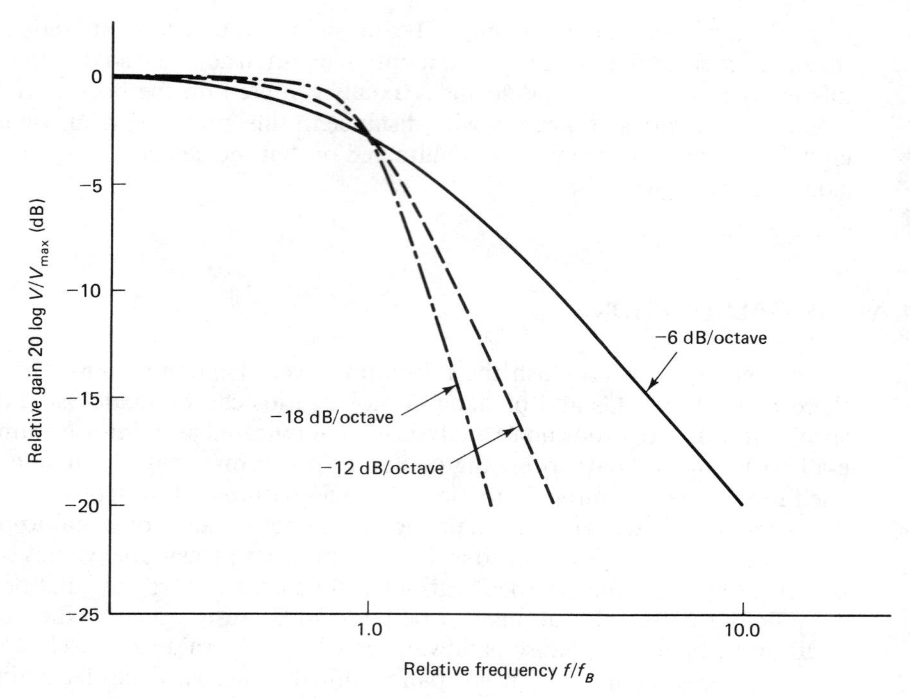

Figure 8-6 Relative low-pass frequency response.

TABLE 8-1 FACTORS FOR CONVERTING ELECTRICAL BANDWIDTH TO NOISE BANDWIDTHS

Slope of Frequency Response (dB)		
Per Octave	Per Decade	Correction Factor KB
−6	−20	1.571 (π/2)
−12	−40	1.222
−18	−60	1
	$\Delta f = K_B f_b$	

Derivation of data sheet NEP and D^* definitions:

$$\text{NEP} = P_0 \qquad \text{(IEEE Standard)}$$

where P_0 is a power level causing a voltage signal V_s that is equal to the noise voltage V_N.

A thermopile generates a voltage V_s that is directly proportional to the incident power P_i. In light of this fact, NEP could be redefined as follows:

$$\text{NEP} = P_0 \qquad V_N = V_s \qquad \text{IEEE Standard}$$

$$\text{NEP} = \frac{V_N P_i}{V_s} \qquad V_s > V_N \qquad \text{alternative definition}$$

The net incident power on the detector is a function of the flux density and the detector area.

$$P_i = HA$$

where P_i is the incident power
H is the incident flux density
A is the effective detector area

Therefore,

$$\text{NEP} = \frac{V_N}{V_s} HA \tag{8-22}$$

This data sheet uses a value of NEP that is normalized to the square root of the bandwidth, which is a common practice. This normalization leads to the final definition for NEP for this data sheet:

$$\text{NEP} = \frac{V_N}{V_s} \frac{HA}{(\Delta f)^{1/2}} \tag{8-23}$$

A similar line of reasoning can be followed to develop the data sheet value of D^*.

$$D^* = \frac{(A\,\Delta f)^{1/2}}{\text{NEP}} \qquad \text{(IEEE Standard)}$$

Therefore,

$$D^* = \frac{(A\,\Delta f)^{1/2}}{\dfrac{(V_N P_i)}{(V_s)}}$$

$$D^* = \frac{V_s}{V_N} \frac{(A\,\Delta f)^{1/2}}{P_i} \tag{8-24}$$

$$D^* = \frac{V_s}{V_N} \frac{(A\,\Delta f)^{1/2}}{HA}$$

Thus

$$D^* = \frac{V_s}{V_N} \left(\frac{\Delta f}{A}\right)^{1/2} \frac{1}{H} \tag{8-25}$$

which is the data sheet form.

The units on the specified values of NEP and D^* also require some consideration. In equation (8-23) for NEP the voltage units cancel, leaving the numerator with units of power and the denominator with units of the square root of hertz. This manufacturer has selected nanowatts as the unit of power to be used. In equation (8-25) for D^* the voltage units also cancel. The denominator units are the result of taking the product of the square root of the area and flux density. The product would have units of power per unit length, as illustrated below.

$$(A)^{1/2}H$$

$$(A)^{1/2}\frac{P}{A}=\frac{P}{(A)^{1/2}}$$

The numerator of equation (8-25) clearly has units of the square root of hertz. The numerator and denominator units are combined as follows:

$$\frac{(\Delta f)^{1/2}}{P/(A)^{1/2}}=\frac{(\Delta f)^{1/2}(A)^{1/2}}{P}$$

The resulting numerator units are the product of the square root of hertz and the square root of area, and the denominator has units of power. Referring to the data sheet of Figure 8-12, this manufacturer uses centimeters as the unit of length and watts as the unit of power. A scaling factor of 10^8 is also employed to obtain easy-to-read numerical values.

Having waded through the definitions for NEP and D^* used on this data sheet, we can now proceed to use these specifications to make some simple calculations.

EXAMPLE 8-3

Estimation of the Voltage Signal-to-Noise Ratio Given Incident Flux Density and Data Sheet Values of D^*

Given: $H = 325\ \mu\text{W/cm}^2$.

Find: Use the maximum data sheet value of D^* in Figure 8-12 to estimate the best possible signal-to-noise ratio.

Solution:

1. Estimate the effective noise bandwidth Δf.

$$T = 28 \text{ ms} \qquad \text{data sheet}$$

$$f_b = \frac{1}{2\pi T} \tag{8-2}$$

$$= \frac{1}{2\pi(28 \text{ ms})}$$

$$= 5.68 \text{ Hz}$$

$$\Delta f = K_b f_b$$

$$= 1.571 \times 5.68 \text{ Hz} \qquad \text{assumes } -6 \text{ db/octave}$$

$$= 8.93 \text{ Hz}$$

2. Estimate the effective surface area, diameter = 0.5 mm.

$$A = \pi\left(\frac{d}{2}\right)^2$$

$$= \pi\left(\frac{0.5\text{ mm}}{2}\right)^2$$

$$= 1.96 \times 10^{-3}\text{ cm}^2$$

3. Use the maximum D^* to estimate the signal-to-noise ratio.

$$D^* = 5.5 \times 10^8 \frac{\text{cm} \cdot (\text{Hz})^{1/2}}{\text{W}} \qquad \text{data sheet}$$

$$D^* = \frac{V_s}{V_N}\left(\frac{\Delta f}{A}\right)^{1/2}\frac{1}{H} \tag{8-25}$$

$$\frac{V_s}{V_N} = H\left(\frac{A}{\Delta f}\right)^{1/2} D^*$$

$$= 325\,\frac{\mu\text{W}}{\text{cm}^2} \times \left(\frac{1.96 \times 10^{-3}\text{ cm}^2}{8.93\text{ Hz}}\right)^{1/2} \times 5.5 \times 10^8 \frac{\text{cm} \cdot (\text{Hz})^{1/2}}{\text{W}}$$

$$= 2648$$

With this level of flux density the signal voltage could be as much as 2648 times larger than the noise voltage. Sometimes the signal-to-noise ratio is expressed in decibels, in which case we would obtain a value of 68.5 dB.

An alternative and perhaps more practical calculation would be one that estimates the minimum flux density needed to achieve a minimum acceptable signal-to-noise ratio. Example 8-4 illustrates a calculation of this type.

EXAMPLE 8-4

Estimation of Minimum Flux Density

Given: Desired minimum signal-to-noise ratio of 25 dB for a model M5 thermopile detector.

Find: The minimum incident flux density.

Solution:

1. Estimate the area and noise bandwidth.

$$\Delta f = 8.93\text{ Hz} \qquad \text{see Example 8-3}$$

$$A = 1.96 \times 10^{-3}\text{ cm}^2$$

2. The poorest NEP is the largest.

$$\text{NEP} = 0.23\ \text{nW/(Hz)}^{1/2}$$

3. Convert the signal-to-noise ratio in decibels to a decimal value.

$$25\ \text{dB} = 20 \log \frac{V_s}{V_N}$$

$$1.25 = \log \frac{V_s}{V_N}$$

$$10^{1.25} = \frac{V_s}{V_N}$$

$$17.78 = \frac{V_s}{V_N}$$

4. Use the poorest NEP value to calculate the flux density value that will yield a 25-dB signal-to-noise ratio.

$$\text{NEP} = \frac{V_N}{V_s} \frac{HA}{(\Delta f)^{1/2}} \tag{8-23}$$

$$H = \text{NEP} \frac{V_s}{V_N} \frac{(\Delta f)^{1/2}}{A}$$

$$= 0.23 \frac{\text{nW}}{(\text{Hz})^{1/2}} \times 17.78 \times \frac{(8.93\ \text{Hz})^{1/2}}{1.96 \times 10^{-3}\ \text{cm}^2}$$

$$= 6.235 \frac{\mu\text{W}}{\text{cm}^2}$$

In this section we have illustrated how NEP and D^* can be used to predict signal-to-noise ratios for a system. A common variant of the definition for NEP was derived from the primary IEEE standard definition. Hopefully, this section will make you cautious about blindly applying "magic" formulas to the solution of noise problems. The units and algebraic definitions of NEP and D^* given on a data sheet should be examined carefully before proceeding with any calculations. Although it may slow you up a little, it is prudent to try and derive the data sheet units from the original IEEE standard definitions.

8-5 BOLOMETERS AND THERMISTORS

Bolometers and thermistors are devices that exhibit changes in resistance with increases in temperature. In optical applications these devices are mounted at the focal point of the optical system.

Bolometers are constructed out of fine pieces of wire, usually platinum or nickel. These devices exhibit an increase in resistance when heated. *Thermistors* are fabricated out of a semiconductor material such as germanium. These materials are called semiconductors because they exhibit a higher resistance than metals. Thermistors exhibit a decrease in resistance when heated. Thermistors tend to be more sensitive than bolometers—that is, they yield a larger change in resistance for the same change in temperature caused by the incident flux. The general equation for bolometer and thermistor operation is

$$R = R_0(1 + \propto \Delta T) \tag{8-26}$$

where R is the resistance of the device at some temperature T
R_0 is the resistance at a reference temperature T_0: $\Delta T = T - T_0$
$\propto$ is a physical constant of the material, called the temperature coefficient of resistance

For wire bolometers the temperature coefficient of resistance will be a positive quantity. For semiconductor thermistors the temperature coefficient of resistance will be negative. The magnitude of the temperature coefficient establishes the detector's sensitivity.

Equation (8-26) is a good equation to use when thinking about thermistor and bolometer operation. If the change in temperature is small and the temperature coefficient of resistance is specified for the temperature measurement range being used, equation (8-26) will yield realistic answers. The unfortunate reality is that detector resistance is not a linear function of temperature, and if detector resistance is plotted as a function of temperature over wide ranges of temperature, very nonlinear curves will result. Therefore, if these devices are to be used as detectors over wide temperature ranges, their nonlinearity will have to be compensated for. For example, an analog meter scale might be calibrated empirically for device response to different levels of radiation. On the resulting meter scale, equal changes in radiation level would result in unequal changes in pointer position.

Figures 8-7 and 8-8 illustrate two circuits capable of converting a change in resistance to an electrical signal. In the circuit of Figure 8-7, resistors R_1 and R_3 would be equal fixed resistors, and R_4 would be the bolometer or thermistor. Resistor R_2 would have the same resistance as the temperature sensitive resistor R_4 at a fixed reference temperature. When the temperature of R_4 is the reference temperature, voltage V_2 would be equal to V_4, and therefore V_x would be equal to zero.

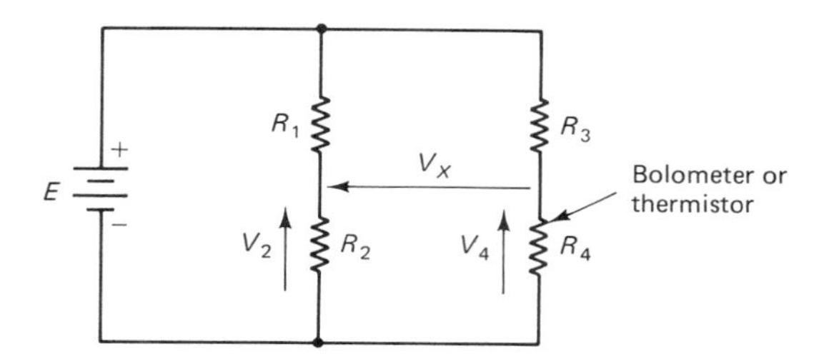

Figure 8-7 Bridge circuit.

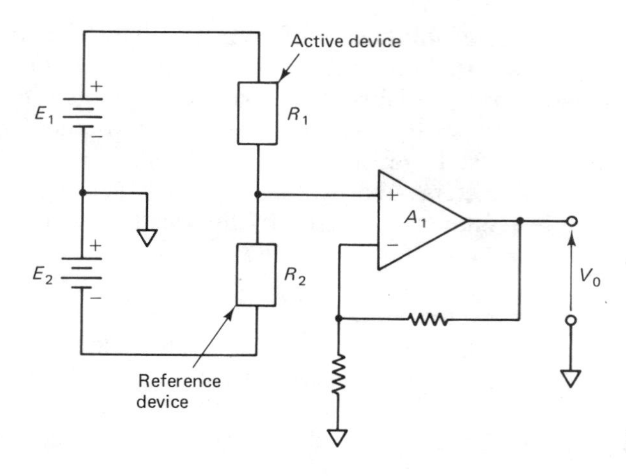

Figure 8-8 Op-aMP thermistor circuit.

If the value of R_4 changed in response to a temperature change, V_2 would no longer be equal to V_4 and V_x would assume some readable value. A refinement of the circuit would be to have R_2 and R_4 be identical thermistors. With R_2 shielded and R_4 irradiated, this configuration provides compensation for room temperature changes.

Figure 8-8 is an example of a similar circuit using an operational amplifier. Voltage sources E_1 and E_2 develop voltages of equal magnitude. Devices R_1 and R_2 are identical thermistors mounted in close physical proximity. R_1 will be exposed to radiation and R_2 will be shielded from radiation. The amplifier A_1 will amplify any change in voltage at the junction of the two resistors that occurs due to an imbalance in the value of R_1 and R_2.

Thermistors and bolometers are examples of devices that accomplish their detection by the conversion of the radiation to heat. Thermistors and bolometers exhibit a change in resistance when heated. These devices always require some external circuitry to convert the resistance change to a usable signal voltage or current.

8-6 THERMOELECTRIC SENSING ELEMENTS

The thermoelectric effect is the physical process used by thermocouples and thermopiles to detect radiation. The thermoelectric effect was discovered by Thomas J. Seebeck and is also called the *Seebeck effect.* Figure 8-9 illustrates how the thermal electric effect caused by the heating of the junction of two dissimilar metals is used. When the two wires are connected at both ends and one junction is heated, a current flow will be measured by the meter. This current flow will be proportional to the temperature difference between the heated and unheated junctions. When the wires are connected at only one end and the junction is heated, an open-circuit voltage will occur between the two wires. The magnitude of this open-circuit voltage will be proportional to the temperature of the heated junction.

When we attempt to measure the open-circuit voltage with a meter, we create

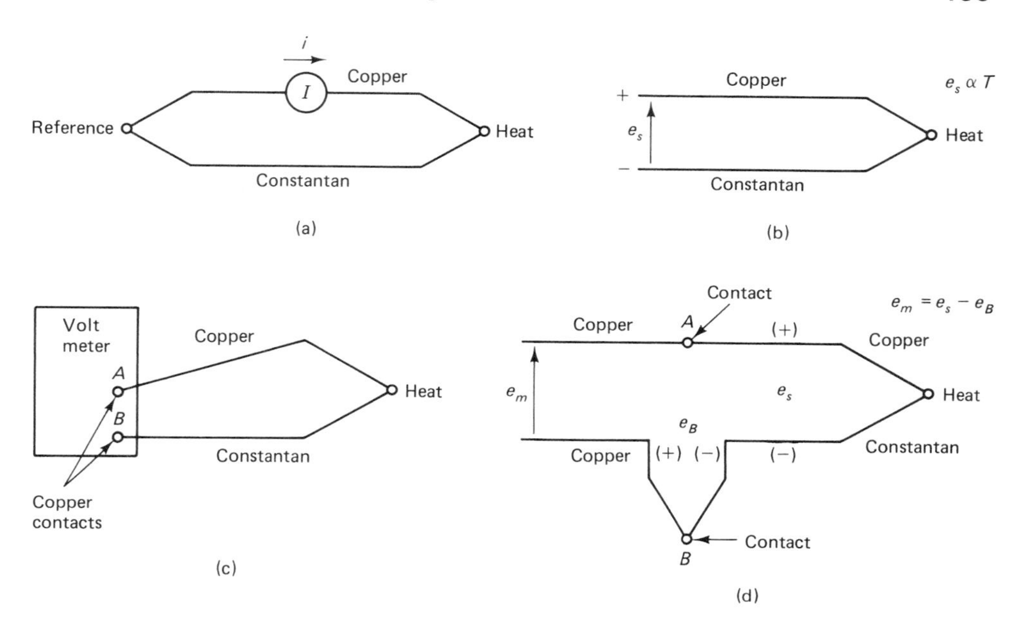

Figure 8-9 Thermocouple operation: (a) Seebeck thermoelectric current, i; (b) Seebeck open-circuit voltage, e_s; (c) measurement of open-circuit voltage; (d) measured voltage is difference between open-circuit voltage, e_s, and contact voltage, e_B.

additional junctions where the thermocouple wires are connected to the meter contacts. The measured voltage will be the difference between the open-circuit voltage and the contact voltages. When the total contact voltage is constant, the measured voltage will be linearly proportional to the open-circuit voltage of the thermocouple.

The magnitude of the open-circuit voltage of a thermocouple junction at a fixed temperature will be established by the metals used. By selecting different metals, the manufacturer can obtain different voltage versus temperature characteristics. Industrial thermocouples actually use lengths of wire welded together to fabricate the thermocouples used in furnaces and ovens. Optical thermocouples are fabricated by depositing very thin layers (thin films) of metal on insulating substrates.

Practical optical detectors that use the thermoelectric effect are *thermopiles.* A thermopile is a device with a number of thermocouples electrically connected in series so that their Seebeck voltages add. A thermopile will have all the active junctions thermally connected to a radiation collecting surface. The interconnecting junctions between the individual thermocouples become reference junctions. All of the reference junctions will be thermally connected to a heat sink, which keeps them at a common temperature that is lower than the temperature of the active junction. Figure 8-10 illustrates a thermopile circuit.

Optical thermopiles often have a circular layout. The active junctions are formed near the geometric center of the device. A coating such as black gold acts as the absorbing surface for radiation, and it is thermally connected to the active junctions. The reference junctions are formed near the outside circumference of the

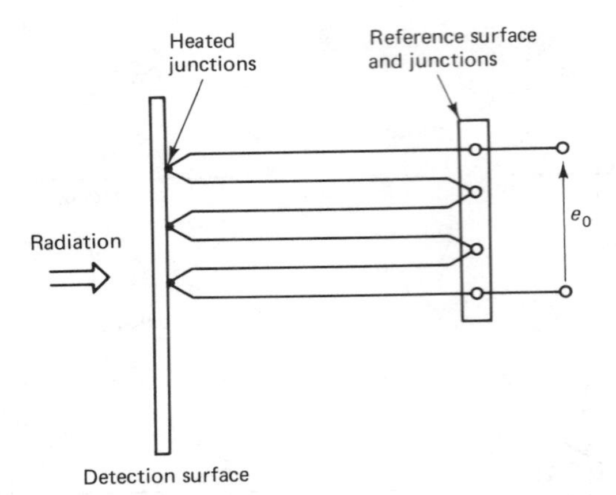

Figure 8-10 Thermopile.

device and are thermally insulated from the detecting area. The device's leads are brought out from the reference junctions. Figure 8-11 illustrates the construction of a circular thermopile. The small drawing in the top left-hand corner of Figure 8-12 is a picture of a thermopile thin-film pattern.

The active junctions of the thermopile are coated with a "black" material that is uniformly energy absorbent from long infrared wavelengths to short ultraviolet wavelengths. In general, these coatings are such excellent radiometric absorbers that any spectral limitations of a thermopile detector are established by the window material used in the detector head.

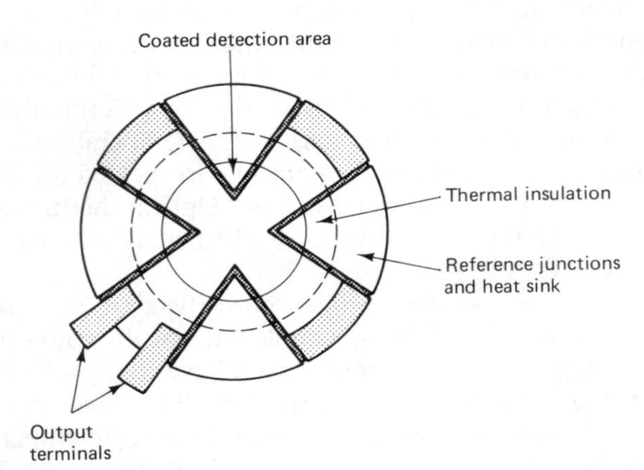

Figure 8-11 Circular thermopile construction.

Figure 8-12(a) and (b) are data sheets for a typical thermopile detector. Figure 8-12(b) lists mechanical characteristics and some maximum operating conditions. Figure 8-12(b) lists some typical spectral characteristics for a particular window material. It should be noted that this typical thermopile is physically small. The case itself is a TO-5 package, which is a standard transistor package with a diameter of about 10 mm. The actual detector is listed as having a sensitive diameter of 0.5 mm. A small detector like this will have limited power-handling capabilities and this device has a listed maximum flux density of 0.1 W/cm^2. The manufacturer claims a linear output voltage for flux densities between 1 μW/cm^2 and 100 mW/cm^2. The table of Figure 8-12(a) lists the dc sensitivity of this device as being 22 μV when irradiated by a flux density of 325 μW/cm^2.

The manufacturer also lists the "responsivity" of the device when illuminated by a constant 500K source as being typically 55 V per incident watt. The units used here could be very misleading considering the maximum incident flux density of 0.1 W/cm^2. Example 8-5 illustrates how the output voltage specification and the responsivity specification can be used to estimate detector open-circuit voltage. The maximum rated incident flux density is used in this example to illustrate the magnitude of the voltages obtained and flux densities used with these devices.

EXAMPLE 8-5

Output Voltage from a Thermopile

Given: The thermopile described by Figure 8-12.

Find: The maximum output voltage that can be achieved by this device at maximum rated incident flux density of

$$0.1\ \frac{\text{W}}{\text{cm}^2} \quad \text{or} \quad 1 \times 10^5\ \frac{\mu\text{W}}{\text{cm}^2}$$

Solution:

1. Calculation using dc output voltage specifications: Let v_o be the output voltage, S_v be the detector-specified response, and H be the incident flux density.

$$\begin{aligned} v_o &= S_v H \\ &= \frac{35\ \mu\text{V}}{325\ \mu\text{W/cm}^2} \times 1 \times 10^5\ \frac{\mu\text{W}}{\text{cm}^2} \\ &= 10.77\ \text{mV} \end{aligned}$$

2. Calculation using responsivity specification: R is the responsivity and A is the area of detector $A = \pi(d/2)^2$.

$$\begin{aligned} V_o &= RHA \\ &= 55\frac{\text{V}}{\text{W}} \times 0.1\ \frac{\text{W}}{\text{cm}^2} \times \pi\left(\frac{0.05\ \text{cm}}{2}\right)^2 \\ &= 10.79\ \text{mV} \end{aligned}$$

August 1986

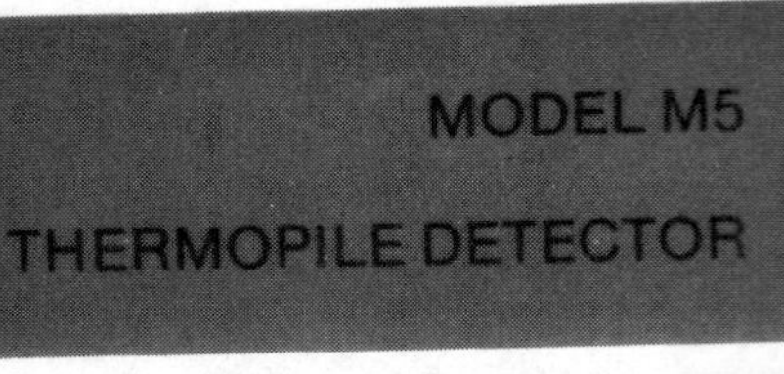

MODEL M5 THERMOPILE DETECTOR

TECHNICAL DESCRIPTION

The Model M5 detector is a miniature multijunction thermopile made of evaporated bismuth and antimony. On the active junction area is deposited an energy absorbing black of either smoke or paint. The element is hermetically sealed in a TO-5 package under a purged atmosphere of Argon or Nitrogen, and then heat treated to insure long term stability. The final package is resistant to both mechanical and temperature shock.

The thermopile is a voltage generating device and therefore requires no bias voltage or current for operation. Since it acts as a pure resistance, it generates no 1/f noise but only the Johnson noise of its resistance.

The spectral absorption of the deposited blacks are essentially flat from the ultraviolet to the far infrared, and thus the spectral sensitivity of the detector depends on the choice of the window material. This spectral band pass may be limited by selecting window materials or filters to replace or augment the standard KBr window.

FEATURES

- LOW COST
- NO COOLING
- NO 1/F NOISE
- HERMETICALLY SEALED
- RUGGED
- HIGH RELIABILITY

ELECTRICAL CHARACTERISTICS: SPECIFICATIONS APPLY AT 25°C WITH KBr WINDOW

Parameter	Conditions	Min.	Typ.	Max.	Symbol	Units	Comments
Resistance		2.0	3.0	4.0	r	KΩ	
Resistance T.C.	0° to 85°C		−0.2			%/°C	Best linear fit (Note 1)
Noise Voltage		5.7	7.0	8.1	V_n	$nV/\sqrt{Hz}$	$V^2_n = 4kTr\Delta f$
Output Voltage	DC 10Hz	22 3.8	35 5.0	45 7.0	V_s	μV RMSμV	H = 325μW/cm² H = 146μW/cm²
Responsivity	(500K,DC,−) (500K,10,1)	34 13	55 17	71 24	R	V/W	$R = V_s/HA$ (Note 2)
Responsivity T.C.	0° to 85°C		−0.4			%/°C	Linear (Note 1)
NEP	(500K,DC,−) (500K,10,1)	.08 .23	.13 .40	.23 .61	NEP	$nW/\sqrt{Hz}$	$NEP = V_nHA/V_s$
D*	(500K,DC,−) (500K,10,1)	1.2 .7	3.5 1.1	5.5 1.8	D*	$10^8 cm\sqrt{Hz}/W$	$D^* = V_s/V_nH\sqrt{A}$
Time Constant	Blackbody		28		τ	mS	Chopped 3dB (Note 1)

NOTE 1: Parameter is not 100% tested, 90% of all units meet these specifications.
2: A is detector area in cm².

DEXTER RESEARCH CENTER, INC.

7300 Huron River Drive/Dexter, Michigan 48130/(313) 426-3921/Telex: 757075

Figure 8-12 Thermopile detector data sheet: (a) specifications; (b) construction and operation. (Courtesy Dexter Research Center, Inc. Dexter, MI.)

PHYSICAL CHARACTERISTICS

Number of Junctions:	10
Sensitive Area:	.5 mm diameter
Package:	TO-5
Window Material:	KBr or to be specified
Encapsulating Gas:	Argon or Nitrogen
Field of View:	80°

OPERATING CONDITION

Temperature Range:	–65°C to 85°C*
Maximum Incidence:	0.1 Watts/cm^2
Spectral Response:	Flat from UV to far IR
Signal Output:	Linear from 10^{-6} to 10^{-1} Watts/cm^2

*Available to 125°C on special order

DETECTOR DIMENSIONS

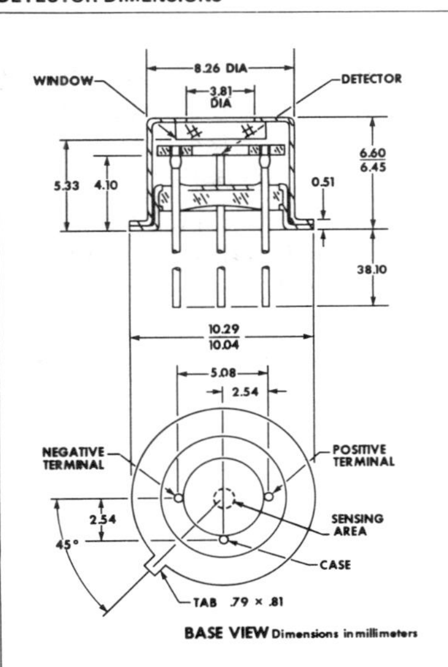

TYPICAL TRANSMISSION OF FIVE WINDOW MATERIALS FOR 1MM THICKNESS

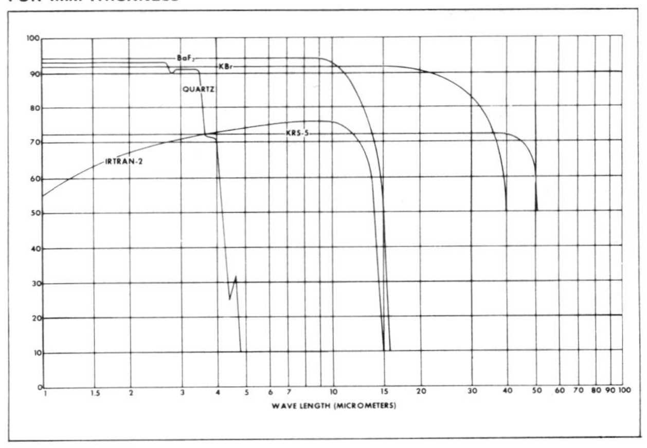

SIGNAL CALCULATION

POWER ON DETECTOR: $\Delta\Phi = \tau_0\tau_1\tau_2\rho(\Delta L)\pi \mathrm{SIN}^2\Theta \mathrm{Ad}$ Watts

$\Theta \simeq \mathrm{TAN}^{-1}\left(\frac{Dm}{2f'}\right)$; $\Delta L = \frac{4\sigma T^3 \Delta T}{\pi}$

Where:

$\tau_1\tau_2$ = Transmission of Windows W_1 & W_2 $\quad \sigma = 5{\cdot}6686 \times 10^{-12}$ W/cm^{2}deg^4

$\tau_0 = 1 - \left(\frac{Dd}{Dm}\right)^2$ $\quad$ T = 273 + °C

ρ = Mirror Reflectance $\quad$ Ad = Detector Area in cm^2

VOLTAGE FROM DETECTOR: $\Delta V = R\Delta\Phi$ Volts

SIGNAL TO NOISE RATIO: $(S/N) = R\Delta\Phi/N$ Where N = Amplifier & Detector Noise

SENSITIVITY:

$$\Delta T = \frac{N(S/N)}{\tau_0\tau_1\tau_2\rho(4\sigma T^3)(R\,\mathrm{Ad})\,\mathrm{SIN}^2\Theta} \quad °C$$

W1 W2 Dm Dd 2Θ f'

(b)

Figure 8-12 *(continued)*

Example 8-5 illustrates that the maximum output voltage that can be expected at the maximum rated flux density is less than 11 mV, a far cry from 55 V!

The time constant of this device is very long, which means that the cutoff frequency is very low. This device has a rated time constant of 28 ms. This corresponds to a cutoff frequency of about 5.7 Hz. This is one reason why the responsivity of the device at a 10-Hz chopping rate is only 17 V/W rather than the 55-V/W dc value. On this data sheet, "dc" means that the flux density is constant, and "10" means that the beam is chopped at a 10-Hz rate. Clearly, these devices are not really suitable for measuring rapidly changing events.

Figure 8-13 is a schematic diagram of a circuit with an operational amplifier connected to a thermopile detector. Note that the detector is connected to the plus input of a noninverting amplifier. The plus input of a noninverting amplifier has a high input impedance, on the order of $10^5\ \Omega$. For a thermopile detector voltage to be amplified effectively it must be connected to a high-impedance load, because the internal resistance of the detector is high; the specifications listed in Figure 8-12(a) indicate that for this device the internal resistance will be in the range 2 to 4 kΩ. If most of the voltage developed by the detector is to be presented to the amplifier input, the amplifier's input impedance must be much greater than the internal impedance of the detector.

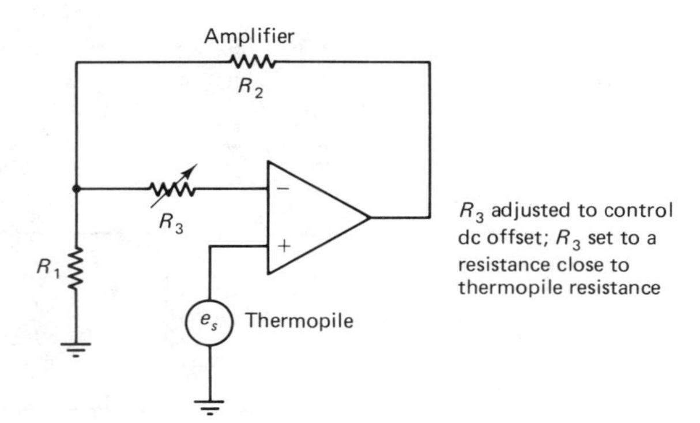

Figure 8-13 Typical thermopile amplifier circuit.

In summary: A thermopile detector is a slow-acting radiometric detector. Thermopile detectors used in optical applications tend to be physically small devices that develop small-signal voltages requiring amplification. The amplifier used with a thermopile detector must have a large input impedance because the output impedance of the detector is high. Engineering estimates of the voltage developed by a detector can be made using the output voltage specification or the responsivity specification. These estimates represent the open-circuit voltage of the detector. A voltage close to this open-circuit voltage can be obtained if the load resistance connected across the detector's terminals is much larger than the detector's internal impedance.

8-7 PYROELECTRIC DETECTORS

Pyroelectric detectors are crystal devices that are used like thermopiles to measure radiant energy uniformly over a wide spectral range. These devices can respond to radiation, which is modulated, chopped, at rates as fast as 100 kHz, but they are not suitable for dc measurements, where "dc" means flux levels that are constant for long periods of time. The crystals used in these devices develop an electric charge on their surfaces when they are heated by incident radiation. Incident radiation is absorbed by the crystal or a coating on the crystal surface and converted to heat. The heating alters the lattice spacing of the crystal and causes a charge differential to occur between the surfaces. This charge differential can be measured as a voltage.

Lithium tantalate ($LiTa0_3$) is the material commonly utilized. The device described by the data sheet of Figure 8-14 is fabricated from lithium tantalate. The crystals used to fabricate pyroelectric detectors are sometimes referred to as ferroelectric crystals. The use of the prefix "ferro" does not imply that these crystals contain iron; rather, it is an attempt to make an analogy between the electrostatic storage characteristics of these crystals and the magnetic storage characteristics of ferromagnetic materials.

The crystal can be thought of as a capacitor. The voltage across a capacitor is directly proportional to the charge on the capacitor and inversely proportional to the capacitance. As the pyroelectric detector is heated, we can envision the capacitor being charged up. At moderate frequencies a circuit model for these devices would be a current source in parallel with a capacitor of 10 to 100 pF and a large resistance. Because the capacitor and the resistor are in parallel, the voltage across the capacitor is also across the resistor. This causes current flow through the resistor, which discharges the capacitor. As the frequency of the radiation pulses is increased, this parallel resistor capacitor network acts like a low-pass filter which establishes an electrical high-frequency cutoff. Above this cutoff frequency, the detector output voltage is attenuated.

At low frequencies the heat capacity and the thermal resistance of the detector establish a low-frequency cutoff. This cutoff frequency is typically about 3 to 5 Hz. Radiation pulses that occur at frequencies less than this low-frequency cutoff are also attenuated. This is the reason that the pyroelectric detector is not suitable for measuring dc flux levels. The combined action of the low-frequency thermal cutoff and the high-frequency electrical cutoff establishes the detector's bandpass characteristic.

Commercially available packages for these devices will often include the pyroelectric detector and a field-effect transistor that is used as the first amplifier for the detected signal. These packages can be used in the three different circuit configurations illustrated on the data sheet of Figure 8-14(b). In two of the circuit configurations, the load resistance connected to the pyroelectric detector is the high input impedance of the FET circuit. In these two circuits the FET amplifier output voltage is proportional to the voltage across the detector circuit. The third circuit configuration is called the current mode of operation, in which the detector is loaded by an apparent short circuit and the amplifier output voltage is proportional to the

P1-70 Ultra-Low Noise Pyroelectric Detector/FET Preamp

Molectron

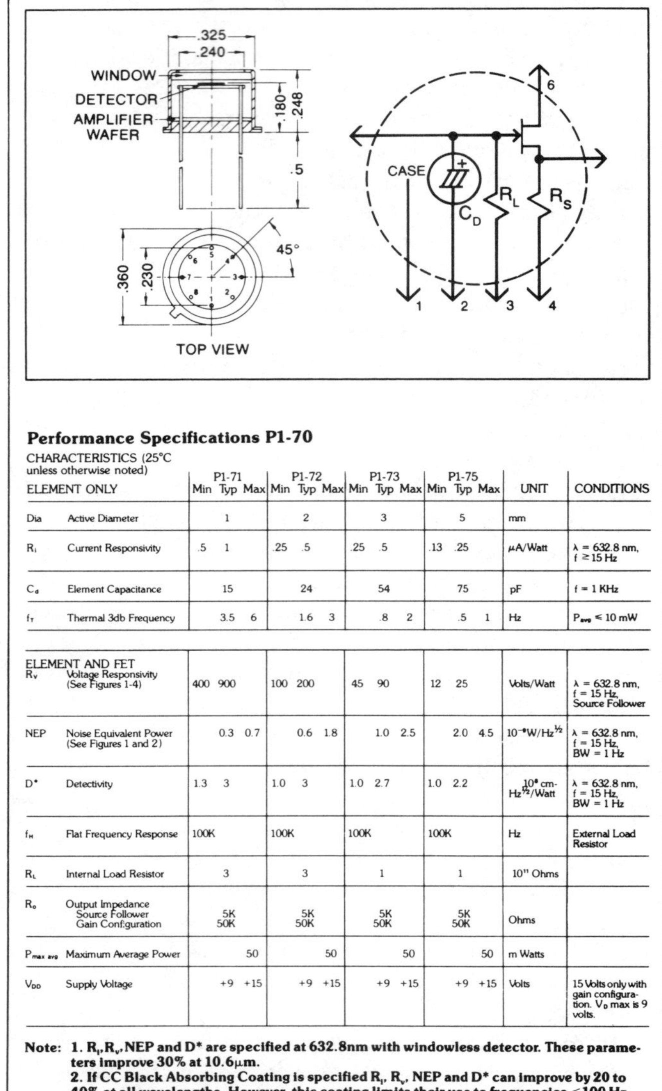

Performance Specifications P1-70

CHARACTERISTICS (25°C unless otherwise noted)

ELEMENT ONLY		P1-71 Min	P1-71 Typ	P1-71 Max	P1-72 Min	P1-72 Typ	P1-72 Max	P1-73 Min	P1-73 Typ	P1-73 Max	P1-75 Min	P1-75 Typ	P1-75 Max	UNIT	CONDITIONS
Dia	Active Diameter		1			2			3			5		mm	
R_i	Current Responsivity	.5	1		.25	.5		.25	.5		.13	.25		μA/Watt	λ = 632.8 nm, f ≥ 15 Hz
C_d	Element Capacitance		15			24			54			75		pF	f = 1 KHz
f_T	Thermal 3db Frequency		3.5	6		1.6	3		.8	2		.5	1	Hz	P_{avg} ≤ 10 mW
ELEMENT AND FET															
R_v	Voltage Responsivity (See Figures 1-4)	400	900		100	200		45	90		12	25		Volts/Watt	λ = 632.8 nm, f = 15 Hz, Source Follower
NEP	Noise Equivalent Power (See Figures 1 and 2)		0.3	0.7		0.6	1.8		1.0	2.5		2.0	4.5	10^{-9}W/$Hz^{1/2}$	λ = 632.8 nm, f = 15 Hz, BW = 1 Hz
D*	Detectivity	1.3	3		1.0	3		1.0	2.7		1.0	2.2		10^{8} cm-$Hz^{1/2}$/Watt	λ = 632.8 nm, f = 15 Hz, BW = 1 Hz
f_H	Flat Frequency Response	100K			100K			100K			100K			Hz	External Load Resistor
R_L	Internal Load Resistor		3			3			1			1		10^{11} Ohms	
R_o	Output Impedance: Source Follower / Gain Configuration		5K / 50K			5K / 50K			5K / 50K			5K / 50K		Ohms	
$P_{max\ avg}$	Maximum Average Power			50			50			50			50	m Watts	
V_{DD}	Supply Voltage		+9	+15		+9	+15		+9	+15		+9	+15	Volts	15 Volts only with gain configuration. V_D max is 9 volts.

Note: 1. R_i, R_v, NEP and D* are specified at 632.8nm with windowless detector. These parameters improve 30% at 10.6μm.
2. If CC Black Absorbing Coating is specified R_i, R_v, NEP and D* can improve by 20 to 40% at all wavelengths. However, this coating limits their use to frequencies <100 Hz.

(a)

Figure 8-14 Pyroelectric detector data sheet: (a) specifications; (b) operating characteristics. (Courtesy Molectron Detector Inc., Sunnyvale, CA.)

Molectron Corporation

Molectron Corporation
177 North Wolfe Road
Sunnyvale, California 94086
Telephone: (408) 738-2661
Telex: 357436, 278780

Eastern Regional Office
118 Crest Lake Drive
Oakridge, NJ 07438
Telephone: (201) 697-3917

Midwest Regional Office
2902 Westmoreland
New Lenox, IL 60451
Telephone: (815) 485-9205

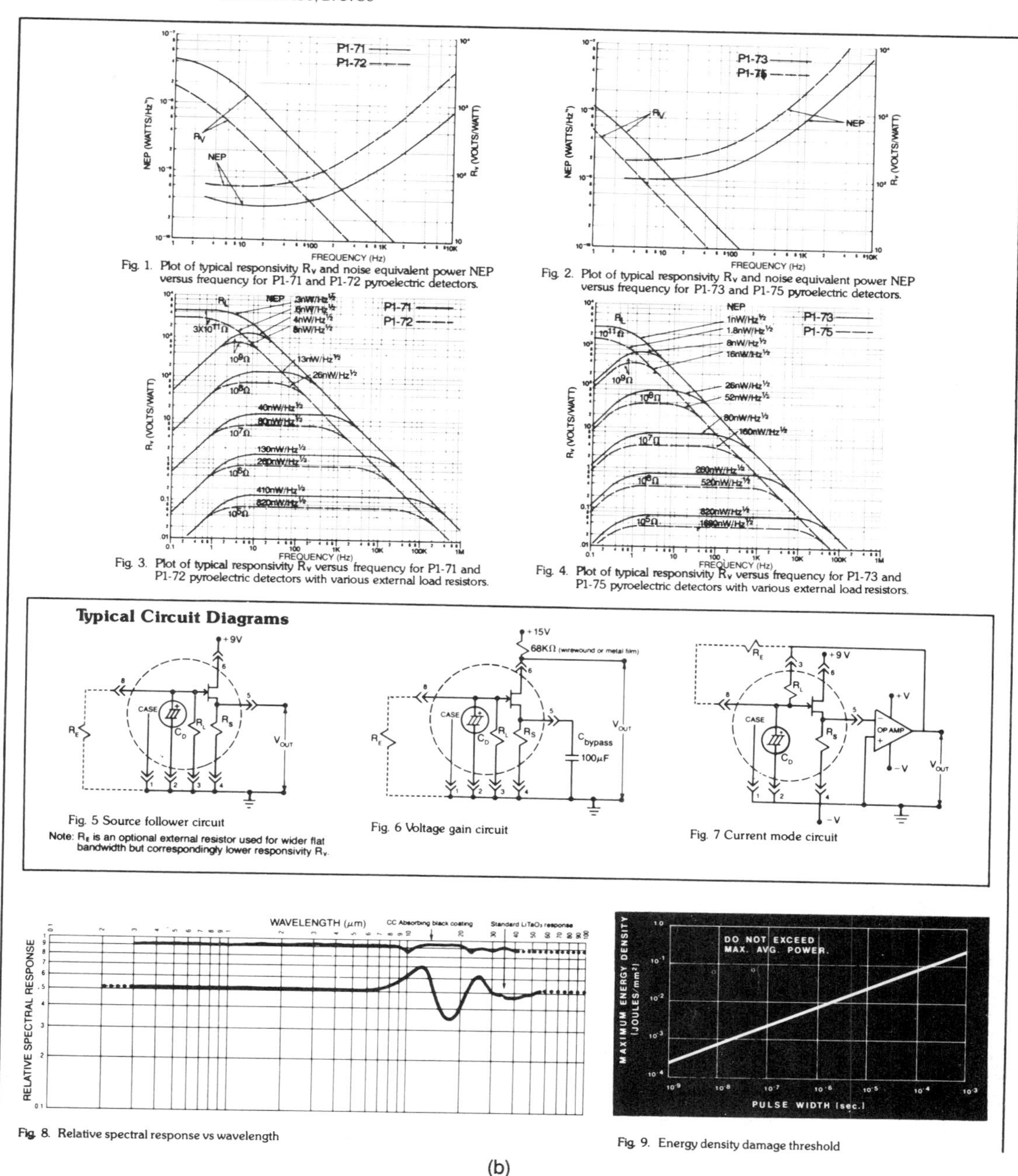

Fig. 1. Plot of typical responsivity R_V and noise equivalent power NEP versus frequency for P1-71 and P1-72 pyroelectric detectors.

Fig. 2. Plot of typical responsivity R_V and noise equivalent power NEP versus frequency for P1-73 and P1-75 pyroelectric detectors.

Fig. 3. Plot of typical responsivity R_V versus frequency for P1-71 and P1-72 pyroelectric detectors with various external load resistors.

Fig. 4. Plot of typical responsivity R_V versus frequency for P1-73 and P1-75 pyroelectric detectors with various external load resistors.

Fig. 5 Source follower circuit
Note: R_E is an optional external resistor used for wider flat bandwidth but correspondingly lower responsivity R_V.

Fig. 6 Voltage gain circuit

Fig. 7 Current mode circuit

Fig. 8. Relative spectral response vs wavelength

Fig. 9. Energy density damage threshold

(b)

Figure 8-14 *(continued)*

detector's short-circuit current. The detector's short-circuit current is proportional to the rate of change of the crystal temperature.

In the voltage mode of operation, external resistors are connected to the circuit package which establish the responsivity of the entire circuit and its high-frequency cutoff. Graphs of responsivity versus frequency for various values of external resistance are shown in Figure 8-14(b). As the size of the external resistor, R_E, is decreased, the responsivity is decreased and the high-frequency cutoff is increased. Radiation pulses with frequencies between the low and high cutoff frequencies can be detected.

The NEP also increases as responsivity is decreased and bandwidth is increased. This means that higher responsivities and narrower bandwidths have better signal-to-noise ratios. The frequency response graphs on the data sheets also list the NEP values for each level of responsivity illustrated. We should note that the NEP specification used on this data sheet is not in the IEEE standard form; rather, it is normalized per unit bandwidth.

The pyroelectric detector can be used to detect the amplitude of radiation pulses which have a pulse width and pulse rate that fall between the high- and low-frequency limits. Figure 8-15 illustrates the detected output voltage pulses that can be expected for radiation pulses of different frequencies. The detected voltage pulse will be most like the square radiation pulse when the pulse frequency is geometrically centered between the low- and high-frequency cutoff points. If pulse wave shape is critical, it will also be necessary to have the ratio of the high-frequency cutoff frequency to the pulse frequency as large as possible. Unfortunately, a large cutoff frequency can be obtained only at the expense of responsivity and increased noise.

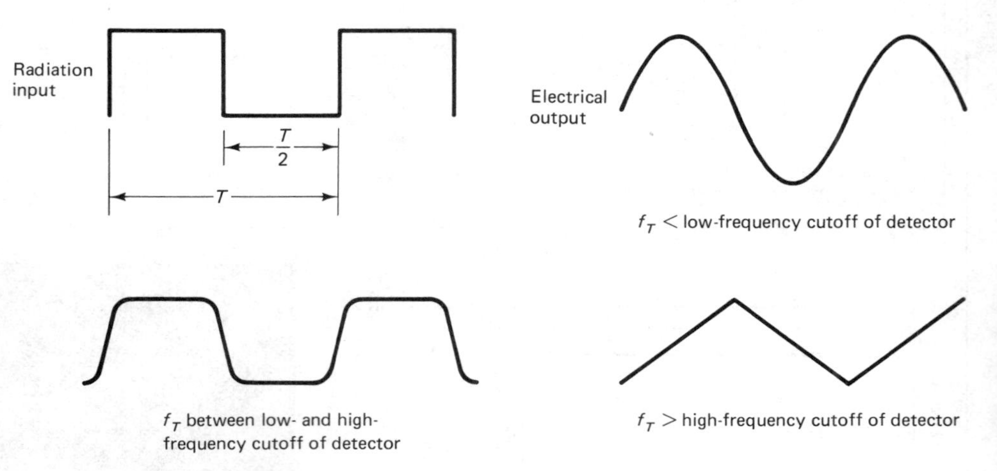

Figure 8-15 Three electrical output wave-shape patterns resulting from three different input radiation pulse frequencies, f_T. Frequency relationships expressed relative to detector amplifier cutoff frequencies. The radiation frequency is expressed in terms of the pulse period T. $F_T = 1/T$.

For any given frequency response, the centered pulse frequency can be calculated as follows:

$$f_T = (f_L f_H)^{1/2} \tag{8-27}$$

where f_T is the pulse frequency
f_L is the low-frequency cutoff
f_H is the high-frequency cutoff

In addition to this "centering" criterion, you would want the ratio of f_H to f_T to be at least 10:1. If the ratio of f_H to f_T is small, say 3:1 or less, the output will look more like a sine wave than a square wave.

EXAMPLE 8-6

Estimation of the Optimum Pulse Frequency for a Given Bandwidth

Given: **(1)** f_L thermal low frequency 3.5 Hz
(2) f_H 10 kHz

Find: **(a)** The optimum pulse frequency f_T
(b) Does f_T meet the 10:1 criterion?
(c) What is the ratio of f_T to f_L?

Solution: **(a)** $f_T = (f_L f_H)^{1/2}$
$= (3.5 \text{ Hz} \times 10\text{kHz})^{1/2}$
$= 187 \text{ Hz}$

(b) $\dfrac{f_H}{f_T} = \dfrac{10 \text{ kHz}}{187 \text{ Hz}} = 53.5$

(c) $\dfrac{f_T}{f_L} = \dfrac{187 \text{ Hz}}{3.5 \text{ Hz}} = 53.5$

It is apparent that we have met the 10:1 frequency ratio without any problem in this case. In fact, this example would indicate that we could probably narrow the bandwidth and increase the gain and still obtain an acceptable pulse shape.

To summarize: The pyroelectric detector is an excellent radiometric detector. The pyroelectric detector is most suited to the detection of pulses with frequencies of several hundred hertz. If some wave-shape distortion is acceptable, pulse rates of

1000 to 10,000 hertz are possible. The pyroelectric detector package illustrated here contains an internal amplifying device called an FET. This FET amplifier requires an external power supply. The responsivity, bandwidth, and NEP of this package are established by the connection of an external resistor to the package terminals.

8-8 SUMMARY

This chapter has discussed the operation of four basic radiometric detectors. In addition, mathematical descriptions of device noise and device optical characteristics have been explored. In an optical laboratory the thermopile will be encountered much more frequently than the pyroelectric detector, the bolometer, or the thermistor.

The discussion of noise as one of the limiting factors on measurement range applies particularly to thermopiles and pyroelectric detectors. In subsequent chapters, noise specifications will be found to be important characteristics of a variety of other detectors.

The discussion of the optical characteristics of the detector package is perfectly general in nature and the terminology and concepts developed apply to any common detector head. The data used for the graph of Figure 8-5 were obtained with a detector head that uses a semiconductor diode as the detector element.

In this chapter the IEEE standard definitions for detector specifications were presented. The typical variations from these standards were also discussed. Some simple example response and noise calculations were presented. The problems that follow will give you a chance to fight your way through this material again.

QUESTIONS

1. What kind of materials are used to construct a bolometer?
2. What kind of materials are used to construct a thermistor?
3. How is an optical thermopile constructed?
4. What physical parameter of a bolometer changes when it is irradiated? Does this parameter increase or decrease?
5. What physical parameter of a thermistor changes when it is irradiated? Does this parameter increase or decrease?
6. What kind of electrical response does a thermopile exhibit when irradiated and short circuited?

7. What kind of electrical response does a thermopile exhibit when irradiated and open circuited?
8. When a pyroelectric detector is open circuited, what is the relationship between input radiation and output voltage?
9. When a pyroelectric detector is short circuited, what is the relationship between input radiation and output current?
10. Why are pyroelectric detectors not suitable for dc operation?

PROBLEMS

1. A detector has an area of 0.6 cm^2. The incident flux density is 0.5 mW/cm^2 and the detector produces a voltage of 12 mV. Calculate the detector's responsivity in units of mV/mW.
2. A thermopile has a −3-dB cutoff frequency of 3.5 Hz. Estimate the time constant of this thermopile.
3. A bolometer has a time constant of 7 s. Estimate the 10 to 90% rise time t_r.
4. Estimate the angle of view of a P1-75 Molectron pyroelectric detector based on detector diameter, window diameter, and their separation. Refer to Figure 8-14. Note that mechanical dimensions of the case are in inches.
5. Estimate the diameter of the spot size for a model M5 thermopile at a distance of 10 cm from the source. Refer to Figure 8-12(b).
6. A square detector with an angle view of 0.2 rad and an aperture of 0.8 cm^2 yields a constant flux reading of 25 μW at distances of 1 to 3 cm from a source. Calculate the radiance of the source.
7. What is the critical distance for a test setup using a source 12 cm in diameter and a circular detector with an angle of view of 56°?
8. A square detector has an angle of view of 80 mrad and an aperture area of 14 mm^2. The instrument connected to the detector indicates a measured power level of 1.8 μW. The source is uniformly illuminated. The area of the source is 7200 cm^2 and the distance from the source to the detector is 120 cm. Estimate the critical distance and the surface radiance of the source.
9. A lab sphere has a circular radiating port that is 5 cm in diameter. The detector used has an angle of view of 56° and a circular aperture with an area of 1 cm^2. Determine the critical distance and then determine the radiance of the port from the data points listed below.

Distance (cm)	Flux (μW)
10	720
20	180
40	45

10. A very small monochromatic source is mounted at the zero location on a lens bench. Power readings are made at three fixed distances from the source. A thermopile with an aperture area of 10 mm^2 is used for the measurements. Use the following data as a basis for calculating the flux density that would be measured at 25 cm and 200 cm.

Location[a] (cm)	Power Readings (nW)
50	800
70	408
100	200

[a] Distance from source.

11. From the following data, determine the angle of view of the circular detector.

Source Diameter (cm)	Critical Distance (cm)
12	22.4
8.5	15.9
6	11.2

12. Estimate the noise bandwidth of a measurement system using a thermopile with a 3.5-Hz cutoff frequency.
13. Two detectors are measured and they yield the data below. Calculate the D^* specification for each device. Which device is best?

	Detector A	Detector B	Units
Modulation frequency	0	0	Hz
Wavelength	632.8	632.8	nm
Bandwidth	10	10	Hz
Radiant power causing a 1:1 signal-to-noise ratio	5	8	μW
Detector area	0.30	0.55	cm^2

14. Your boss has asked you to compare the data sheet noise performance specifications of three thermopiles and to select the "best" one. The data obtained for the three devices tested at the same wavelength and at room temperature is listed below.

Device 1	Device 2	Device 3
NEP $0.18 \frac{\text{nW}}{(\text{Hz})^{1/2}}$	NEP 0.12 nW	D^* $5 \times 10^6 \frac{\text{m(Hz)}^{1/2}}{\text{W}}$
−3 dB 5.6 Hz	−3 db 4.8 Hz	−3 dB 5 Hz
Area 2×10^{-3} cm^2	Area 0.19 mm^2	Area 0.22 mm^2

15. Two different bead thermistors are used in your laboratory to monitor heat sink temperatures. From the data below determine the temperature coefficient of resistance for each of these devices. Use 20°C as the reference temperature.

Temperature (°C)	Device Resistance (kΩ)	
	A	B
20	13.15	6.04
25	11.70	5.38
30	10.25	4.71
35	8.81	4.05
40	7.36	3.38

16. A thermopile detector has a rated maximum flux density of 250 mW/cm^2. The specified responsivity is 42 mV/mW. The diameter of the active area is 0.07 cm. Estimate the maximum "safe" output voltage.

9

Semiconductor Concepts and Photoresistors

9-1 INTRODUCTION

Photoconductors, photodiodes, and phototransistors are all optoelectronic devices fabricated from semiconductor materials. A basic understanding of the atomic and subatomic properties of these materials will help in understanding the functional characteristics of these devices. The Bohr model for atomic and subatomic particle behavior can be used as a starting point for understanding the electric and photoelectric behavior of materials. In this chapter we develop some general material concepts with an emphasis on the electrical property of conductivity. Photoconductive devices and their specifications are discussed.

9-2 THE BOHR MODEL

The Bohr model of an atom describes the arrangement of subatomic particles in a single isolated atom. This model was described in some detail in Chapter 5. It was pointed out in that discussion that at any specific distance r from the center of the nucleus, an electron must have a specific kinetic energy to maintain orbital position. If the electron's energy is decreased, the electron will decrease its orbital radius and move closer to the nucleus. If the electron's energy is increased, the electron will increase its orbital radius and move farther away from the nucleus. Electrons are only permitted to assume specific discrete levels of energy.

Figure 9-1(a) illustrates a silicon atom that has 14 electrons. These electrons are distributed over three principal energy levels. As is the case with all atoms, there is a maximum number of electrons that can exist in each permitted level. Principal levels have fixed sublevels capable of holding one or more pairs of electrons. For

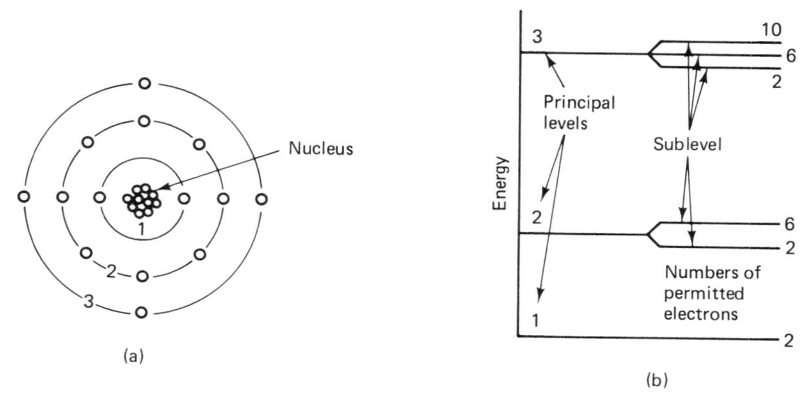

Figure 9-1 (a) Schematic of Bohr atom; (b) energy levels and ground-state electron distribution.

the first three levels of an atom the maximum number of electrons is indicated in Figure 9-1(b). These levels fill from the lowest level to the highest. When an energy level reaches the maximum number of permitted electrons, it is said to be *full.*

The first two levels of a silicon atom are full; the third level is not full. The last level to receive electrons, the outermost level, is called the *valence level.* A silicon atom has four valence electrons in the valence level. Electrons cannot exist in the regions between the permitted principal and sublevel energy levels. When an electron changes energy levels, it can move only between permitted levels. This means that the electrons of the atom can lose or gain only the amount of energy required to make the changes between permitted levels of energy. These specific quantities of energy are called *quanta.* Photons of radiation are the carriers of energy quanta.

9-3 CONDUCTION IN SOLIDS

When the individual atoms come together to form a solid, they interact with each other, causing changes in the energy-level structure. In general, the major effect is that the outermost energy levels widen into continuous energy bands, and in some cases the bands overlap. Energy bands can be thought of as being composed of many closely spaced energy levels or as being continuous regions of permitted energies.

Figure 9-2 illustrates what will happen to the energy levels of a representative single atom as it is incorporated into an increasingly dense material. As the number of atoms per unit volume gets very large, the final "solid" state is reached. The solid has energy characteristics that are different from those of the single atom. First, the magnitude of the quanta of energy required to cause electrons to change energy levels is reduced, and second, the three discrete energy levels of the atom have become three energy bands.

The energy-band characteristics of a material can be used to explain its elec-

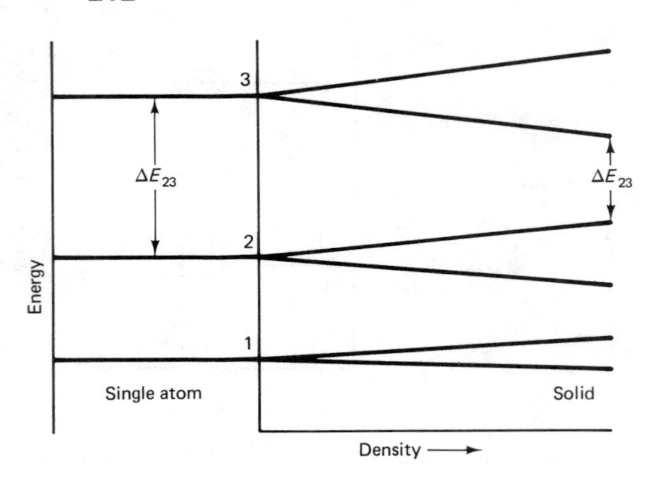

Figure 9-2 Changes in the energy diagram of an atom as a function of the proximity of the atom to other atoms.

trical conduction characteristics. The permitted band of energy immediately above the valence band is called the *conduction band.* The energy difference between the valence band and the conduction band is a major factor in determining the conductivity of a material and is called the *gap energy.* The energy region between the valence band and the conduction band is sometimes referred to as the *forbidden gap* because electrons are forbidden to exist within it. A material will conduct electric current when a voltage is applied if there are electrons in the conduction band. There are two ways electrons can enter the conduction band.

1. When electrons absorb photons and jump over the forbidden gap between the valence band and the conduction band, an external voltage can act on the electrons in the conduction band and cause current flow within the conduction band. These migrating electrons leave vacancies in the valence band which also contribute to current flow. These vacancies are called *holes.*

   Hole A mobile vacancy in the electronic valence structure of a semiconductor that acts like a positive electron charge with a positive mass. (*Source:* ANSI/IEEE Standard 100-1984. Reprinted by permission of the IEEE Standards Department.)

2. If the conduction and the valence bands overlap, the holes and electrons in this composite band can easily be moved by an external voltage. Metals are materials that have this overlapping band structure.

Materials can be grouped into three broad electrical conduction classifications: insulators, semiconductors, and conductors. An ideal *insulator* will not support conduction; it exhibits zero conductivity. *Semiconductors* are poor conductors; they exhibit low conductivity. Ideal *conductors* would offer no opposition to the flow of electrical current; they would have infinite conductivity. Practical conductors exhibit small amounts of opposition to current flow. Figure 9-3 illustrates typical energy-band patterns for an insulator, a semiconductor, and a conductor.

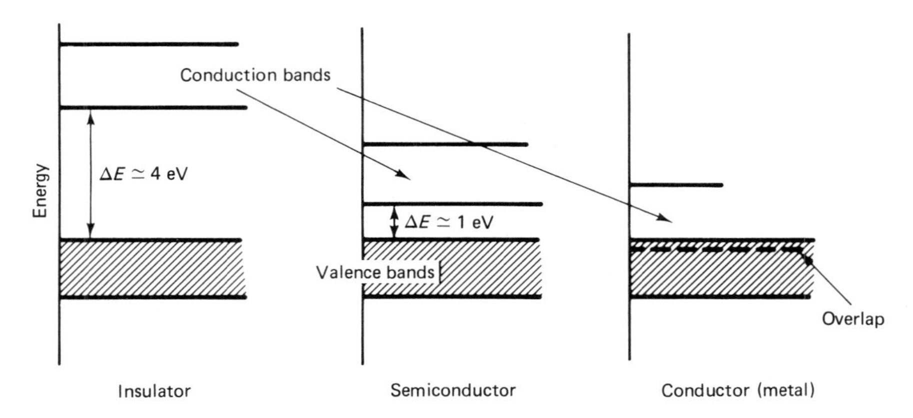

Figure 9-3 Conduction-band and valance-band relationships for three classes of materials.

The *insulator* is characterized by a filled valence band separated from the conduction band by a large energy difference. Under normal temperature conditions and at moderate applied voltage levels there are no electrons in the conduction band and no holes in the valence band, so no current flow takes place. If a very high voltage is applied to the insulator, valence-band electrons can be accelerated and they will jump to the conduction band. A large breakdown current will flow when this takes place, causing permanent damage to the material.

An *intrinsic (pure) semiconductor* material is composed of a single element and it has a valence band separated from the conduction band by a small energy difference. At room temperature the energy quanta, the photons, available from the environment will cause some electrons to move into the conduction band. Because there are a few electrons in the conduction band and a few holes in the valence band, a small current will be produced when a moderate voltage source is applied. The number of current flow electrons available in the conduction band of the semiconductor depends on the number of quanta of energy available to move electrons from the valence band to the conduction band.

When a valence electron's energy is increased so that it moves from the valence band to the conduction band, it leaves behind a hole; this process is called electron–hole pair formation. The application of an electric field to the material will cause holes to flow from positive to negative in the valence band and electrons will flow from negative to positive in the conduction band.

A *conductor,* such as gold, has a valence band that overlaps the conduction band. This means that electrons in the valence band can contribute to current flow without requiring an energy band change to the conduction band. The current flow electrons in metals flow in this overlapping energy band.

In summary: Insulators do not conduct; because they have empty conduction bands and large energy differences between the conduction bands and the valence bands, there are no electrons free to contribute to a current flow. Metals conduct readily because the valence band overlaps the conduction band; thus the applied voltage will cause electron current flow within this overlapping band. Intrinsic semiconductors conduct because hole–electron pairs are formed when quanta of energy are absorbed moving electrons to the conduction band. An applied voltage (electric field) will cause electron current flow within the conduction band and hole current flow within the valence band of an intrinsic semiconductor.

9-4 EXTRINSIC SEMICONDUCTORS

In Section 9-3 we discussed the atomic and subatomic characteristics of intrinsic semiconductors, which are solid materials composed of only one element. Most semiconductor devices have other materials added to the pure material. These added materials are called *impurities* or *dopants.* The process of adding impurities is called *doping.* Materials are doped in order to change the number of charge carriers in the resulting solid and to establish the bandgap energy. This process of combining elements is also referred to as *alloying.*

Silicon is a common intrinsic semiconductor-based material used in electronic devices and optical detectors. When pure silicon is grown as a crystalline solid, the four valence electrons of the individual atoms form symmetrical bonds with adjacent atoms. The crystal's valence band is very stable and well defined. Intrinsic silicon exhibits about six times more resistance than an identically shaped piece of copper. When an element with five valence electrons, such as phosphorus, is introduced into the silicon, the gap energy and the resistivity of the resulting combination will be different from that of intrinsic silicon.

Figure 9-4 illustrates how the silicon atoms (Si) share their four valence electrons with the adjacent silicon and phosphorus atoms (P). The straight lines represent these shared valence-band electrons. The extra phosphorus electron, called a *donor electron,* resides in a donor energy level just below the conduction band. Only a very small amount (quanta) of energy is required to raise an electron from the donor energy level to the conduction band, so the number of electrons in the conduction band of the doped material is greater than that of pure silicon at normal temperatures. Because there are more electrons in the conduction band after doping, the conductivity of the material is increased. The phosphorus atoms have increased the number of available electrons, which have a negative charge, so the resulting material is called N-type material.

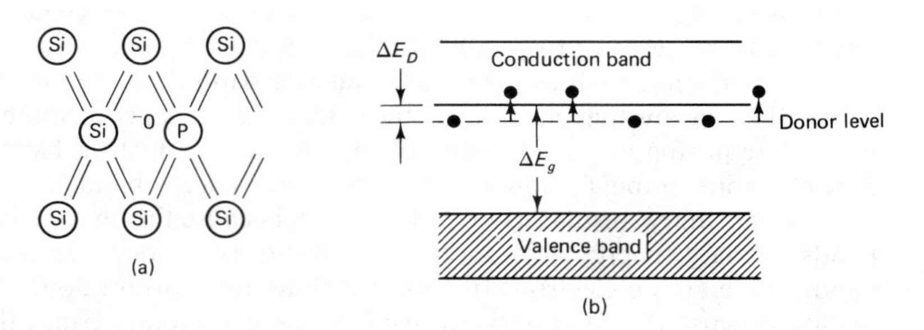

Figure 9-4 N-doped silicon: (a) bonded atoms; (b) energy levels.

Silicon can also be doped with atoms of boron, which has only three valence electrons. This will cause the creation of an energy level just above the valence band called the *acceptor level.* When a silicon valence electron bonds itself to a boron atom, it moves to the acceptor energy level. This process creates a hole in the silicon valence band. Figure 9-5 illustrates the energy-level configurations that will occur when silicon is doped with boron. The silicon boron combination has more holes than pure silicon. Because holes are considered positive, the resulting material is called P material.

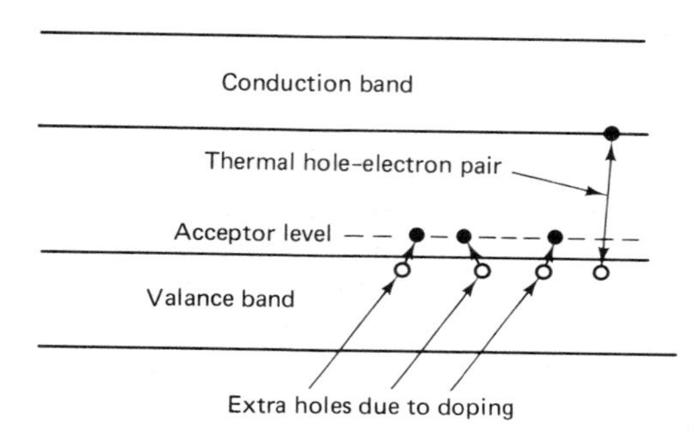

Figure 9-5 P-type semiconductor energy levels.

When fabricating semiconductor devices, dopants are added to the basic material, thus altering the material's conductivity and energy-band structure. By controlling the relative concentration of the dopants, the manufacturer can achieve a variety of spectral sensitivities and conductivities. The number of valence electrons of the dopant material will cause the resultant solid to be either P material or N material. P material will have more holes in the valence band than the basic material. N material will have more electrons in the conduction band than the basic material.

9-5 SPECTRAL CHARACTERISTICS OF PHOTOCONDUCTIVE CELLS

A photoconductive cell is an example of a detector whose conductivity is established by the process of electron–hole pair formation. Photoconductive cells are sometimes called *photoresistors.* Resistivity and conductivity are reciprocals of each other. The resistivity of a cell is established by the material from which the cell is fabricated and the number of quanta of the appropriate wavelength incident on the cell.

$$p = \frac{1}{g} \tag{9-1}$$

where p is the resistivity
g is the conductivity

Table 9-1 lists the resistivity of some common metals and semiconductor materials at room temperature.

TABLE 9-1 ELECTRICAL RESISTIVITY AT 20°C OF ELEMENTS USED IN SEMICONDUCTOR TECHNOLOGY

Element	Resistivity ($\mu\Omega \cdot$ cm)	Element	Resistivity ($\mu\Omega \cdot$ cm)
Aluminum	2.7	Indium	8.4
Cadmium	7.6	Lead	20.6
Copper	1.7	Selenium	12.0
Gallium	17.4	Silicon	50×10^9
Germanium	46×10^6	Silver	1.6
Gold	2.4	Tungsten	5.7

Cell characteristics are expressed either in terms of units of resistance or units of conductance. Resistance is measured in ohms and is a function of the resistivity of the material as well as its length, thickness, and width. Conductance is measured in siemens and is equal to the reciprocal of resistance. The relationship that exists between resistivity and resistance is illustrated in Figure 9-6.

The cell manufacturer establishes the cell's nominal resistance at a given level of illumination when they establish the thickness, width, and length of the conducting path and selects the cell's material composition. When the illumination level falling on the cell increases, more quanta are absorbed by the cell, causing the resistivity of the material to decrease and decreasing the cell's resistance.

$$G = \frac{1}{R} \qquad \text{where } R = \frac{\rho l}{tw} \tag{9-2}$$

$$G = \frac{1}{\rho l / tw}$$

$$G = g\,\frac{tw}{l} \tag{9-3}$$

where G is the conductance, in siemens
R is the resistance, in ohms
l is the length, the distance between the electrical terminals
t is the thickness

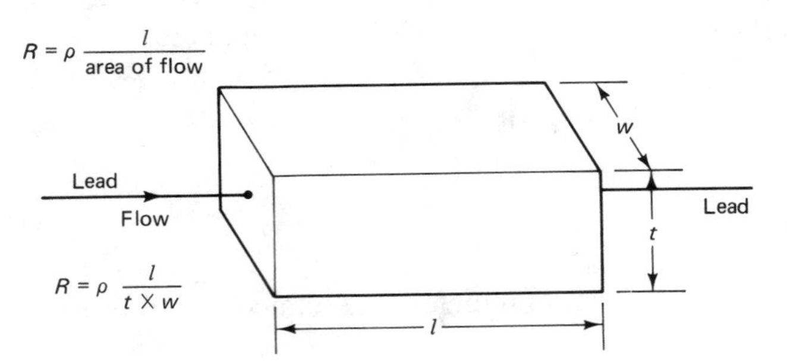

Figure 9-6 Resistance and resistivity of a block of material.

w is the width
ρ is the resistivity of the material, which varies with the wavelength and intensity of the incident radiation

Photoconductive detectors are constructed from intrinsic semiconductor materials, extrinsic semiconductors, and polycrystalline materials formed by alloying various rare earths, metals, and semiconductors. The longest wavelength that a photoconductive cell will respond to is established by the bandgap between the valence band and the conduction band of its material. The magnitude of this bandgap is a function of the materials used and the fabrication process. Photoconductors exhibit a decrease in conductivity for wavelengths significantly shorter or longer than the wavelength that corresponds to the bandgap energy. The surface characteristics of the material and any coverings also limit the wavelength response. Photoconductive cells will exhibit the greatest increase in conductance when exposed to radiation containing wavelengths at or near the wavelength that corresponds to the bandgap energy. The wavelength corresponding to the bandgap energy can be estimated by

$$\lambda = \frac{hc}{E_g} \tag{9-4}$$

$$f = \frac{c}{\lambda}$$

where λ is the wavelength, in meters
h is the Planck's constant, 4.135×10^{-15} eV · s
c is the speed of light, 3×10^8m/s
E_g is the bandgap energy, in electron volts
$hc = 1241$ eV · nm
f is photon frequency, in hertz

Two of the materials listed in Table 9-2, germanium and silicon, are intrinsic semiconductors. The remaining materials are alloys, mixtures of elements. The

TABLE 9-2 MATERIALS AND TYPICAL BANDGAP ENERGIES AND MAXIMUM WAVELENGTHS

Material	Symbol	λ (nm)	E_g (eV)	f (Hz $\times 10^{12}$)	Region
Indium antimonide	InSb	7000	0.18	43	Infrared
Lead sulfide	PbS	2000	0.62	150	Infrared
Germanium	Ge	1852	0.67	163	Infrared
Silicon	Si	1128	1.1	266	Infrared
Gallium arsenide	GaAs	886	1.4	338	Infrared
Cadmium telluride	CdTe	856	1.45	351	Infrared
Cadmium selenide	CdSe	713	1.74	421	Infrared
Gallium phosphide	GaP	539	2.3	556	Visible
Cadmium Sulfide	CdS	506	2.45	591	Visible

listed values of wavelengths are for specific ratios of the individual elements in these alloys. Manufactured devices will exhibit operating ranges of wavelengths around the values listed for a given material. Cadmium selenide and cadmium sulfide are commonly used in the fabrication of photoconductive cells.

Figure 9-7 illustrates the spectral response characteristics for two photoresistor material combinations: cadmium sulfide and cadmium selenide. These graphs illustrate the relative response of these devices versus wavelength in nanometers. Table 9-2 lists cadmium sulfide as having a typical peak wavelength response at 506 nm; the curve for the CdS material on the graph has a peak response at about 565 nm. Table 9-2 lists cadmium selenide as having a peak response at 713 nm; the

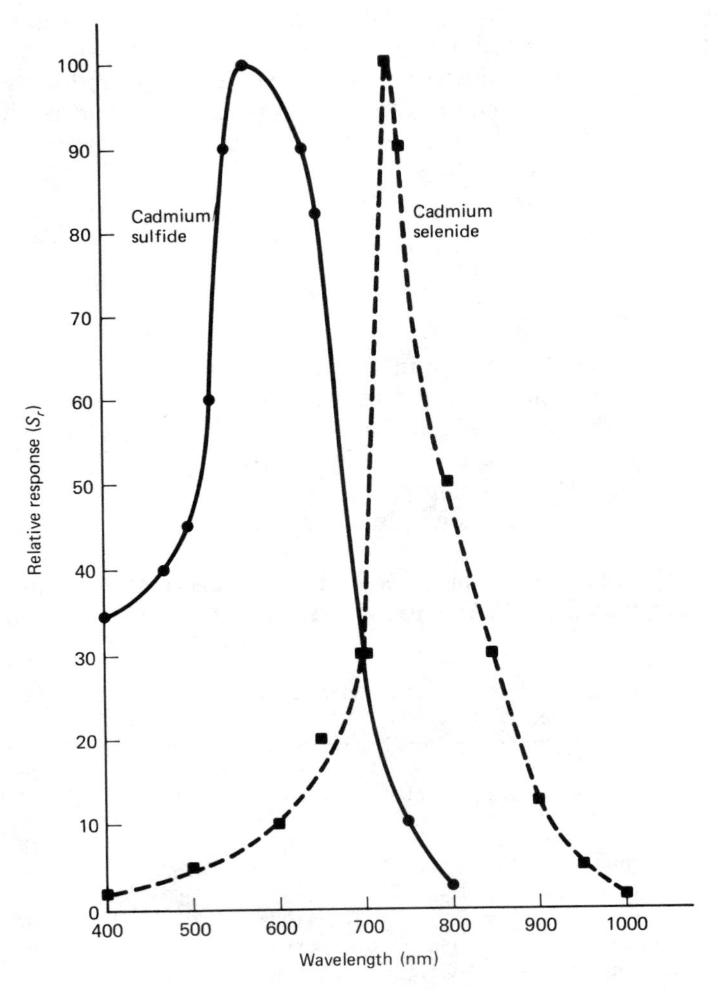

Figure 9-7 Relative response of two photoconductive materials as a function of wavelength.

curve for the CdSe material on the graph has a peak response at about 735 nm. These differences illustrate how the manufacturer is able to engineer the spectral response of a material by altering material formulation. Relative response is used for the vertical axis of the spectral response curves because it allows one curve to describe all the devices manufactured from a given material combination.

Relative response, sometimes called *relative sensitivity,* is the ratio of cell conductance at the test wavelength to the cell conductance at the wavelength of peak response.

$$\frac{G_\lambda}{G_P} = S_r \tag{9-5}$$

where G_λ is conductance at any wavelength
G_P is conductance at the wavelength of peak response
S_r is the relative sensitivity

Because conductance is the reciprocal of resistance, equation (9-5) can be rewritten

$$\frac{R_P}{R_\lambda} = S_r \tag{9-6}$$

where

$$R_\lambda = \frac{1}{G_\lambda}$$

$$R_P = \frac{1}{G_P}$$

Figure 9-8 illustrates the physical appearance of a typical photoconductive cell. The active material is deposited on an insulating substrate. Lead wires are inserted through the substrate and bonded to the photoconductive material. Some type of protective coating will cover the active material. An opaque masking material may be used to create a regularly shaped active area. Sometimes the device is placed in a metal package with a lens.

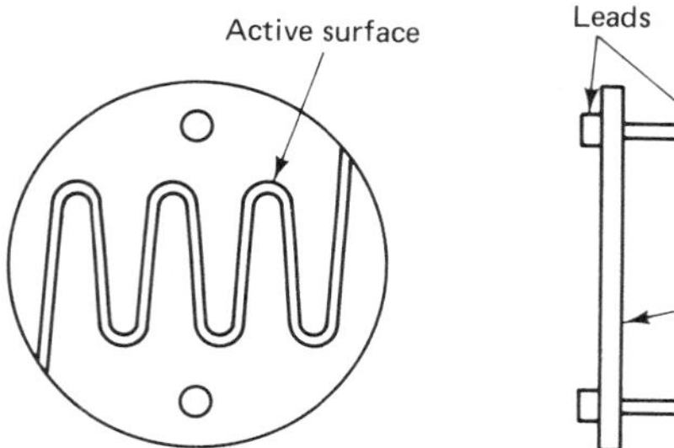

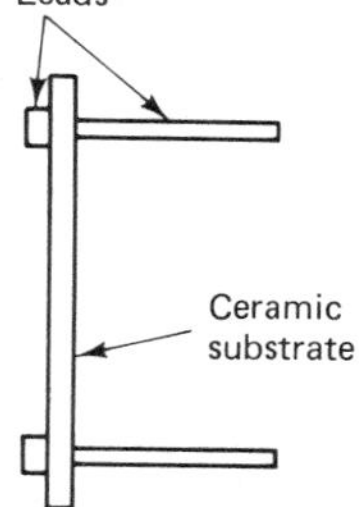

Figure 9-8 Construction of a typical photoresistor.

EXAMPLE 9-1

Estimation of Cell Resistance at a Specific Wavelength

Given:

Test Source		Resistance	
λ (nm)	He (mW/m^2)	Dark (Ω)	Illuminated (Ω)
640	1	504 k	4 k

Find: The resistance of a CdS cell described by the data below and the graph of Figure 9-7 when the cell is illuminated by a 450-nm source with a flux density of 1 mW/m^2.

Solution: From Figure 9-7 and the data above:

λ (nm)	Relative Response S_r	R (Ω)
640	0.90	4k
450	0.38	R

$$\frac{R_P}{4\ \text{k}\Omega} = 0.90$$

$$\frac{R_P}{R} = 0.38$$

$$4\ \text{k}\Omega \times 0.90 = R \times 0.38$$

$$R = 9.47\ \text{k}\Omega$$

The spectral response can be used to predict the response of the cell to monochromatic sources, but many photoconductive cell applications use an incandescent (polychromatic) source whose color temperature will affect cell response. Figure 9-9 illustrates the responses of four cells tested with an incandescent source whose color temperature was 2800 K. This figure is a typical plot of cell resistance versus illumination for cells illuminated by a source of constant color temperature. If a source with a higher or lower color temperature were used, slightly different curves would be obtained.

If you will page ahead to the data sheet of Figure 9-15, you will find a graph that illustrates the effect of source color temperature on cell sensitivity with flux density held constant. The cells described by this graph are CdS and CdSe cells and

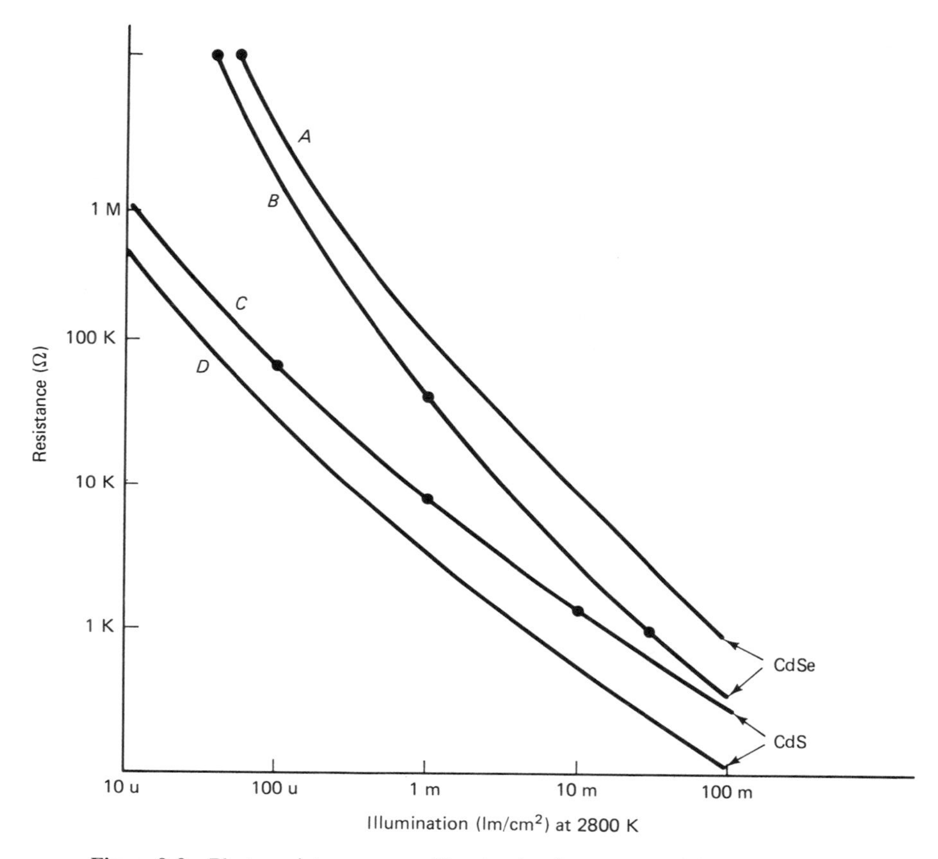

Figure 9-9 Photo resistance versus illumination for two materials.

their sensitivity is rated at 100% when they are illuminated by a source with a color temperature of 2854 K. When a source with a lower color temperature is used, the cells exhibit a greater sensitivity, which means a greater conductance and a lower resistance. An incandescent source with a lower color temperature will have a higher percentage of its radiated energy in the red and infrared region than the standard source. When a source with a higher color temperature than the standard source is used, all but one of the cell types exhibit a reduced sensitivity, which means a lower conductance and a higher resistance. A source with a higher color temperature will have a higher percentage of its radiated energy in the blue and ultraviolet region than will the standard source.

Notice that the type 5H cell has a sensitivity that is relatively immune to changes in source color temperature. This cell type has a peak spectral sensitivity near 520 nm, which is in the center of the green region of the visible spectrum, so that this cell responds roughly equally to red and blue wavelengths. Cell type 3 exhibits the greatest change in sensitivity of all the cell types described by this graph. This cell type has a peak spectral sensitivity at 735 nm, which is in the red region of the visible spectrum near the infrared boundary. The other cells described by this graph have peak spectral sensitivities at wavelengths between these two wavelength values. Starting with cell type 5H and moving down the graph to cell type 3, each successive cell type has a longer wavelength of peak sensitivity than the cell above it. The cells with wavelengths of peak sensitivity at the red end of the spectrum can be expected to exhibit increased resistance as the color temperature of the source is increased, while flux levels are held constant, because the percentage of the incident energy at the longer wavelengths decreases as the color temperature of the source is increased.

It is possible, however, to formulate CdS cells, which have spectral responses which extend through the shorter wavelengths and into the UV region. This type of cell would exhibit an increase in resistance, hence a decrease in conductance, as the color temperature of the source is decreased, because the cell is more sensitive to blue than to red wavelengths.

A majority of photoconductive cells are constructed using a cadmium-based polycrystalline material. A particular material combination is chosen to give a spectral response match between the available light source and the material. Manufacturers also offer material combinations which have a spectral response that approximates the response of the human eye for photometric work. These cells are formed by combining cadmium sulfide, which has a strong blue response, with cadmium selenide, which has a strong red response. Table 9-3 lists three common cadmium-based material combinations and their spectral characteristics.

You may recall that most clear glass tungsten incandescent lamps have color temperatures in the range 2000 to 3000 K, with 2800 K being a representative nominal value. Over this 1000 K range the type 3 cell of Figure 9-15 exhibits a 12% change in sensitivity, and all the other cells exhibit a smaller change in sensitivity. We can conclude that for the photoresistive cells described by this graph, almost any clear glass incandescent tungsten lamp operated at its rated voltage can be used as a test source. This tends to be true for most cadmium-based photoresistive cells intended for use with incandescent sources. When using photoconductive cells in a system design where the source color temperature is expected to be less than 2000 K or greater than 3000 K, the cell resistance should be measured when the cell is illuminated by a source that has the same color temperature as the source that will be used in the system.

The change in cell response caused by source color temperature or wavelength are two ways of describing a fundamental characteristic of these devices, which is that these devices are photon sensors, not power sensors. In other words, the

TABLE 9-3 CADMIUM CELL MATERIAL CHARACTERISTICS

Material	Characteristics
Cadmium sulfide	Peak blue (low red), low sensitivity when compared to other materials
Cadmium sulfo-selenide	Approximates the spectral response of the human eye
Cadmium selenide	Peak red (low blue), fast response time

response of these devices depends on the number of photons of the appropriate wavelength present in the incident radiation, not the total incident power. This is a fundamental difference between these devices and the thermal detectors discussed in Chapter 8.

9-6 SENSITIVITY OF PHOTORESISTORS

Assigning a resistance sensitivity specification to photoresistive cells presents a number of problems. The primary problem is that a change in resistance by itself is not terribly useful. The change in resistance must be used to cause a change in a measured current or voltage. Circuit design parameters such as the circuit's resistance and supply voltage will determine how the measured current or voltage changes in response to the cell resistance change. In the final analysis it is circuit sensitivity that is the item of interest.

A second problem arises from the nonlinear mathematical relationship that exists between cell resistance and incident illumination. A sensitivity specification should allow us to predict how much the detector's characteristic will change in response to a change in illumination. For example, a linearly responding current detector might have a current sensitivity specification of 10 μA/lm. This specification implies that each one lumen increase or decrease in illumination will cause a 10 μA change in current.

Since this sort of linearity does not exist with photoresistors, it is not possible to come up with a single sensitivity number which when multiplied by a change in illumination will yield the corresponding change in resistance. Thus the techniques for evaluating or ranking sensitivity of devices will have to be based on a different approach.

Resistance versus illumination graphs are one way of assessing device sensitivity. The devices with the steepest curves are the ones with the greatest sensitivities. In Figure 9-10 the line illustrating the response of cell 5 is clearly steeper than the line illustrating the response of cell 2. Cell 5 is therefore more sensitive than cell 2. A useful mathematical indicator of slope would be one that has a constant value independent of the range of illumination used to determine it. A method for obtaining such a slope factor is illustrated below.

Actual photoresistor characteristics are not nice straight lines like those found in Figure 9-10. If you glance at the photoresistor curves of Figure 9-9, you will find that the lines are noticeably curved. The straight lines on Figure 9-10 represent simplified engineering approximations of actual cell characteristics. Straight-line approximations of cell characteristics are used to develop cell sensitivity specifications. To obtain an equation that is an accurate mathematical description of the straight-line approximation of a cell, you must take into account the fact that the cell characteristics are plotted on a log-log graph. The equation for a straight line on a log-log resistance versus illumination graph is given by

$$\ln R = K + N \ln E_v \tag{9-7}$$

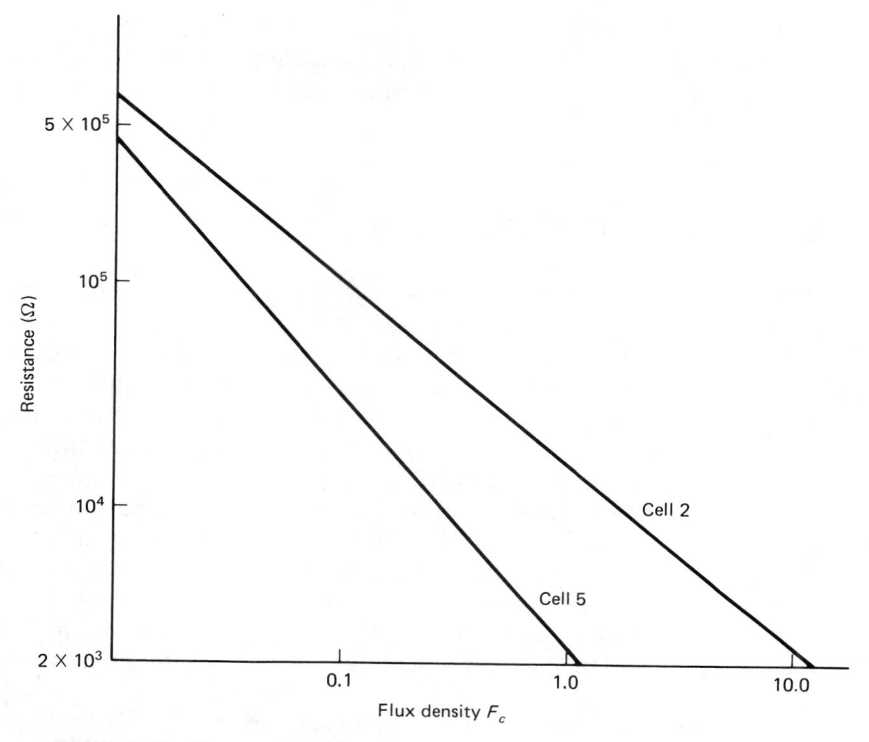

Figure 9-10 Two straight-line approximations of photoresistor characteristics illustrating slope differences.

where R is the resistance at E_v
K is a reference constant
N is a dimensionless number that determines the line's slope
E_v is the illumination level

Notice that equation (9-7) has the same form as the more familiar equation for a straight line plotted on Cartesian graph paper. This more familiar form is illustrated by

$$y = b + mx \tag{9-8}$$

On the resistance versus illumination log-log graph, ln R corresponds to y and ln E_v corresponds to x. By selecting two values of E_v and their corresponding values of R, it is possible to solve for the unknown factors K and N in equation (9-7). Let R_1 be the resistance caused by E_1 and R_2 be the resistance caused by E_2. Then

$$\ln R_1 = K + N \ln E_1$$

$$\ln R_2 = K + N \ln E_2$$

Subtracting yields

$$\ln R_1 - \ln R_2 = N \ln E_1 - N \ln E_2$$

Invoking the properties of logarithms gives

$$\ln \frac{R_1}{R_2} = N \ln \frac{E_1}{E_2}$$

Therefore,

$$N = \frac{\ln(R_1/R_2)}{\ln(E_1/E_2)} \tag{9-9}$$

The slope factor N meets our criterion for a good sensitivity factor. It has a constant value that is independent of the range of illumination. The calculated value for N can be substituted back into either of the two original equations, and a value of K can then be calculated. An alternative approach for obtaining K would be to observe that the term "$N \ln E_v$" will be zero when E_v has a magnitude of 1 because the log of 1 to any base is zero. Therefore, K is equal to the $\ln R$ when E_v is equal to 1 unit of illuminance. The value of R caused by an illumination level of 1 unit of illuminance is given the symbol R_0.

Example 9-2 illustrates the derivation of the value of K, R_0, and N for the line representing cell 2 on Figure 9-10.

EXAMPLE 9-2

Determination of the Values of the Constants K, R_0, and N in Equation (9-7) for Cell 2 of Figure 9-10

Data read from the graph of Figure 9-10:

$$\begin{aligned}
E_1 &= 2 \text{ fc} \\
R_1 &= 9.8 \times 10^3\ \Omega \\
E_2 &= 0.01 \text{ fc} \\
R_2 &= 1 \times 10^6\ \Omega \\
R_0 &\cong 1.8 \times 10^4\ \Omega \qquad \text{when } E_v = 1 \text{ fc} \\
K &= \ln R_0 \\
&\cong \ln 1.8 \times 10^4\ \Omega \text{ at } 1 \text{ fc} \\
&\cong 9.798
\end{aligned}$$

Calculations:

$$\begin{aligned}
N &= \frac{\ln(R_1/R_2)}{\ln(E_1/E_2)} \\
&= \frac{\ln(9.8 \times 10^3\ \Omega / 1 \times 10^6\ \Omega)}{\ln(2 \text{ fc}/0.01 \text{ fc})} \\
&= -0.873
\end{aligned}$$

As a check on the values calculated above, calculate K by substituting the calculated value of N and the data values of R_1, E_1 and R_2, E_2 into equation (9-7) and compare the result with the value determined using R_0 from the graph:

$$K_1 = \ln R_1 - N \ln E_1 = 9.795$$

$$K_2 = \ln R_2 - N \ln E_2 = 9.795$$

These two values for K are essentially the same as the value for K determined using R_0.

Equation (9-7) is cumbersome to work with if all you desire is to calculate values of R for corresponding values of E_v. The properties of logarithms can be used to modify equation (9-7) to a more useful form once N and K are known. This derivation is illustrated below.

$$\ln R = K + N \ln E_v \tag{9-10}$$

Therefore,

$$\ln R = K + \ln E_v^N$$

$$\ln R = \ln R_0 + \ln E_v^N$$

where R_0 is the resistance of the cell when E_v is 1 unit. R_0 is the antilog of K.

$$\ln R = \ln R_0 E_v^N$$

Finally,

$$R = R_0 E_v^N \tag{9-11}$$

Equation (9-11) can then be used to calculate values of R corresponding to different values of E_v on a straight-line approximation of a cell.

EXAMPLE 9-3

Comparison of Calculated and Plotted Values of Photocell Resistance

Find: The resistance of cell 2 of Figure 9-10 for various values of illumination and compare the calculated values with the graphed values.

$R = R_0 E_v^N$

$R_0 = 1.8 \times 10^4\ \Omega$ when $E_v = 1$ fc (read from graph)

$N = -0.873$ from Example 9-2

$R = 1.8 \times 10^4\ \Omega \times E_v^{-0.873}$

As Table 9-4 illustrates, this approach provides us with a tool to predict cell resistance. The exponential equation (9-11) could be written into a circuit analysis computer program to predict circuit performance. All of the discussion above assumes a straight-line approximation for the actual cell data, and in many cases an "eyeball fit" of a straight line to the line on the data sheet will be close enough. Occasionally, the more structured approach discussed below will be required.

TABLE 9-4 COMPARISON OF GRAPHED AND CALCULATED VALUES OF PHOTOCELL RESISTANCE *R*

E_v (fc)	R_{graphed} (Ω)	$R_{\text{calculated}}$ (Ω)
0.01	1×10^6	1×10^6
0.05	2.45×10^5	2.46×10^5
0.10	1.35×10^5	1.34×10^5
0.50	3.25×10^4	3.29×10^4
1.00	1.8×10^4	1.8×10^4
2.00	9.8×10^3	9.8×10^3

Figure 9-11 illustrates three easy-to-obtain straight-line approximations for a photoconductive cell. All three lines are parallel with a slope factor of -0.784. Line A is drawn from one end of the desired illumination range to the other above the cell's resistance curve. Calculated values for cell resistance based on line A will be greater than the plotted values over the entire range. Line B is drawn tangent to the cell's curve parallel to line A; all calculated cell resistance values based on this line will be less than the plotted values over the entire range. Line C is a compromise line. When using an equation based on line C, calculated cell resistance values near the end of the range will be less than the plotted values. Calculated cell values near the center of the range will be greater than the plotted values. A compromise line such as line C is usually the most useful approach. The only difference between the equations for the three lines is the R_0 value used. The slope factors of the three lines are the same because they are parallel.

The following procedure steps are a systematic way of achieving a line such as line C:

1. Draw line A and use its values to calculate the slope factor N.
2. Calculate R_0 or read its value off the graph for line A.
3. Draw line B parallel to line A and read or calculate the R_0 value for line B.
4. Select a value for R_0 between the two values obtained above. A centered line such as line C would have an R_0 value equal to the square root of the product of these two values.

In this section a method for determining cell sensitivity was presented, and a method for calculating cell resistance was derived. More elaborate and detailed approaches could be developed, but generally the approximations presented here are

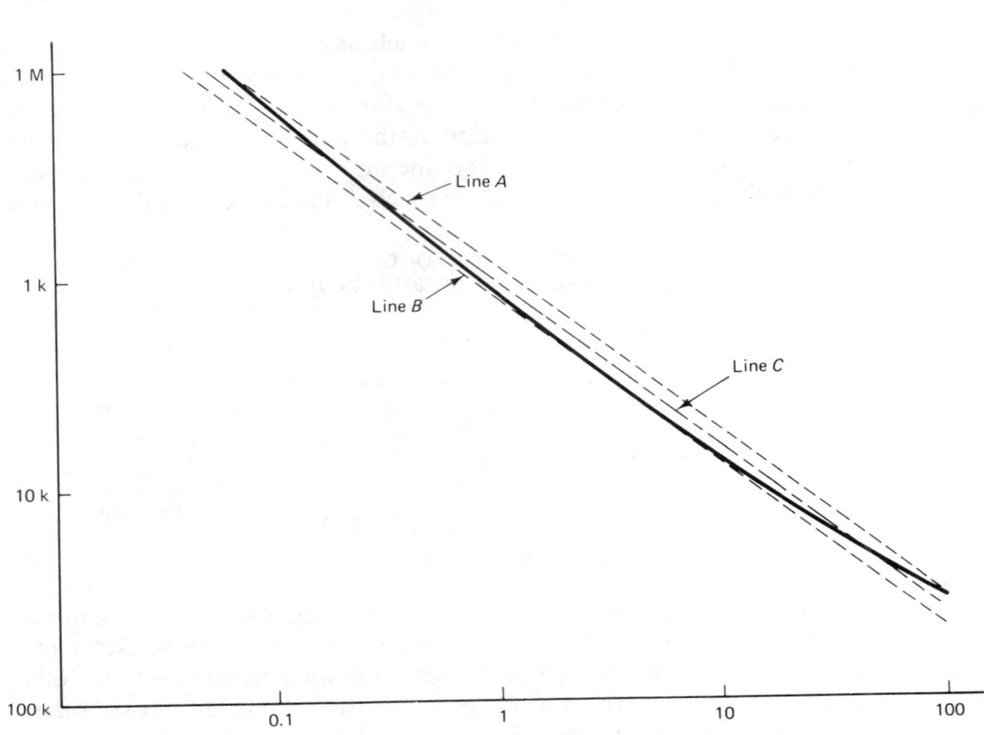

Figure 9-11 Three possible straight-line approximations of a photoresistor characteristic.

more than adequate. Photoconductive cells can have tolerances of ±30%. Because of this wide tolerance range, scrupulous attention to mathematical precision is not warranted. It is hoped that these equations and graphs drive home the point that the resistance of these cells is exponentially, rather than linearly, related to the illumination level.

9-7 PHOTOCONDUCTIVE CELL SPECIFICATIONS

In addition to the spectral, color temperature, and resistance characteristics of photoconductive cells that have already been discussed in detail, there are six other common cell specifications: (1) change in resistance due to temperature, (2) response time, (3) resistance tolerance, (4) light history effect, (5) maximum voltage, and (6) maximum power. Characteristics 1 through 6 will now be detailed in order.

1. Change in Resistance Due to Temperature

The number of charge carriers available for conduction in a photo semiconductor material is a complex function of illumination and device temperature. In general, elevated temperatures yield lower resistances. The specification used to describe the temperature-related resistance change is the "percent change in resistance" from a standard value, often the resistance at 25°C. This specification is computed as illustrated below.

$$\text{SPEC} = \frac{R_n - R_t}{R_t} \times 100\% \tag{9-12}$$

where R_n is the resistance at the temperature and illumination level of interest
R_t is the specified resistance at 25°C and the same level of illumination

$$R_n = R_t \times \frac{\text{SPEC}}{100\%} + R_t$$

This specification is illustrated in Figure 9-12. Note that temperature changes have a much greater effect on resistance at low levels of illumination.

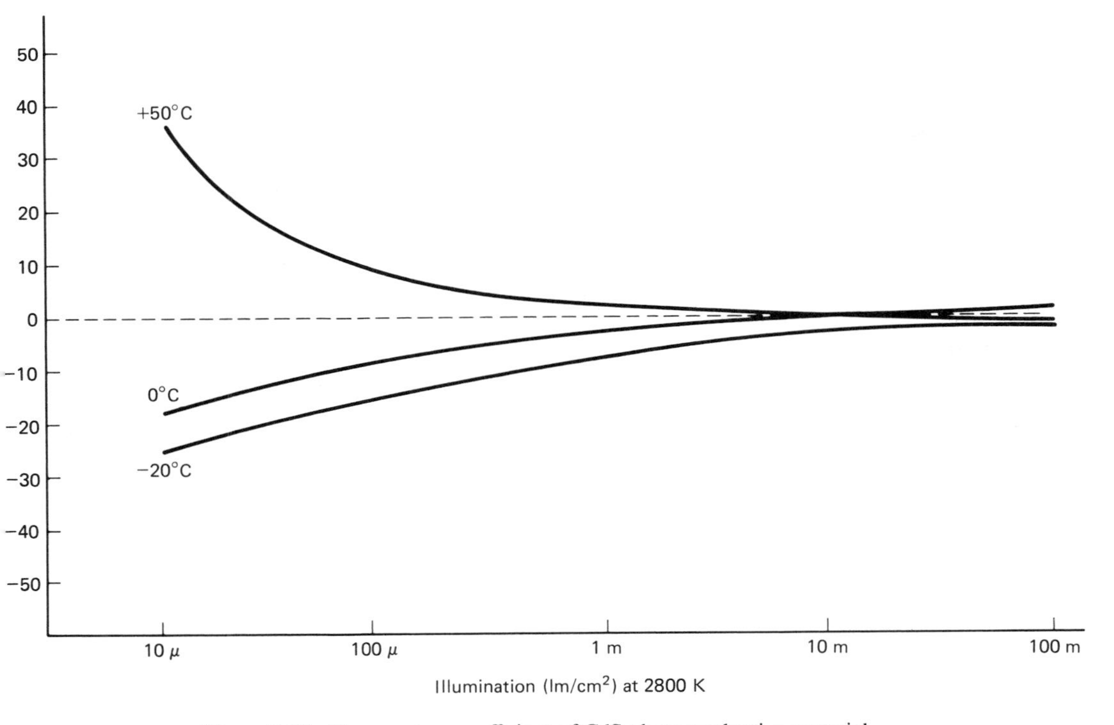

Figure 9-12 Temperature coefficient of CdS photoconductive material.

EXAMPLE 9-4

Cell Resistance and Temperature

Find: The resistance of a CdS cell. Refer to Figure 9-12 for an illumination of 1 mlm/cm^2, 0°C, and 50°C. Assume that the resistance at 25°C for 1 mlm/cm^2 is 8 kΩ.

Solution: From Figure 9-12, when the temperature is at 0°C, the SPEC is −0.3%.

$$R_n = R_t \times \frac{\text{SPEC}}{100\%} + R_t$$

$$= 8\ \text{k}\Omega \times \frac{-0.3\%}{100\%} + 8\ \text{k}\Omega$$

$$= 7760\ \Omega$$

The temperature spec at 50°C is +2%.

$$R_n = R_t \times \frac{2\%}{100\%} + R_t$$

$$= R_t = 8160\ \Omega$$

---

2. Response Time

Response time T is defined by the graph of Figure 9-13 as the $1 - 1/\varepsilon$ time (that is, the time required for the resistance to reach 63% of the final value) from dark to light. This is the time constant specification defined in Chapter 8. In general, higher levels of illumination cause faster response times. Decay time is a measure of the time required to go from light to dark resistance. Decay time is usually faster than rise time, as illustrated.

EXAMPLE 9-5

Cell Rise Time and Response Time

Find: The rise time for a CdS cell at 10 mlm/cm^2. That is, convert the response time to rise time.

Solution:

1. The illustrated response time is $T = 0.007$ s (reference: Figure 9-13).
2. The rise time $t_r = 2.2T$, $t_r = 2.2 \times 0.007\ \text{s} = 0.015$ s.

The rise time, or response time, specifications can be used to predict how fast a cell will change from high dark resistance to low illuminated resistance. The decay time specification can be used to predict how fast the cell will change from a low illuminated resistance value to a high dark resistance value. The rise and fall times of such a cell is a function of the cell material, the difference between the initial and final levels of illumination, and to a small extent, temperature.

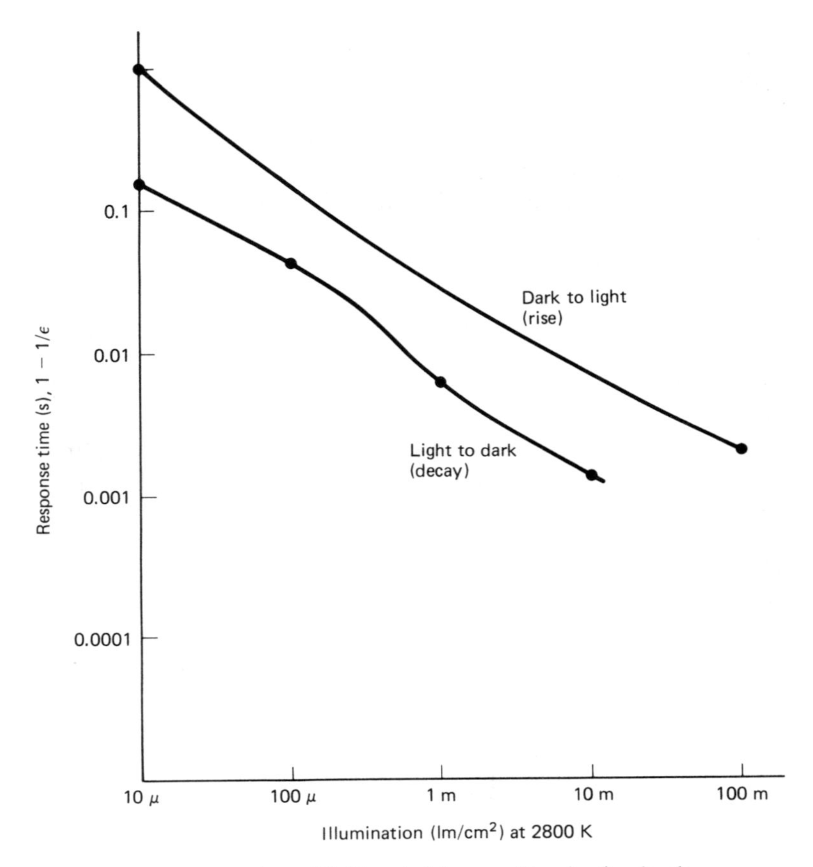

Figure 9-13 Response time of CdS material versus illumination levels.

3. Resistance Tolerance

Resistance tolerance describes the acceptable range of resistance values for a particular set of device part numbers. Values of tens of percent around the nominal value are usual. The tolerance will be established at a specified level of illumination from a source of a stated color temperature and with either device temperature or ambient temperature held constant.

4. Light History Effect

The various cell specifications may have footnotes referring to different periods of light or dark adaptation. This is because the resistance value that a cell will assume when the light level changes is related in part to the cell's light history. Although *light history* is the preferred term for this effect, other names used are *fatigue, hysteresis,* and *light memory.* The phenomenon of light history affects only the initial value of resistance that a cell will exhibit when the illumination level changes. Each light level will cause a final equilibrium resistance that is independent of the cell's light history.

When a cell is maintained at a fixed level of illumination for a long time, it assumes a fixed equilibrium value of resistance. When the illumination changes, the initial value of resistance that the cell assumes will be different from the equilibrium resistance. The magnitude of this difference will depend on the magnitude of the difference between the initial and final illumination levels.

The variation of resistance due to light history is a function of the initial and final illumination levels and the specification is often tabulated. The illumination levels listed in the table will be the final illumination levels. The numbers in the body of the table will represent the ratio obtained by dividing the initial resistance that occurs at the operating illumination level after a long period at a higher illumination level by the initial resistance that occurs at the operating illumination level after a long period in the dark. The high illumination level used to obtain these data will be on the order of 30 fc.

The left half of Figure 9-14 illustrates the change in resistance that will occur when the cell has a light history. The right half of the figure illustrates the change in resistance that will occur when the cell has a dark history. When a cell has been illuminated at the high light level E_1 for a long time, it will assume the equilibrium resistance R_1. At time T_1 the illumination level is reduced to the operating level E_0

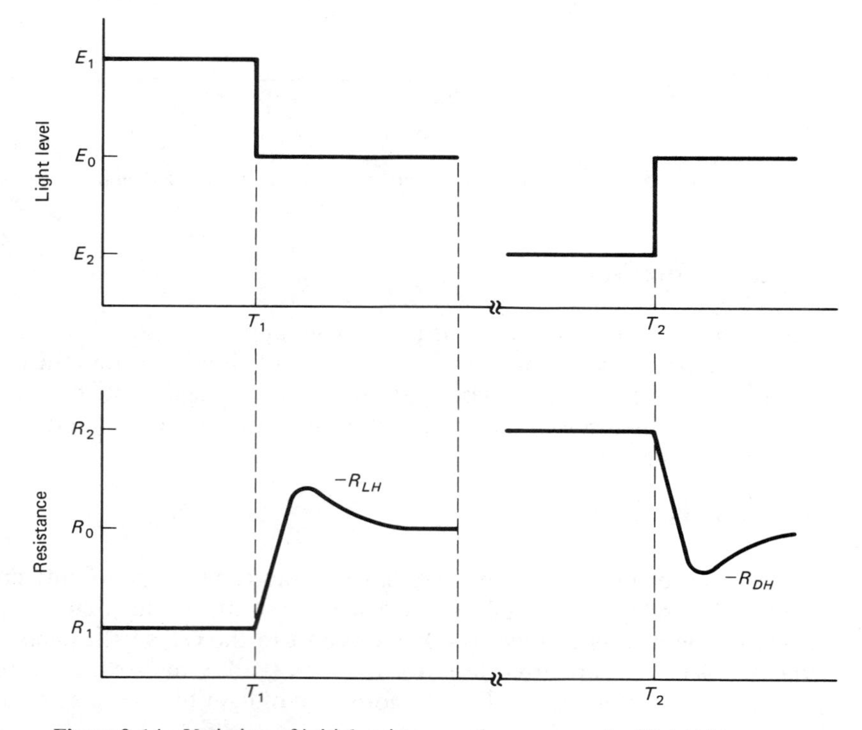

Figure 9-14 Variation of initial resistance values as a result of light history.

and the cell resistance rises to the light history resistance R_{LH}. This value of resistance then decays asymptotically to the operating equilibrium resistance R_0.

When a cell has been in the dark, represented by illumination level E_2, for a long time it assumes the equilibrium resistance value R_2, which is the dark resistance. At time T_2 the illumination level is increased to E_0 and the cell resistance decreases to the dark history resistance R_{DH}. This value of resistance then rises asymptotically to the operating equilibrium resistance R_0.

As a general rule, decreasing illumination levels will cause initial resistances that are greater than the operating equilibrium resistance. Increasing illumination levels will cause initial resistances that are less than the operating equilibrium resistance.

5. Maximum Voltage

Maximum cell voltage is a specification that is often misunderstood. It might be thought that it is related to cell power dissipation, but this is not true. The maximum cell voltage is the voltage that will cause a breakdown in cell insulation or perhaps electrical arcing between cell leads or resistance paths. The maximum cell voltage is typically 100 V or more, and it is measured when the cell is in the high-resistance dark state. Measurement of maximum cell voltage is a device-destructive test. Allowable cell voltages at lower resistance values are determined by the cell's power limitations. These power-limited voltages will be much lower than the rated maximum cell voltage.

6. Maximum Power

A cell's useful maximum power level is limited by the maximum safe cell temperature. Since this relationship is common to all electronic devices, it is discussed in depth in the next section.

9-8 THERMAL MANAGEMENT

All electronic devices have a maximum temperature which if exceeded will cause the device to fail. The temperature that a device reaches is related to the temperature of the ambient in which the device is used and the amount of electrical power supplied to the device.

When electrical power is supplied to a device, the device's temperature becomes elevated above the ambient temperature. If the device can release heat readily, the magnitude of temperature rise will be small. If the device does not release heat readily, the temperature rise will be large. There is in effect a resistance to the flow of heat from a device to the ambient. This resistance to the flow of heat is called the *thermal resistance* and it is expressed in °C of temperature rise per watt of heat.

$$\theta_{DA} = \frac{\Delta T}{P} \tag{9-13}$$

where θ_{DA} is the thermal resistance from device to ambient
ΔT is the temperature rise above the ambient of the device
P is thermal power, which is usually taken as equal to the electrical power applied to the device

To use the thermal resistance specification, it is useful to know the device's maximum permitted operating temperature. Example 9-6 explores one method of dealing with thermal specifications.

EXAMPLE 9-6

Determination of Device Temperature and Power Limits

Given: **(1)** Maximum power in a 25°C ambient = 80 mW
(2) Thermal resistance = 200°C/W

Find: **(a)** The device temperature at rated power when the device is in an ambient of 25°C.
(b) The maximum permissible power in an ambient of 30°C.

Solution:

(a)
$$\begin{aligned} \Delta T &= P\theta_{DA} \\ &= 80\text{ mW} \times 200°\text{C/W} \\ &= (80 \times 10^{-3}\text{W}) \times 200°\text{C/W} \\ &= 16°\text{C} \\ &= T_D - T_A \end{aligned}$$

where T_D is the device temperature
T_A is the ambient temperature

$$\begin{aligned} T_D &= 16°\text{C} + 25°\text{C} \\ &= 41°\text{C} \end{aligned}$$

(b) Assume that the device temperature obtained above is the maximum permitted temperature, and solve for maximum power permitted in a 30°C environment.

$$\begin{aligned} P &= \frac{\Delta T}{\theta_{DA}} \\ &= \frac{41°\text{C} - 30°\text{C}}{200°\text{C/W}} \\ &= \frac{11°\text{C}}{200°\text{C/W}} \\ &= 55\text{ mW} \end{aligned}$$

Example 9-7 illustrates a method for determining the maximum power that can be applied to a cell. Often this information is given as a *derating factor*. This kind of specification would usually read something like the following:

> Maximum power in a 25°C ambient is 80 mW
> Derate at 5 mW/°C

The derating factor is the decrease in power required to maintain a safe operating temperature as the ambient temperature increases. A relationship between the derating factor and the thermal resistance is derived below.

$$\begin{aligned} T_D &= \theta_{DA}P_1 + T_{A1} \\ T_D &= \theta_{DA}P_2 + T_{A2} \\ \hline 0 &= \theta_{DA}(P_1 - P_2) + (T_{A1} - T_{A2}) \end{aligned}$$

The derating factor is equal to $\Delta P/\Delta T$:

$$\frac{P_1 - P_2}{T_{A2} - T_{A1}} = \frac{\Delta P}{\Delta T} = \frac{1}{\theta_{DA}} \qquad (9\text{-}14)$$

The derating factor is equal to the reciprocal of the thermal resistance. Often a *derating curve,* a plot of safe permitted power versus ambient temperature, is given. The slope of this plot can then be used to derive either the derating factor or the thermal resistance, as illustrated in Example 9-7.

EXAMPLE 9-7

Determination of Thermal Resistance

Given: The power derating curve of Figure 9-15.

Find: The thermal resistance of series 700 cells.

Solution:

$$\begin{aligned} T_D &= 126\text{ mW} \times \theta_{DA} + 25^\circ\text{C} \\ T_D &= 0 \qquad\qquad\qquad\quad + 75^\circ\text{C} \\ \hline 0 &= 126\text{ mW} \times \theta_{DA} - 50^\circ\text{C} \\ \theta_{DA} &= 50^\circ\text{C}/126\text{ mW} \\ &= 397^\circ\text{C/W} \end{aligned}$$

The free air power rating curve provides two pieces of information:

1. Maximum safe device temperature, which is the temperature where the curve intersects the zero watts axis.
2. Thermal resistance is equal to the absolute value of the reciprocal of the slope of the curve.

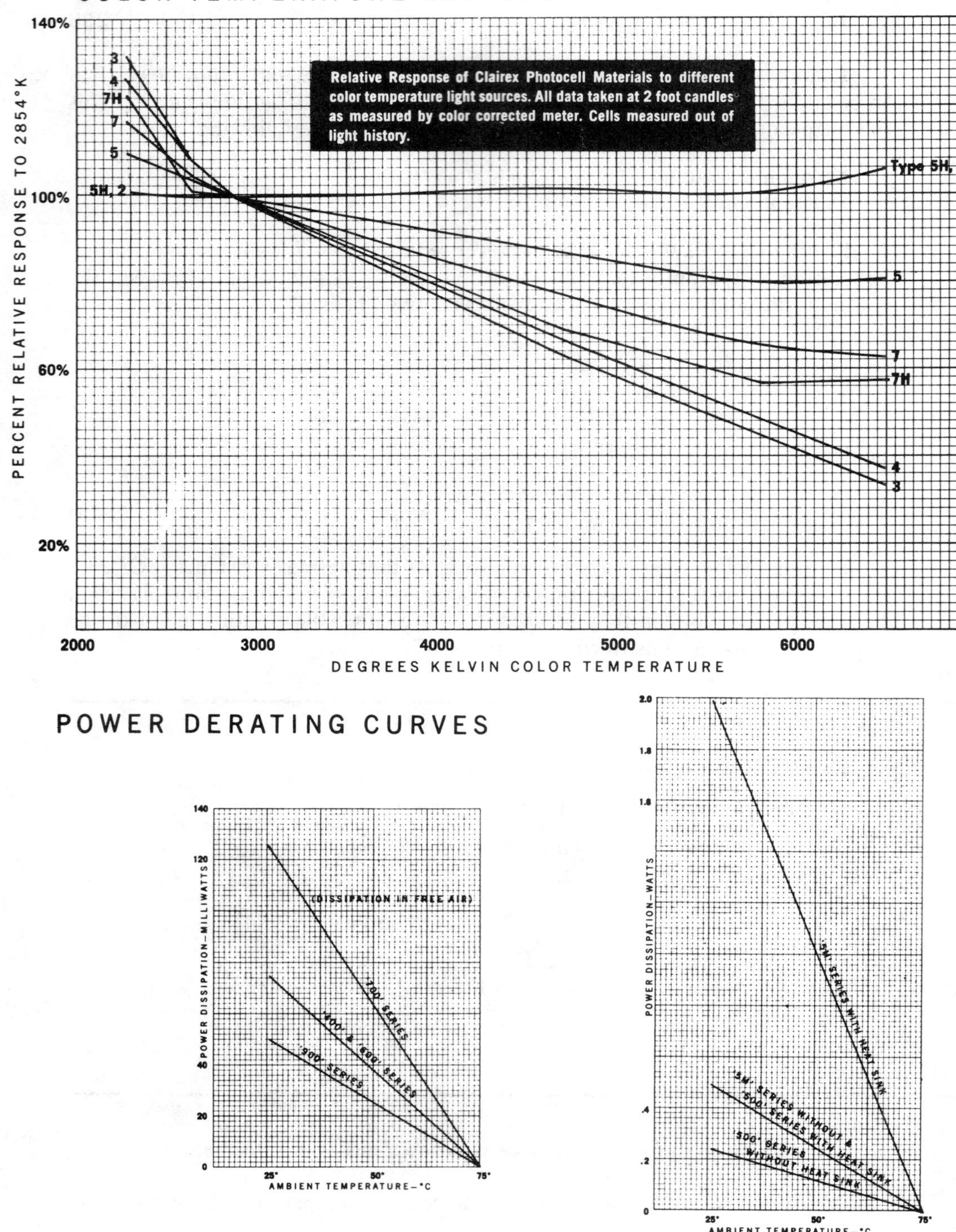

Figure 9-15 Color temperature and power derating curves. (Courtesy Clairex Corporation, Mt. Vernon, NY.)

The point of this section is to emphasize that safe operating power levels, voltage levels, and current levels are not constant values that can be read from a data table, but rather that they are dependent on device temperature. Device temperature depends in turn on the electrical power applied to the device, the ambient temperature, and the device's thermal resistance. Once a safe power level is calculated based on the highest expected ambient temperature and the device's thermal resistance, the maximum permissible voltage and current levels can be determined.

9-9 PHOTOCONDUCTOR APPLICATIONS

Sometimes photoconductors are used to sense the presence or absence of a light beam. At other times they are used to develop a signal proportional to the light level. Three circuits are analyzed in this section. Figure 9-16 illustrates a variable current photoconductor circuit. The current I_T in the circuit would be given by

$$I_T = \frac{E}{R_\lambda + R_m}$$

where R_λ is the resistance of the photoconductor
R_m is the resistance of the current meter

If the resistance of the photoconductor is always much larger than the resistance of the meter, the circuit current will be inversely proportional to the photoconductor's resistance. Because photoconductor resistance decreases as illumination increases, the current will increase as illumination increases.

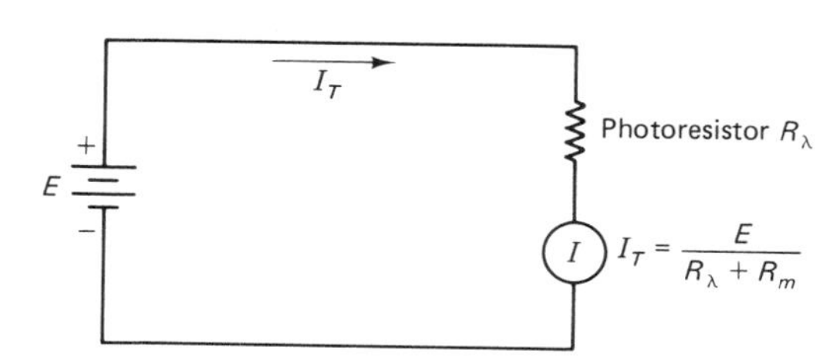

Figure 9-16 Variable-current photoresistor circuit.

Instrument current meters typically have a resistance between several hundreds of ohms and 2000 Ω. The meter resistance limits the choice of photoresistors to cells that have resistances of several thousand ohms at high illumination levels. Unfortunately, these cells will have resistances of several megohms at low illumination levels. These large resistances may decrease circuit current below a level that will cause a useful meter indication.

A typical current meter is specified as having a full-scale deflection of 100 μA and a meter movement resistance of 1800 Ω. The resolution of this type of meter would be 2 μA. Cell resistance values for two typical cells are listed in Table 9-5.

TABLE 9-5 RESISTANCE OF TWO PHOTOCONDUCTIVE CELLS

Illumination (fc)	Resistance (Ω)	
	Cell X	Cell Y
100	250	2,500
50	400	3,500
10	1,050	10,500
5	1,935	19,446
2	4,342	43,916
1	8,003	81,332
0.9	14,749	150,627
0.1	61,000	630,000
0.09	110,000	1,500,000
0.01	800,000	8,000,000

$$\text{Cell X: } R_0 = 8003\ \Omega,\ N = -0.8821$$

$$\text{Cell Y: } R_0 = 81333\ \Omega,\ N = -0.8891$$

The resistance values of these cells, the battery voltage and the meter resistance, can be used to calculate circuit currents. These current values are listed in Table 9-6. The useful measurement range of each cell will be limited by the maximum and minimum values that can be read from the meter. Cell X can be used over the range 0.01 to 0.5 fc and Cell Y can be used from 0.05 to 10 fc. In each case, since current is a nonlinear function of illumination, a nonlinear scale will have to be created.

TABLE 9-6 CURRENT VALUES FOR TWO PHOTOCONDUCTIVE CELLS IN A CIRCUIT USING A 1.2-V NICKEL–CADMIUM BATTERY AND THE 1800-Ω MICROAMMETER WITH A 100-μA RANGE

Illumination (fc)	Current (μA)	
	Cell X	Cell Y
10	421	98
5	321	57
2	195	26
1	122	14
0.5	73	8
0.1	19	2
0.05	11	1
0.01	3	0.3

The photoconductive cell can also be used to create a variable potential output. Two circuits of this type are illustrated in Figure 9-17. If the value of R_s used in Figure 9-17(a) is much less than the smallest value of R_λ, the output voltage is

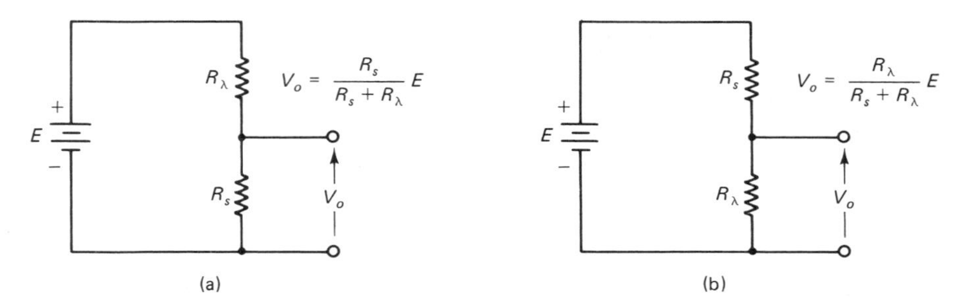

Figure 9-17 Two variable-potential photoresistor circuits.

directly related to the illumination. If the circuit of Figure 9-17(b) is used, the output voltage will be inversely proportional to the illumination.

Accurate measurement of a circuit's voltage without changing it when a meter is connected requires a voltmeter with a resistance that is much larger than the circuit resistance it is placed across. This is easy to achieve for the circuit of Figure 9-17(a) because R_s will have a fixed small value. Measuring the output voltage of Figure 9-17(b) is much more difficult because R_λ is variable and will assume some very large values. Thus, in general, a circuit like that of Figure 9-17(a) is preferred.

In both the current meter circuit of Figure 9-16 and the preferred potential circuit of Figure 9-17(a) the photoconductive cell creates a circuit output by varying circuit current. A photoconductive cell can also be used with an operational amplifier circuit which creates an output voltage that is proportional to the input current to yield an output voltage that is related to illumination. A circuit of this type is illustrated in Figure 9-18. This circuit has a number of features to recommend it:

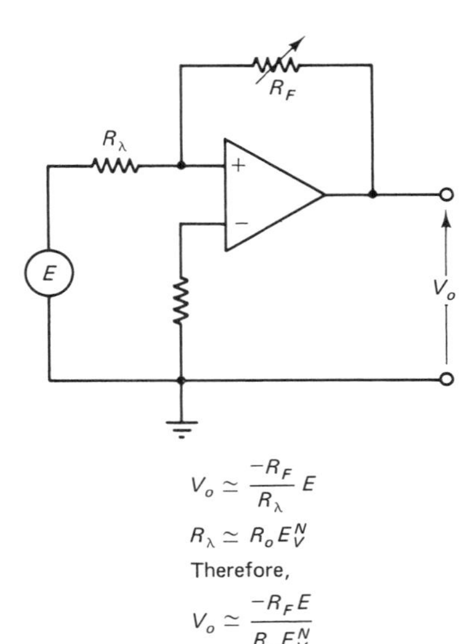

$$V_o \simeq \frac{-R_F}{R_\lambda} E$$

$$R_\lambda \simeq R_o E_V^N$$

Therefore,

$$V_o \simeq \frac{-R_F E}{R_o E_V^N}$$

Figure 9-18 Photoresistor in an operational amplifier circuit.

(1) the output voltage will increase as the illumination increases, (2) the magnitude of the output voltage can be adjusted by adjusting the value of R_F, (3) the output impedance of the circuit is very low and almost any meter can be driven, and (4) the signal source voltage E can be a sine-wave source that will avoid dc drift problems. Any circuit that can convert a change in resistance to a change in voltage or current can be utilized with a photoconductive cell.

This discussion of circuits brings us back full circle to the topic of cell sensitivity discussed in Section 9-6. It is obvious that the circuit components and power supply voltage have a significant effect on circuit sensitivity. In addition, you must keep in mind the useful limits of the meter. In the series meter circuit, for example, too much current and the meter is pegged, too little current and the meter displacement is too small to observe. Practical instruments will usually have a range switch that will place fixed resistors in parallel with the meter to shunt off excess current at high illumination levels. At low illumination levels the range switch will increase the size of the shunt resistor and the switch might also connect a higher-voltage supply to the circuit.

Yes, but what does this have to do with cell sensitivity? Assuming that the meter circuit is designed to be compatible with the cell characteristics, the cell with the highest resistance sensitivity will yield a full-scale deflection with the smallest change in illumination. This will allow you to measure smaller increments of illumination. This is illustrated in the example below.

EXAMPLE 9-8

Full-Scale Deflection for a Meter Circuit with Two Different Photoresistive Cells

Given: **(1)** $R_m = 1.2\ \text{k}\Omega$
(2) $I_{max} = 100\ \mu\text{A}$
(3) $E = 1.2$ V dc

Find: The range of illumination that will cause the meter to go from zero deflection ($I < 2\ \mu\text{A}$) to full-scale deflection ($I = 100\ \mu\text{A}$). Use each of the two cells, cells 2 and 5, described by the graphs of Figure 9-10.

Solution:

1. Calculate the cell resistance that will cause full-scale deflection.

$$E = IR_m + IR_\lambda$$

$$R_\lambda = \frac{E}{I} - R_m$$

$$= \frac{1.2\ V}{100\ \mu\text{A}} - 1.2\ \text{k}\Omega$$

$$= 10.8\ \text{k}\Omega$$

2. Read the value of illuminance from the graph for each cell, and calculate N the photoresistor slope factor.

Cell Number	E_v (fc)	N
2	1.80	−0.873
5	0.325	−1.304

3. Calculate the ratio of the required illuminance values.

$$\frac{1.80}{0.325} = 5.538$$

We can conclude that in this application the circuit using cell 5 is about 5.5 times as sensitive as the circuit using cell 2. The cell resistance values used in these calculations were obtained from the graph of resistance versus illumination. As might be expected, the circuit using the cell with the largest slope factor N has the greatest sensitivity. There is no quick and dirty way to estimate how much more sensitive one cell and circuit combination will be when compared to the possible alternative combinations. Sometimes circuit calculations cannot be avoided.

9-10 SUMMARY

Photoconductive cells are devices whose conductance increases as illumination increases. These cells are constructed from a variety of semiconductor materials and material combinations. The bandgap energies of these materials determine the wavelengths to which these cells will respond. The difference in the response to various wavelengths is called the relative sensitivity, and the relative sensitivity specification can be used to estimate the resistance at various wavelengths if the resistance at one wavelength is known. The response of the cell to changes in illumination can be modeled mathematically using an approximate exponential equation relating resistance to illumination; this equation is valid as long as source wavelength or color temperature is constant. Cells with the steepest resistance versus illumination slopes will have the largest exponent value and exhibit the greatest sensitivity.

The resistance of the completed photo cell is determined by the conductance of the material used and the width, thickness, and length of the material path deposited on the device's substrate. Different cells fabricated out of the same material will have the same resistance sensitivity, N, and spectral characteristics even though each cell may exhibit a different resistance. When the resistance versus illumination curves of different cells made of the same material are plotted on log-log graph paper, the plotted lines will be parallel.

These cells exhibit a wide tolerance range on their nominal resistance and resistance will also change with temperature. Like all detectors, these cells exhibit a rise-time and fall-time characteristic. Also, a long time constant effect called the light history can affect measurements. Maximum voltage ratings are based on breakdown conditions when the cell is darkened. Safe operating voltage current and power levels are temperature dependent.

A simple thermal model can be used to estimate safe operating power, voltage, and current levels for a given ambient temperature. The electrical circuit in which the cell is used is a major determinant of the voltage and current sensitivity that can be achieved. Some simple circuits were analyzed in Section 9-9.

PROBLEMS

1. Estimate the resistance of a CdS photoconductor at the wavelengths listed below. Assume that the resistance measured when irradiated by a 565-nm source is 10 kΩ and that flux density is constant at all wavelengths. Refer to Figure 9-7.
   (a) 475 nm
   (b) 550 nm
   (c) 660 nm
   (d) 700 nm
2. Refer to Figure 9-10. Determine the values of R_0 and the slope factor N for cell 5.
3. Refer to Figure 9-9. Determine the values of R_0 and N for cell A and cell B over the illumination range 100 μlm/cm^2 to 10 mlm/cm^2. Which cell is most sensitive?
4. Which of the following compounds would make the best photometric detector: Cadmium selenide, Cadmium sulfide, or Cadmium sulfo-selenide?
5. A CdS cell is illuminated by 1 mlm/cm^2 at 25°C: the cell exhibits its nominal rated resistance of 200 kΩ. Estimate the cell resistance at −20°C. Refer to Figure 9-12.
6. A CdS cell when illuminated with 100 μlm/cm^2 has a resistance of 40 kΩ at 25°C. What will the resistance of this cell be at the following temperatures?
   (a) −20°C
   (b) 0°C
   (c) +50°C
7. A photoconductive cell has a data sheet specified response time of 0.008 s. Convert this value to a pulse rise time.
8. Refer to Figure 9-15.
   (a) What is the thermal resistance of a 5M series cell without a heat sink?
   (b) What is the thermal resistance of a 5M series cell with a heat sink?
9. A certain cell has a thermal resistance of 150°C/W. Its rated failure temperature is 75°C. What is the maximum power that can be applied to the cell in a 30°C ambient?
10. When a cell is placed on a heat sink, the resultant thermal resistance is approximately equal to the parallel combination of the heat sink thermal resistance and the cell thermal resistance. What thermal resistance heat sink is required if you want to apply 60 mW to a cell with a thermal resistance of 200°C/W in a 35°C ambient? Assume a maximum permitted cell temperature of 41°C.

11. A CdS photo resistor has the following specifications:
    - (1) Maximum power at 25°C: 250 mW
    - (2) Thermal resistance: 100°C/W
    - (3) Resistance tolerance: $\pm 33\frac{1}{3}\%$ at 2 mlm/cm^2 and 25°C
    - (4) R dark: 1 MΩ
    - (5) R at 2 mlm/cm^2: 300 V
    - (6) Maximum voltage: 2.6 kΩ
    - (7) Material: CdS

    (a) Estimate the minimum acceptable resistance at +50°C and 2 mlm/cm^2 (refer to Figure 9-12).

    (b) Estimate the maximum voltage that can be applied to the cell at this resistance and under these test conditions.

12. Why is the maximum voltage rating of a cell always much greater than the actual voltage that can be applied to the cell in use?

13. It is desired to use a cell in series with a current meter to build a light meter that will measure up to 50 mlm/cm^2. The meter movement available has a resistance of 1000 Ω and a full-scale deflection of 500 μA. A 1.2-V battery will be used. The meter discrimination is 2 μA. Assuming nominal cell values:

    (a) At 50 mlm/cm^2, which cell on Figure 9-9 will cause a current closest to the current required for full-scale deflection?

    (b) What will the minimum measurable illumination be with this cell?

14. A type B cell (Figure 9-9) is to be used in a meter circuit where the maximum illumination will be 70 mlm/cm^2 and the maximum cell current will be 7 mA. What power will be delivered to the cell? What will the maximum safe ambient temperature be? Assume that the maximum safe cell temperature is 41°C and that the thermal resistance is 200°C/W.

15. What would the resistance of a type 3 (Figure 9-15) cell be if the color temperature of the lamp is increased to 5000°C and the measured illumination remains constant? Assume that the resistance at the standard color temperature of 2854 K is 1 MΩ.

10

PN Junction Detectors

10-1 INTRODUCTION

In Chapter 9, P- and N-type semiconductor materials were introduced. It was pointed out that different semiconductor material combinations would result in devices with different bandgaps of energy and therefore different spectral responses. In this chapter the properties of a junction of P- and N-type semiconductor materials are discussed. These PN junctions are the basic building blocks of semiconductor devices. A single PN junction can be fabricated into a two-lead device called a diode. Diodes are used directly in electronic circuits, often under the name rectifier. The light-emitting diode is a fundamental optoelectronic light source built from a PN junction. Photodiode detectors are also PN junctions. Combinations of PN junctions are used to fabricate more complex electronic devices like transistors. In this chapter the operation of the single junction diode in the forward-biased, reversed-biased, and photovoltaic modes will be discussed.

Photodiodes are detectors that develop a current as a function of radiation. Four basic forms of semiconductor photodiode construction are utilized: planar PN junction, Schottky barrier photodiode, PIN photodiode, and avalanche photodiode. These photodiodes are used in three typical circuit configurations:

1. Reverse biased with a series load resistor.
2. Short circuited, usually connected to an operational amplifier.
3. Photodiode terminals connected across a load resistor with no external bias voltage.

Circuit configurations 1 and 2 are called the photoconductive mode, and configuration 3 is called photovoltaic operation. Actually, the short-circuit mode is the boundary between photoconductive and photovoltaic operation.

10-2 OPERATION OF PN JUNCTIONS

The PN junctions are fabricated in or on a semiconductor substrate material. The fabrication of junctions is a highly technical process whose exact steps are closely guarded secrets. Two techniques commonly used are diffusion and deposition. In a diffusion system the basic material is exposed to gas and heat. The gas penetrates the hot material and introduces impurities into it. In a deposition process the basic material is exposed to chemicals and/or gases and heat, and a layer of material is deposited on the surface. By careful control of these processes, alternate regions of P and N material can be obtained. When a diode intended for use as an electronic rectifier is fabricated, the junction is encapsulated to shield it from radiation and to protect it from the environment. When a diode intended for use as a photodiode is fabricated, the junction is encapsulated to protect it from the environment, and the geometry of the device is arranged so that radiation can reach the PN junction region.

When the PN junction is formed, there will be a force of attraction between the electrons and holes near the junction. Electrons move across the junction into the P material and neutralize holes, and holes move into the N material and neutralize electrons. This creates an electrically neutral zone near the junction called the *depletion region.* The length of this depletion region and the amount of forward voltage required to force charge carriers across it are a function of the bandgap energy of the material, material temperature, and doping levels. At 25°C the dominant factor that determines this required external potential is the type of material used. Typical values of forward voltage for various materials are listed in Table 10-1. At low forward current levels the diode forward voltages will be near the low end of the indicated forward voltage range. If the basic material used in the fabrication of PN junction is known, a good estimate of the forward voltage required to cause conduction can be made.

Silicon and germanium are used to fabricate electronic devices and photodiode detectors. Gallium is used in high-speed, high-frequency electronic devices and in the fabrication of light-emitting diodes, laser diodes, and photodiode detectors. Selenium is used primarily to fabricate photodiodes used in photometric detector applications.

When the more positive terminal of an external voltage source is connected to the P material of a junction and the less positive terminal is connected to the N material, conduction takes place. The junction is said to be forward biased. The

TABLE 10-1 TYPICAL VALUES OF DIODE FORWARD VOLTAGE

Base Material	Voltage (V)
Germanium	0.3–0.4
Silicon	0.5–0.7
Gallium alloys	1.4–1.7
Selenium	0.5

current that flows when a PN junction is forward biased is called the forward current. When the external voltage is applied in the opposite direction, the junction is said to be reverse biased, sometimes called back biased. Ideally, zero current would flow when the junction is reverse biased. As a practical matter a small amount of current on the order of a few nanoamperes to not more than 10 μA will flow; this current is called the reverse leakage current.

Figure 10-1 illustrates a forward- and a reverse-biased junction. When the junction is forward biased the (+) potential of the battery pushes the holes across the depletion region while the (−) battery potential pushes electrons across the depletion region. Under these conditions the forward current I_F will flow. When the junction is reverse biased, the (−) potential of the battery attracts the holes in the P region and the (+) potential of the battery attracts the electrons in the N region. These effects widen the depletion region and cut off forward current flow. A small reverse current I_R will still flow under these conditions. The forward and reverse voltage and current characteristics of a typical rectifier diode are plotted on a four-quadrant graph such as the one in Figure 10-2.

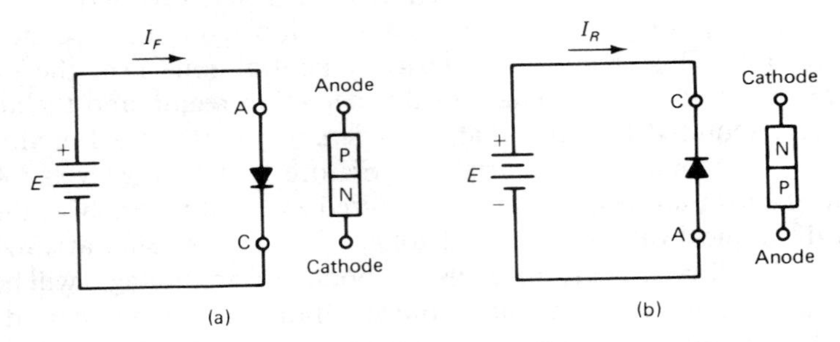

Figure 10-1 (a) Forward- and (b) reverse-biased PN junction diodes.

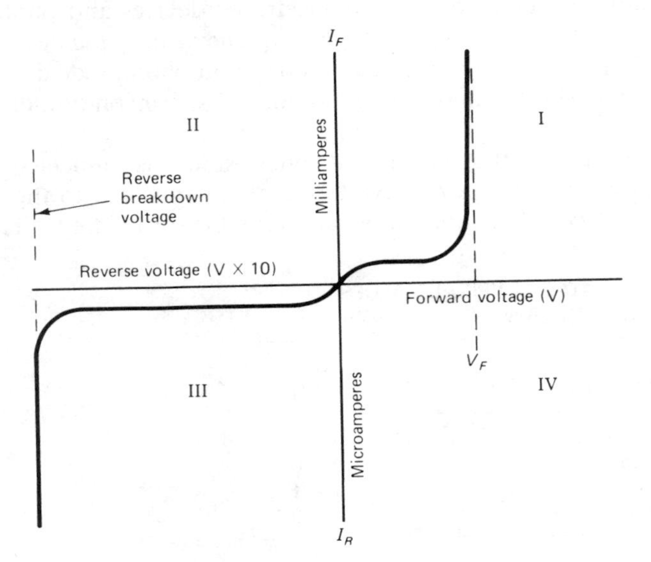

Figure 10-2 Typical PN junction characteristics.

As is illustrated in Figure 10-2, the vertical current axis and the horizontal voltage axis have different scaling magnitudes in the first and third quadrants. The first quadrant contains the plot of the forward characteristics. As this plot illustrates, the voltage drop across the diode is essentially constant for a wide range of forward current values. The third quadrant contains the plot of the reverse bias characteristics. A relatively constant value of reverse current will be observed until the reverse breakdown voltage is reached, at which time current flow increases dramatically. There are some diodes, called regulator diodes, that are designed to operate in the reverse breakdown voltage region; however, most diodes fail if the reverse breakdown voltage is exceeded. Even regulator diodes will fail if the current through is not limited to safe levels by the external circuit. Diodes intended for use as rectifiers in electronic circuits are operated in the first and third quadrants. The rectifier diode can be thought of as a polarity sensitive switch. When forward biased the switch is closed (on) and current flows; when reverse biased the switch is open (off) and current flow stops.

An understanding of the rectifier diode characteristics is useful because this understanding provides a platform from which other diode structures can be discussed. Radiating diodes, such as light-emitting diodes (LEDs) and injection laser diodes (ILDs), for example, are operated forward biased and have voltage and current characteristics similar in form to those of a rectifier diode. As we shall see in this chapter, detector diodes are normally operated reverse biased. If a detector diode is incorrectly installed in a circuit so that it is forward biased (something I'm sure my intelligent readers would never do), the detector will behave like any other forward-biased junction as long as its power dissipation limits are not exceeded. In addition, rectifiers are often used as components in optoelectronic circuits.

The fundamental characteristics of the unilluminated junctions used in electronics applications can be summarized as follows: When these junctions are forward biased, the voltage drop across them will be relatively constant and equal to the forward voltage characteristic of the material used to fabricate the diode. When a diode is reversed biased, the current through it is very small, essentially constant, and independent of the magnitude of reverse voltage. If the value of reverse breakdown voltage is exceeded, the current through the diode will increase dramatically and the diode will probably fail.

Diode forward voltage and current characteristics are temperature dependent. If the temperature of the diode is increased significantly, the forward voltage drop will decrease. If the diode is cooled, its forward voltage will increase. The value of reverse current is also temperature dependent and will increase as junction temperature increases.

10-3 PHOTODIODE CONSTRUCTION AND PHYSICS

Photodiodes are fabricated with one side of the semiconductor structure open to radiation through a window or other protective covering. Planar PN junction photodiodes are very thin devices with a large surface area and an N material substrate much thicker than the surface P material top layer through which intercepted radiation passes to the junction. The construction of a conventional planar PN photodiode is shown in the cross-sectional drawing of Figure 10-3. Most conventional planar PN photodiodes are fabricated by gas diffusion into silicon. These devices are called *planar diffused photodiodes.*

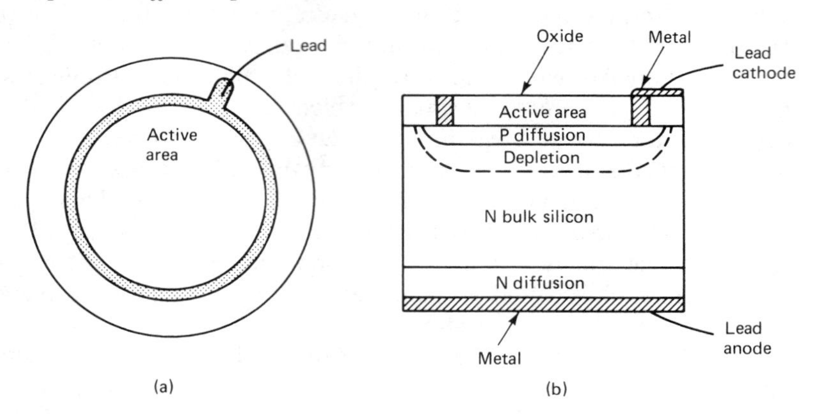

Figure 10-3 Planar diffused photodiode construction: (a) top view; (b) cross section.

The *Schottky barrier photodiode* is an alternative structure, constructed as an N-type semiconductor material with no P semiconductor layer. Instead, a very thin layer of gold film is deposited on the surface, and the interface between the gold layer and the N material acts as the junction. Schottky barrier photodiodes have a flatter spectral response in the infrared and visible region than the conventional PN junction photodiode and they are also more sensitive in the ultraviolet region. On the other hand, Schottky barrier photodiodes are more sensitive to heat than PN junction photodiodes, and they do not always operate dependably at high radiation levels. Figure 10-4 illustrates the spectral response of a typical planar diffused photodiode and a typical Schottky photodiode.

A third very important photodiode type is the PIN photodiode. The PIN photodiode is constructed with a top layer of P material followed by a layer of intrinsic material followed by a layer of N material. The layering of the PIN photodiode gives it its name: P for P material, I for intrinsic material, and N for N material. The intrinsic material widens the depletion region, creating a greater separation between the depletion region's positive and negative surfaces. This increased separation of these charged surfaces reduces device capacitance, thus decreasing photodiode response time.

The physical size of a photodiode and its package will be a function of its intended application. A photodiode designed for use in a fiber optics system will only be a few millimeters on a side, and it will be mounted in a similarly small package. A photodiode or photodiode array intended for solar diode power applications will be in a large package several inches in diameter. Photodiodes designed for use as detectors in radiation measuring instruments often have areas on the order of 1 cm^2. A system designer can select from a wide range of device sizes and packages between these extremes.

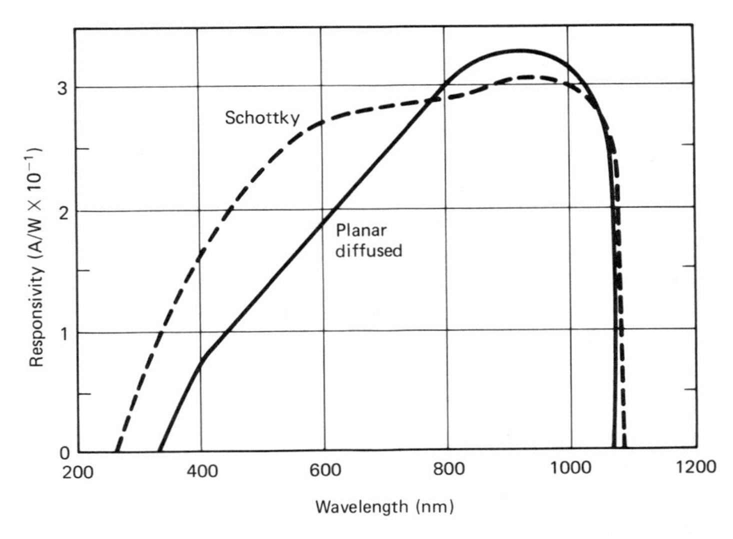

Figure 10-4 Comparison of the responsivity of a Schottky photodiode and a planar diffused photodiode.

The bulk semiconductor photoresistors discussed in Chapter 9 exhibit increased conduction when exposed to radiation. The incident radiation is absorbed, causing an increase in the number of free charge carriers. The incident photons with appropriate energies cause electrons to make the permitted transition to the conduction band, where they can contribute to the device current.

In photodiodes the incident radiation causes an increased reverse current flow across the photodiode junction. When the depletion region of the photodiode receives radiation there will be a net increase in reverse current. If an external voltage source forward biases the junction, the observed change in current flow will be negligible. Hence photodiodes are not operated with forward bias; photodiode terminals are connected directly to a load, short circuited together or reverse biased. The change in reverse current flow caused by the absorption of photons will be quite noticeable under these conditions. Photons that reach the photodiode depletion region successfully are absorbed, causing hole–electron pairs to be formed; this results in more electrons in the conduction band and more holes in the valence band. The internal field of the depletion region causes the holes and electrons to separate, with the electrons migrating toward the N material (the cathode) and the holes migrating toward the P material (the anode). We can observe the result of this process in four ways:

1. If the photodiode is open circuited, a voltage that is logarithmically related to the incident flux density will be observed.
2. If a resistor is placed across the photodiode, a voltage rise and a current flow will be observed.
3. If the photodiode is short circuited, a current flow will be observed through the short circuit which is directly proportional to the incident flux density.

4. If the photodiode is reversed biased, a current flow that is directly proportional to the incident flux density will be observed.

The open-circuit connection and the unbiased load resistor connection are both called the photovoltaic mode. The short-circuit connection and the reverse-biased connection are both called the photoconductive mode. Be careful not to confuse the PN junction photo conduction mode with photo resistor conductivity. Solar diodes are photodiodes operated in the photovoltaic mode, while communication and instrument detectors are typically operated in the photoconductive mode.

In either the photovoltaic or the photoconductive mode, the incident photons will cause the anode (P material) to become more positive than the cathode (N material). When an external path (circuit) for current flow is established between the anode and the cathode, the photon-generated conventional current will flow from the anode to the cathode in the external circuit. Since this direction of conventional current flow is opposite to the current direction defined as forward current for a diode, the photon-generated current is a reverse current!

In addition to the desired photon current, a practical photodiode will also exhibit a leakage current or dark current. This current is the result of the formation of hole–electrons pairs that are not related to the intercepted radiation, and generally increases with temperature and with reverse bias voltage. When this leakage current is measured with intercepted radiation blocked, it is called the dark current. To be measured reliably, the photon generated current must be larger than the leakage current.

At a fixed wavelength or color temperature the photon-generated current of a photodiode operated in the photoconduction mode will be directly proportional to the incident flux density. The effectiveness of any given wavelength is described by the device's spectral response curve. Photodiode current is also directly proportional to the active area of the device. If you have two photodiodes fabricated by the same techniques, the photodiode with the larger area will yield the larger detected current. These photodiode characteristics are related to the quantum efficiency of these devices as is described below.

The purpose of a photodiode is to absorb photons and to yield up charge carriers that can contribute to current flow. The ratio of the number of emitted electrons to the number of absorbed photons is called the *quantum efficiency.* The quantum efficiency is given the symbol η. A perfect device would have a quantum efficiency of 1 but actual photodiodes will have quantum efficiencies of less than 1. The quantum efficiency of actual diodes also varies with wavelength. Quantum efficiency can be calculated by dividing the average photodiode current by the current that an ideal quantum detector would generate under equivalent conditions of wavelength and flux density.

$$\eta = \frac{I_p}{i_p} \tag{10-1}$$

where η is the quantum efficiency
I_p is the average photodiode current
i_p is the current of the ideal quantum detector

The charge flow rate in a quantum detector is caused by the radiant energy flow rate into the detector; an ideal quantum detector will yield one electron of charge for every photon of incident energy, so the charge flow rate will be equal to the photon flow rate. When you divide the incident energy flow rate (the incident power) by the energy of one photon (one quantum of energy), you obtain the photon flow rate, and therefore the electron flow rate that will be produced in an ideal quantum detector. In practical circuits we are usually more interested in current magnitudes than in the number of charges present. Current is the charge flow rate. This thought process is used to develop an equation for the photon-generated current in an ideal quantum detector.

$$P = H_0 A \tag{10-2}$$

where P is the energy flow rate, in joules per second (watts)
H_0 is the flux density, in watts per unit area
A is the detector's active area

$$E_p = \frac{hc}{\lambda} \tag{10-3}$$

where E_p is the photon energy
h is Planck's constant, 6.625×10^{-34} joule second
c is the speed of light, 3×10^8 m/s
λ is the free-space wavelength of the photon

$$i_p = \frac{P}{E_p} e \tag{10-4}$$

where i_p is the photon-generated current in the ideal quantum detector
P/E_p is the photon flow rate
e is the charge of one electron, 1.602×10^{-19} C

Substituting the fundamental qualities into equation (10-4) yields

$$i_p = H_0 A \frac{e\lambda}{hc} \tag{10-5}$$

Equation (10-5) describes the current response of an ideal quantum detector. As equation (10-5) illustrates, the amount of photon current created will be directly proportional to flux density, detector area, and the wavelength of radiation. Quite often the ideal current equation is converted to the responsivity equation, where the responsivity is defined as the number of amperes per watt.

$$R = \frac{i_p}{P} \tag{10-6}$$

where R is the responsivity
i_p is the photon current of the ideal quantum detector
P is the incident power

Therefore,

$$R = \frac{i_p}{H_0 A} = \frac{e\lambda}{hc} \tag{10-7}$$

$$= \lambda (8.06 \times 10^{-4}\ \text{A/W} \cdot \text{nm})$$

(*Note:* λ must be in nanometers to use the numerical constant shown.) The responsivity of the quantum detector is directly proportional to the wavelength. The graph of Figure 10-5 illustrates the responsivities of an ideal quantum detector with an efficiency of 1 and a conceptual quantum detector with an efficiency of 0.8; also illustrated is the responsivity of a typical planar diffused silicon detector.

Equations (10-5) and (10-7) can be rewritten, including the quantum efficiency term to represent the detected average current and responsivity of a photodiode. The quantum efficiency of a photodiode is not a constant value but is a function of wavelength. This variation of the quantum efficiency with wavelength is discussed below.

$$I_p = \eta H_0 A \frac{e\lambda}{hc} \tag{10-8}$$

$$I_p = \eta H_0 A\ (8.06 \times 10^{-4}\ \text{A/W} \cdot \text{nm})\ \lambda$$

Therefore,

$$R = \frac{I_p}{H_0 A} = \eta \frac{e\lambda}{hc} \tag{10-9}$$

$$= \eta\ \lambda\ (8.06 \times 10^{-4}\ \text{A/W} \cdot \text{nm})$$

where I_p is the average photodiode current
- η is the ratio of electrons to photons for that device (quantum efficiency)
- e is the charge of one electron, 1.602×10^{-19} C
- λ is the free-space wavelength of the photon (λ must be in nanometers to use the numerical constant shown)
- h is Planck's constant, 6.625×10^{-34} J · s
- c is the speed of light, 3×10^8 m/s

The responsivity curve of the photodiode has a shape similar to that of the ideal quantum detector in the sense that there is a range of wavelengths for which responsivity increases as the incident energy is changed from short wavelengths to longer wavelengths. There are long and short wavelength values outside of which responsivity of the photodiode decreases. Four physical conditions cause a photodiode's responsivity to be different from that of an idealized quantum detector. These four conditions are the reasons that a photodiode's quantum efficiency varies with wavelength.

1. *Reflection.* The physical process that limits the quantum efficiency of a photodiode detector over the range of wavelengths where its responsivity parallels the responsivity of the quantum detector is reflection. Semiconductors have an

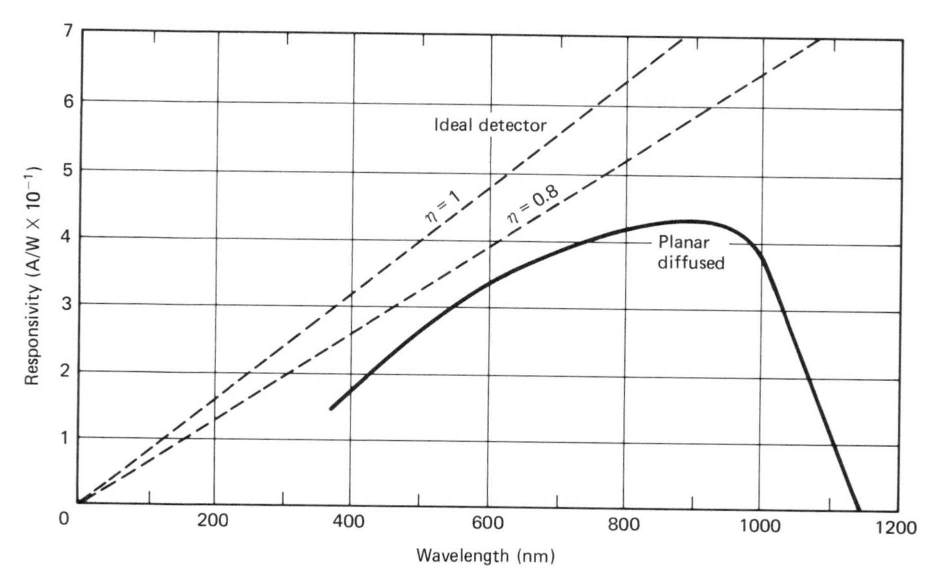

Figure 10-5 Comparison of the responsivity of a planar diffused photodiode to that of two ideal quantum detectors with different quantum efficiencies.

index of refraction of about 3.5. This causes a reflection of about 30% at the semiconductor surface. Photons that are reflected can not contribute to conduction; thus the quantum efficiency of a photodiode will always be less than 1 because of reflection loss. The peak responsivity of a photodiode is always limited by reflections from the detector's surface and losses and reflections associated with the optics covering the detector.

2. *Decreasing absorption of long wavelengths.* The probability that a photon will cause an energy-level transition in the depletion region is related to its wavelength. Short wavelengths are more likely to be absorbed and are said to have a higher absorption coefficient. The exact manner in which the absorption coefficient varies with respect to wavelength is different for each material, but in general the absorption at the shortest useful wavelength is 10^4 times as large as the value of absorption coefficient at wavelengths near the bandgap limit. The absorption coefficient is a measure of how readily a photon will be absorbed per unit distance traveled in the material. The absorption of long wavelengths can be enhanced by making the depletion region longer. The responsivity of the planar diffused photodiode illustrated in Figure 10-5 begins to decrease above 900 nm because of the decreasing absorption of the longer wavelengths.

3. *Short-wavelength absorption.* A decreasing responsivity to short wavelengths is also illustrated by Figure 10-5. This occurs because the shorter-wavelength photons are absorbed by the top layer of the photodiode before they ever reach the depletion region, so that they do not contribute to detected current. Response to the shorter wavelengths can be improved by making the top layer thinner. A Schottky barrier photodiode has an apparent upper P region that is much thinner than that of a planar diffused photodiode. Figure 10-4 illustrates that a Schottky barrier photodiode has more responsivity than a planar diffused photodiode at short wavelengths.

4. *Bandgap limit.* The incident photons must have enough energy to cause electrons to make bandgap transitions. As wavelengths are increased, the photons have less energy. For any given material there will be some wavelength beyond which the photons have too little energy to cause the electrons to make bandgap transitions. For silicon this wavelength is about 1100 nm. The bandgap limit is the upper boundary of λ for long-wavelength absorption.

Some measures can be taken to partially overcome these four effects:

1. The detector surfaces can be coated with materials that will reduce the reflections. So-called "blue-enhanced photodiodes" receive this type of coating.
2. The depth of the depletion region is established by device construction and reverse bias voltage. Therefore, increasing the depth of the depletion region by construction, as is done in a PIN structure, or by increasing the reverse bias voltage can extend the long-wavelength response by a small amount.
3. Reducing the thickness of the top layer of the photodiode will increase the short-wavelength responsivity.

The idealized quantum detector responsivity equation serves as a tool that predicts the maximum possible responsivity that can be achieved at any wavelength. Photodiode detectors can never reach these ideal values of responsivity. The quantum efficiency of a detector at any given wavelength is a figure of merit or quality that allows comparison of the responsivity of the photodiode with those of the idealized quantum detector and other photodiodes.

The *avalanche photodiode* (APD) is a device which has a design that attempts to circumvent the quantum efficiency limitations of the simple PN junction structures. Simply stated, the APD is a detector that generates an electron flow rate greater than the incident photon flow rate. The process by which this is achieved is conceptually similar to that employed by the vacuum photo multiplier discussed in Chapter 7. During the absorption process the APD performs like a PN junction. In other words, the number of photon-generated electrons is similar for an APD and the basic junction device.

The process that takes place after the photon-generated electrons are produced causes the APD to deliver increased current. A very large reverse voltage is used to bias the junction of the APD. The photon-generated electrons are accelerated by this voltage to high velocities. These high-speed particles have large kinetic energies, so that on impact with the atomic structure of the depletion region they cause additional hole-electron pairs to be formed. This process increases the net current flow, causing currents hundreds of times greater than those of a simple junction photodiode. The avalanche photodiode is said to exhibit current gain, where the magnitude of the gain is calculated by dividing the APD current by the current that would be detected by an idealized quantum detector with a quantum efficiency of 1. Voltages of several hundreds of volts are required to achieve this gain. At the present time all "off-the-shelf" avalanche photodiodes are fabricated from silicon. The gain of these devices is very sensitive to changes in the voltage across the photodiode and to temperature changes. Practical circuits using these detectors are more complicated than those required for the other photodiode types. These circuits usually employ techniques that provide voltage regulation and temperature compensation in an attempt to keep gain constant.

The gain of the APD represents the average number of electrons generated per incident photon. The actual number of electrons generated by any particular photon may be higher or lower than the average number. This number varies in a random manner. The net effect of this random process is a noise signal superimposed on the average value of current. This type of noise is different from the thermal noise discussed previously because it is related to absorption and multiplication phenomena, rather than to temperature and bandwidth criteria. At very low power levels, where the number of incident photons is small, the advantage of the avalanche photodiode's gain may be offset by this increase in noise level.

10-4 PHOTODIODE CIRCUITS

The simplest photodiode circuit would be a photodiode connected across a load resistance. The load resistance could represent a meter or an amplifier input. Figure 10-6 illustrates this simple circuit. This circuit is the self-biased or photovoltaic mode of operation. The output voltage across the resistor is also the voltage across the photodiode. The voltage across the resistor and the current through the resistor are governed by Ohm's law, whereas the voltage across the photodiode and the current through the photodiode are governed by the photodiode's photovoltaic characteristic curve. The operating point of the circuit will be where the current versus voltage curve of the resistance intersects the current versus voltage curve of the photodiode. In essence the circuit must satisfy two independent equations relating current and voltage simultaneously. The solution can be found by drawing the straight line that represents the resistor's current and voltage characteristic on the photodiode characteristic curve. This method of circuit analysis illustrated in Figure 10-7 is called *load-line analysis* and the resistor's line is called a *load line.* This approach to problem solution has also been used in Chapter 7.

The three curved lines of Figure 10-7 represent the photodiode's current and voltage characteristics at three different light levels: dark, H_1, and $2H_1$. The straight dashed line represents the characteristics of the load resistor R_L and it has a slope equal to the reciprocal of the load resistance. The points where the dashed line intersects the photodiode curves are called the *operating points.* When the

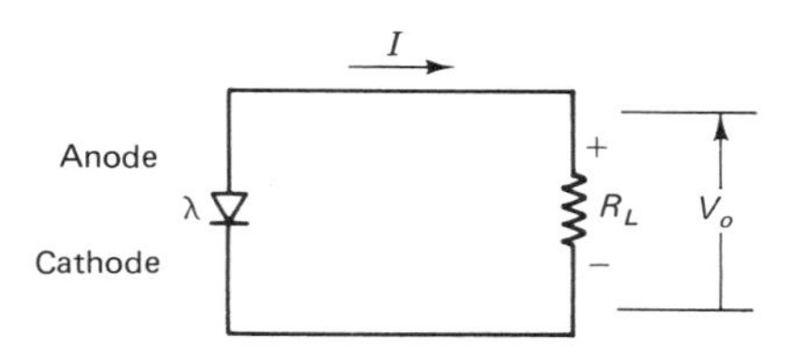

Figure 10-6 Photodiode in the photovoltaic mode.

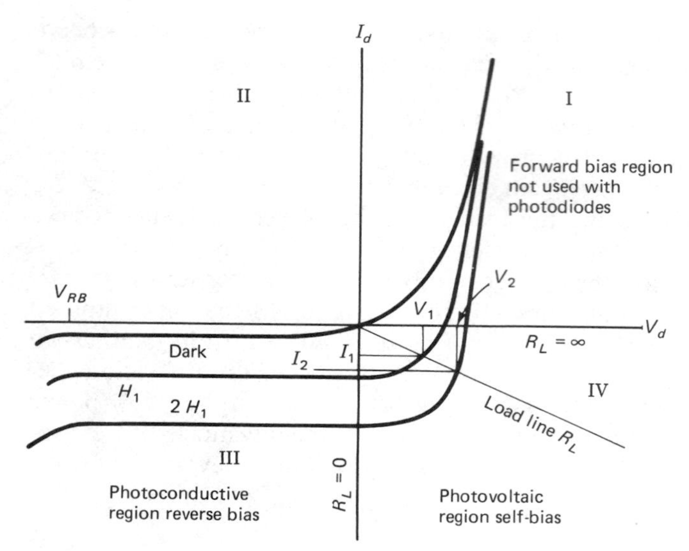

Figure 10-7 Photodiode four quadrant characteristic current and voltage curves illustrating a load line in the photovoltaic mode.

photodiode is illuminated by flux density H_1, current I_1 will flow in the circuit and the voltage V_1 will exist across the resistor. When the flux density is increased to $2H_1$, the current increases a small amount to I_2 and the voltage across the resistor increases a small amount to V_2.

In the photovoltaic mode with the photodiode loaded by a large resistor, relatively large changes in flux density will result in only small changes of output voltage and current. In addition, and much worse from a measurement standpoint, the changes in output voltage and current are not linearly related to the incident flux density. The maximum voltage that can be reached is limited to the photodiode's forward voltage. As load resistance approaches zero ohms, the current's response to changes in flux density becomes linear and the voltage change with changing flux density becomes very small.

The photovoltaic curves lie in the forth quadrant of the photodiode characteristic curves. The horizontal voltage axis values are the open-circuit values that occur when R_L is equal to infinity. The vertical axis current values are the short circuit currents that occur when R_L is equal to zero. The physics of photodiodes is such that the external voltage across the photodiode, when open circuited, is logarithmically related to the incident flux density with a maximum value essentially equal to the photodiode's forward voltage.

$$V_o = K_1 \ln K_2 H_0 \tag{10-10}$$

where V_o is the open-circuit voltage across the photodiode on the horizontal axis
K_1 is a physical constant related to temperature
K_2 is a physical constant related to wavelength and photodiode construction
H_0 is the flux density

For simplicity, assume for a moment that the only current flowing through the photodiode is this average photon-generated current. As was shown previously, the photodiode's photon-generated reverse current is linearly related to incident flux density and active surface area.

$$I_p = \eta H_0 A(8.06 \times 10^{-4}\ \text{A/W} \cdot \text{nm})\,\lambda \tag{10-8}$$

The most useful detection devices are those whose response is linearly related to the detected quantity. Referring to Figure 10-7, observe that the change in current with respect to the change in flux density is linear near and to the left of the zero volts axis. When the photodiode is operated on the zero volts axis, a linear relationship exists between detected current and incident flux density. The photodiode will operate on the zero volts axis if it is short circuited. This can be achieved by connecting the photodiode to a current-to-voltage converter amplifier circuit like that illustrated in Figure 10-8. The apparent input resistance of this type of amplifier is very close to zero ohms. This type of circuit is sometimes called a *transimpedance amplifier.*

The photodiode connected to the apparent short-circuited input of the amplifier will develop a current that is linearly proportional to the incident flux density. The transimpedance amplifier develops a voltage V_o proportional to the input current, in this case the photodiode current. Therefore, the output voltage of the amplifier will be proportional to incident flux density. Actually, the photodiode will have some small amount of leakage current even when darkened, which will cause some small initial offset voltage at the output of the amplifier. This offset voltage can be subtracted from the measurement values or it can be zeroed out with adjustments built into the amplifier.

The photodiode can also be used as a linear detector of flux if it is operated in the reverse-bias mode. This mode of operation also results in a decrease in device capacitance and therefore a decrease in response time. A reverse-bias voltage of about 10 V is typical for silicon devices. A reverse-biased circuit is illustrated in Figure 10-9. The Kirchhoff voltage law equation for this circuit is given below.

$$V_o = V_B - V_D$$

$$I_p R_L = V_B - V_D$$

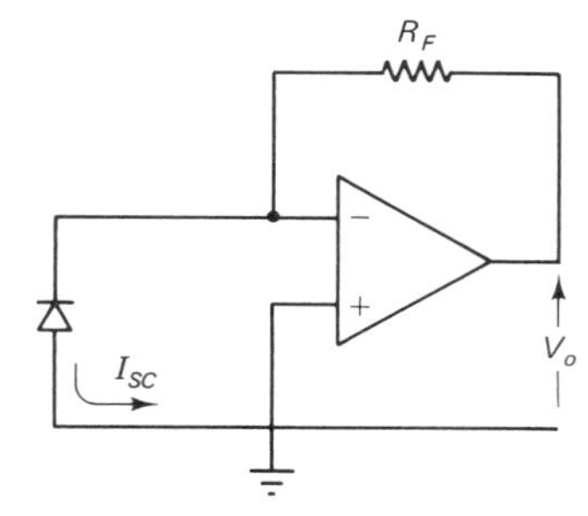

Figure 10-8 Photodiode connected to a current-to-voltage converting amplifier. The photodiode is in the short-circuit mode of operation.

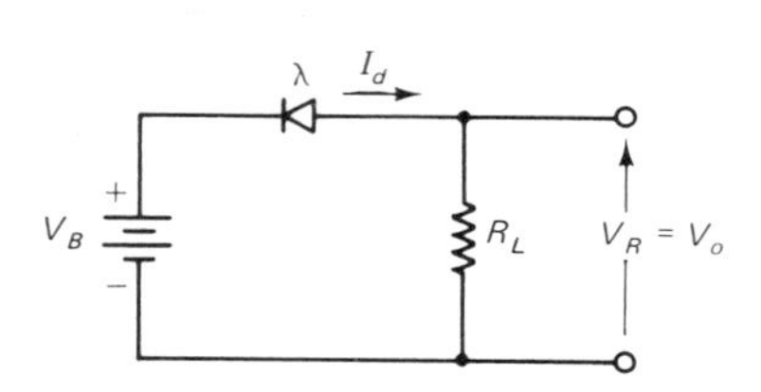

Figure 10-9 Photodiode in a reverse-biased, photoconductive mode circuit.

When I_p, the circuit current, is equal to zero, the voltage across the photodiode will be equal to the bias voltage. To prevent photodiode failure under these conditions the bias voltage must be less than the photodiode reverse breakdown voltage. The theoretical maximum value of current will flow when the voltage across the photodiode is zero.

$$V_D = V_B \qquad \text{when } I_p = 0$$

When $V_D = 0$,

$$I_p = I_{\text{short circuit}} = I_{\max} = \frac{V_B}{R_L}$$

The theoretical maximum value of current is called the *saturation current,* I_{SAT}. This KVL analysis can be plotted on the reverse-bias characteristics of the photodiode, as illustrated in Figure 10-10. Reverse-bias characteristics are plotted in the third quadrant of the characteristic curves.

Figure 10-10 graphically illustrates reverse-bias circuit operation. The resistor load line is plotted between the two points V_B and the saturation current, I_{sat}. An illustrative operating point is also drawn. When flux density H_2 falls on the photodiode, it causes current I_2 to flow. Current I_2 causes voltage drop V_R to appear across the load resistor and voltage V_D to appear across the reverse-biased photodiode. If the flux density were to increase to H_4, the circuit current would increase to I_4. This increased current flow would result in a greater voltage drop across the load resistor and a decreased voltage drop across the photodiode. The maximum flux density that could be detected would be the flux density corresponding to the saturation current. Increasing the flux density above the value that causes the saturation current would not cause any appreciable change in circuit current or resistor

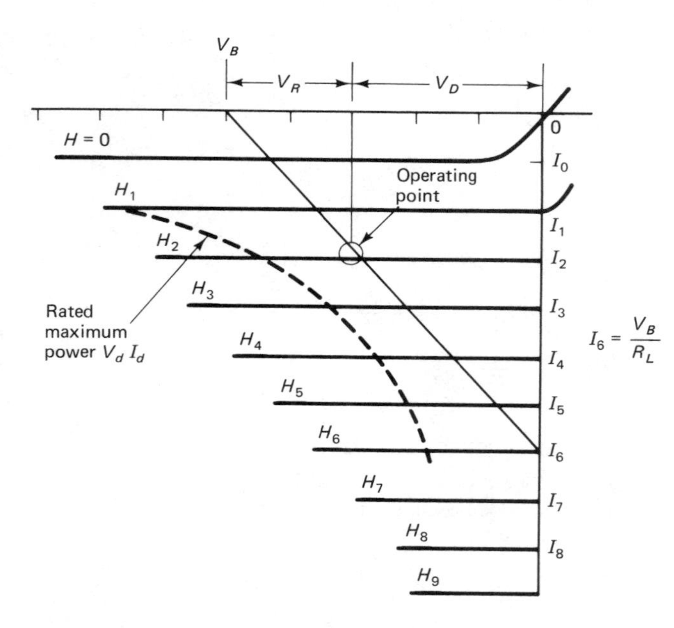

Figure 10-10 Photodiode third-quadrant, reverse-biased characteristic current and voltage curves illustrating a load-line plot and the maximum power rating curve.

voltage. The circuit is said to be *current limited* or *saturated* at this level of flux and current.

If a greater current range is required, either the bias voltage will have to be increased or the resistor size decreased. Care must be taken to select a bias voltage less than the reverse breakdown voltage. The product of the voltage across the photodiode and the current through it must always be less than the power dissipation capability of the photodiode. A curve delineating the maximum power limits can be plotted against the voltage and current axis of the characteristic curves by calculating the maximum permitted current corresponding to each value of diode voltage. The load line must always fall inside this rated maximum power curve as shown in Figure 10-10, to ensure safe photodiode operation. The maximum photodiode dissipation will occur when the flux density causes a current to flow that is equal to one-half of the saturation current. At this current level the voltage drop across the photodiode will be equal to one-half the bias voltage V_B. For you math buffs the calculus derivation of the conditions for maximum dissipation is given below. That is followed by a numerical example for everyone else.

Calculus derivation of the conditions that will cause maximum diode dissipation in the circuit of Figure 10-10:

V_D = photodiode voltage
P_D = photodiode dissipation
I_p = circuit current, average photodiode current
R_L = series resistance
V_B = bias voltage

Derivation:

$$P_D = I_p V_D \qquad \text{definition of photodiode dissipation}$$

$$V_D = V_B - I_p R_L$$

Therefore,

$$P_D = I_p(V_B - I_p R_L)$$

$$P_D = I_p V_B - I_p^2 R_L$$

$$\frac{dPD}{dI_p} = V_B - 2I_p R_L$$

At a peak or maximum value, $dP_D/dI_p = 0$ and

$$0 = V_B - 2I_p R_L$$

Therefore,

$$I_p = \frac{V_B}{2R_L} \qquad \text{at maximum dissipation} \tag{10-11}$$

and

$$V_D = V_B - \left(\frac{V_B}{2R_L} R_L\right) = \frac{V_B}{2}$$

Finally,

$$P_{max} = I_p V_D = \frac{V_B^2}{4R_L} \tag{10-12}$$

EXAMPLE 10-1

Variation of Photodiode Dissipation in a Reverse-Bias Circuit with Changes in Current

Given: **(1)** $R_L = 10\ \text{k}\Omega$
(2) $V_B = 10\ \text{V}$

Find: **(a)** The maximum possible current, I_{sat}.
(b) The maximum power dissipation.
(c) The power dissipation at five values of device current.

Solution: **(a)** $I_{sat} = \dfrac{V_B}{R_L}$

$$= \frac{10\ \text{V}}{10\ \text{k}\Omega} = 1\ \text{mA}$$

(b) $P_{max} = \dfrac{V_B^2}{4R_L}$

$$= \frac{(10\ \text{V})^2}{4 \times 10\ \text{k}\Omega}$$

$$= 2.5\ \text{mW}$$

(c) $V_D = V_B - I_p R_L$
$P_D = V_D I_p$

I_p (mA)	V_D (V)	P_D (mW)
0.1	9	0.9
0.2	8	1.6
0.5	5	2.5
0.8	2	1.6
1.0	0	0

In conclusion, we can say that practical linear photodiode detector circuits will have very small load resistors if they are operated in the photovoltaic mode, or the detector may be short circuited or operated with a reverse-bias voltage. The reverse-biased and short-circuit configurations utilize the linear flux density-to-current relationships of the third quadrant of the characteristic curves. In the reverse-biased circuit the bias voltage and resistor size will establish the maximum power dissipation that can occur. Care must be taken to keep this maximum dissipation within the operating limits of the device.

Unlike detector photodiodes, *solar diodes,* sometimes called *solar cells,* are always operated in the photovoltaic mode. A solar cell is often actually an array of solar diodes electrically connected together. In this application obtaining maximum power output from the solar diode rather than linearity is the major concern. An obvious example would be solar power energy conversion. For any given solar diode there will be an optimum load resistance for maximum power transfer. This resistance can be determined graphically from the fourth-quadrant photodiode characteristic curves. The conventional solar diode data sheet shows the fourth-quadrant characteristics rotated into the first quadrant. This conventional presentation is illustrated in Figure 10-11.

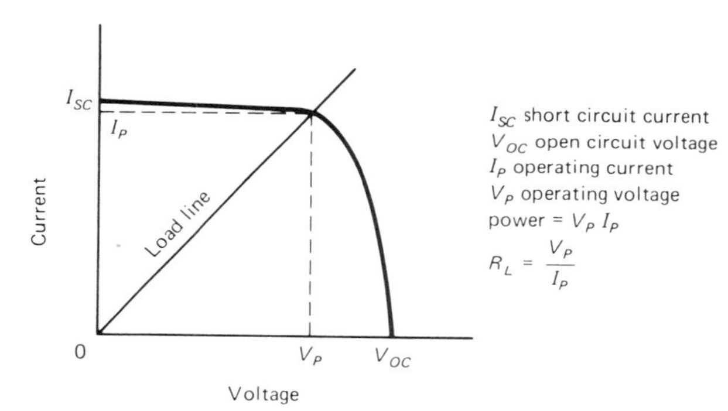

Figure 10-11 Typical solar cell photovoltaic voltage and current plot with load line.

When the solar diode's photovoltaic curve is available, a few trial values of R_L can be evaluated near the knee of the curve that will yield the required load resistance for maximum power transfer. After this value is known, an attempt can be made to make the system resistance appear to be this value. The location of the knee point resulting in maximum power transfer depends on the assumed illumination level, so a representative value of illumination must be used.

Photodiodes are operated in the third or fourth quadrant of the diode characteristic curves. Reverse-biased and short-circuited photodiodes that operate in the third quadrant are used as linear detectors. Solar diodes operate in the fourth quadrant. It is possible to obtain relatively linear responses in the self-biased, fourth quadrant if the load resistances are small. Photodiodes are not operated forward biased!

10-5 PHOTODIODE SMALL-SIGNAL MODELS

A photodiode circuit model, sometimes called an *equivalent circuit,* is composed of idealized components. The circuit equations that describe the model can be used to predict the behavior of the photodiode for the range of operating conditions over which the model is valid. The equivalent-circuit element values are established from the photodiode's measured characteristics. Some of these measured values depend on environmental conditions. In other words, some of the equivalent circuit's element values depend primarily on device construction and materials, while others are controlled or affected by temperature, bias voltage, and incident radiation levels. Nominal element values are sometimes quoted on data sheets for particular test conditions. These nominal values represent averages for a given part number, and often individual devices of a given part number will exhibit noticeably different values; variations of 2:1 or 3:1 can occur and variations of 10:1 are not unheard of.

You could ask: "Why are models important if their circuit element values can vary so much?" First, the model can be used as a basis for calculated estimates of the "average circuit's" performance. Keep in mind the fact that the actual values of the model's components may vary by as much as an order of magnitude from data sheet values. In general, elaborate algebraic or numerical analysis is not warranted! The second and perhaps more important use of models is to predict trends. In other words, when you know how the model's element values are going to change and you know something about the illumination levels and the load resistance, you can predict how the circuit performance will change as these variations take place. A good designer will attempt to construct the circuit so that the effects caused by changes in photodiode characteristics are minimized.

The photodiode model that will be discussed in this section is illustrated in Figure 10-12. Each component of the model is discussed in turn. The "ideal diode" of the circuit model is often omitted from the model because photodiodes are usually operated with reverse bias and therefore, an "ideal diode" would be an open circuit.

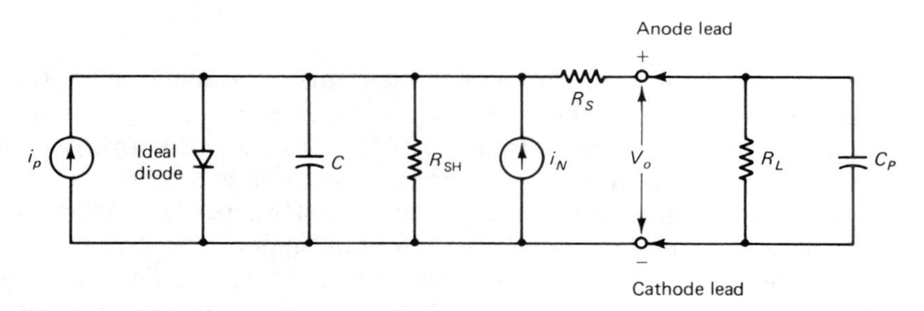

Figure 10-12 Small-signal photodiode model.

Current Generator

The current generator I_p represents the average current caused by the incident radiation.

$$I_p = RP \tag{10-13}$$

where I_p is the average photo current
R is the responsivity of the photodiode, in amperes/watt
P is the net incident radiant power, in watts

The value of R used depends on the spectral content of the source radiation and the spectral response of the photodiode.

Shunt Capacitance

The shunt capacitance, C, is the junction capacitance of the reverse-biased photodiode. Junction capacitance is directly proportional to photodiode area. Its value is also affected by the bulk resistivity of the photodiode material. Finally and perhaps most important from a circuit design standpoint, capacitance is inversely proportional to the square root of the bias voltage. The value of the junction capacitance can be decreased by increasing the reverse-bias voltage that appears across the photodiode.

$$C = \frac{KA}{[p(V_0 - V_D)]^{1/2}} \tag{10-14}$$

where C is junction capacitance, in picofarads
- K is constant of proportionality of about 19,000 for the units used here
- A is active area, in square centimeters
- p is resistivity in $\Omega \cdot$ cm; values range from 10 to 10,000
- V_0 is the depletion voltage of the material: about 0.6 V for silicon
- V_D is reverse-bias voltage in volts that appears across the photodiode; V_D is negative

Shunt Resistance

The shunt or current source resistance R_{SH} is a function of the device's surface area, material resistivity, and temperature as well as the bias voltage that appears across the photodiode. It is used in predicting noise performance. The shunt resistance is sometimes called the *channel resistance.* Typical values of R_{SH} are in the range 100 kΩ to several megohms or more. The shunt resistance is usually ignored when making rise-time calculations. In practical circuits the load resistance will be much smaller than the shunt resistance. The shunt resistance decreases as the device temperature or device area are increased. The shunt resistance generally increases as the bias voltage increases. The shunt resistance is specified as being equal to the slope of the diode voltage and current curve at zero volts or as the ratio of the reverse-bias voltage across the diode to the dark current.

$$R_{SH} = \frac{\Delta V}{\Delta I_d} \tag{10-15}$$

or

$$R_{SH} = \frac{V_D}{I_d} \text{ at } V_D$$

where R_{SH} is the source resistance
- ΔV is the voltage change corresponding to the current change ΔI_d on the diode curves at zero volts
- I_d is the dark current at V_D
- V_D is the reverse-bias voltage across the diode

Noise Current

The noise current source I_n accounts for all the photodiode noise effects. In Chapter 8 thermal noise was discussed as a limiting factor on the performance of thermal detectors. Noise characteristics also limit the performance of photodiode detectors. Photodiodes have three noise-producing processes: quantum noise, shot noise, and thermal noise. Quantum noise is caused by the photon absorption process, and shot noise is a characteristic of semiconductor junction operation.

Quantum noise is a noise process that becomes important at low incident power levels. It occurs in all quantum detectors, including vacuum photodiodes and photoresistors. Quantum noise does not occur in thermoelectric detectors. In the photodiode detection process some of the incident photons are absorbed, causing a current, and some are not. This means that at each instant of time the actual detected current may have a value that is greater or less than the average detected current. When the total incident energy is large, this random process results in a very small current amplitude variation around the average current level. As the average incident energy decreases, the total number of available photons decreases, so the change in current caused by a nonabsorbed photon becomes more significant. This can cause a noise current level that is significant compared with the average current level. The noise current caused by this process is called the *quantum noise current.*

The flux density intercepted by the photodiode must fall to very low levels before the *quantum noise* becomes a significant factor. The intercepted power level at which the quantum noise power is equal to the signal power generated by the average detected current is given approximately by

$$P_m = (H_0 A)_m \simeq \frac{hc}{\lambda} \frac{\Delta f}{\eta} \tag{10-16}$$

where P_m is the incident power level that will cause a signal-to-noise power ratio of 1:1
hc is 6.625×10^{-34} J · m
λ is the photon wavelength
Δf is the noise bandwidth
η is the quantum efficiency

Example 10–2 is presented to illustrate how small the incident power level must be for quantum noise to be a problem.

EXAMPLE 10-2

Calculation of the Incident Power Level at Which Quantum Noise Power is Approximately Equal to Signal Power

Given: **(1)** $\lambda = 800$ nm
(2) $\Delta f = 100$ kHz

(3) $\eta = 0.6$

(4) $P_m \simeq \dfrac{hc}{\lambda} \dfrac{\Delta f}{\eta}$

$\simeq 4.14 \times 10^{-14}$ W

Solution: If we desire our system to have a 100:1 (20-dB) signal-to-noise power ratio, the incident power would have to fall to about 4 pW. Such a low power level would be unusual in an optical bench setup.

In all active electronic devices, both semiconductor and vacuum, there is a noise process called *shot noise.* The current flow through these devices is not a continuous process. A better conceptual model would be to treat the current flow as resulting from the addition of many small current pulses caused by the charge carriers moving in discrete jumps. This pulsing process causes a current variation around the average value of the current. This current variation is called the *shot noise current;* its magnitude depends on the average current through the photodiode, which is the sum of the dark current and the average photon-generated current. In a reverse-biased PIN photodiode circuit at normal levels of flux density, the shot noise is the major noise mechanism. Equation (10-17), which follows, illustrates the relationship that exists between shot noise current, I_s, and the average device current. Shot noise current can be reduced by operating the photodiode in the short-circuit mode, which reduces the dark current.

$$I_s = [2e\,\Delta f(I_d + I_p)]^{1/2} \tag{10-17}$$

where I_s is the average shot noise current
e is the charge of the electron 1.6×10^{-19} C
Δf is the noise bandwidth
I_d is the dark current
I_p is the average photo current

Thermal noise will also make a small contribution to the total noise current of a reverse-biased photodiode. Thermal noise may be the dominant noise current when the photodiode is operated in the short-circuit mode. The magnitude of the thermal noise depends upon device temperature, shunt resistance, and noise bandwidth:

$$I_t = \left(\frac{4KT\,\Delta f}{R_{\text{SH}}}\right)^{1/2} \tag{10-18}$$

where I_t is the average thermal noise current
K is Boltzmann's constant, 1.38×10^{-23} J/K
T is the temperature, in kelvin
$4KT = 1.65 \times 10^{-20}$ at 25°C
Δf is the noise bandwidth
R_{SH} is the photodiode shunt resistance

The photodiode load resistance R_L will cause an increase in the thermal noise current.

$$I_t = \left[4KT\,\Delta f\left(\frac{1}{R_{SH}} + \frac{1}{R_L}\right)\right]^{1/2} \tag{10-19}$$

where R_{SH} is the photodiode shunt resistance
R_L is the load resistance

The net noise current, I_n, is given by the following equation, which assumes that the incident flux is large enough to make the quantum noise negligible, which is usually the case.

$$I_n = (I_s^2 + I_t^2)^{1/2} \tag{10-20}$$

An estimate of the noise equivalent power (NEP) can be made using the noise current calculated with equation (10-20) and the detector responsivity.

$$\text{NEP} = \frac{I_n}{R} \tag{10-21}$$

where I_n is the net noise current
R is the photodiode responsivity

The value of the net noise current, I_n, from the noise current generator depends on the thermal noise current and the shot noise current. Thermal noise current is affected by temperature, bandwidth, and load resistance. The normally dominant shot noise current is affected by bandwidth and dark current. Dark current increases with bias voltage and temperature. The designer affects the noise current when he selects the device operating temperature, the system bandwidth, the load resistance, and bias voltage.

EXAMPLE 10-3

Illustration of the Noise Contribution of the Three Noise Effects in a Planar Silicon Photodiode

Given: **(1)** $H_0 = 1.0$ mW/cm^2
(2) $A = 4$ cm^2
(3) $\lambda = 930$ nm
(4) $R = 0.40$ A/W
(5) $R_L = 1000\ \Omega$
(6) $\Delta f = 10$ kHz
(7) $R_{SH} = 1$ MΩ
(8) $I_d = 1.3\ \mu$A

Find: The noise contribution of the noise effects in the photodiode.

Solution:

1. Estimate the incident power level.

$$\begin{aligned}P &= H_0 A \\ &= 1.0 \text{ mW/cm}^2 \times 4 \text{ cm}^2 \\ &= 4 \text{ mW}\end{aligned}$$

2. Estimate the average, dc, photocurrent.

$$\begin{aligned}I_p &= RP \\ &= 1.6 \text{ mA}\end{aligned}$$

3. Estimate the quantum efficiency.

$$\begin{aligned}R &= \eta\lambda(8.06 \times 10^{-4}) \\ \eta &= 0.53\end{aligned} \qquad (10\text{-}9)$$

4. Determine the incident power level at which the quantum noise is significant.

$$P_m \simeq \frac{6.625 \times 10^{-34}\,\Delta f}{\lambda} \qquad (10\text{-}16)$$

$$\simeq 1.3 \times 10^{-23} \text{ W}$$

Conclusion: The quantum noise can be ignored because P_m is about 10^{20} times smaller than the available incident power.

5. Estimate the shot noise current.

$$\begin{aligned}I_s &= [2e\,\Delta f(I_d + I_p)]^{1/2} \\ &= 220 \text{ nA}\end{aligned} \qquad (10\text{-}17)$$

6. Estimate the thermal noise current.

$$I_t = \left[4KT\,\Delta f\left(\frac{1}{R_{SH}} + \frac{1}{R_L}\right)\right]^{1/2} \qquad (10\text{-}19)$$

$$= 203 \text{ pA}$$

Conclusion: Clearly, the shot noise is dominant in this example because I_s is approximately 1000 times larger than the thermal noise current I_t.

Series Resistance

The series resistance component (R_S) of the photodiode model represents the resistance of the photodiode material and the lead connections. Series resistance has a pronounced effect on device rise time, and linearity. Series resistance can be as small as 0.1 Ω in a large photodiode, and it may be as large as several hundred ohms in a small photodiode. Series resistance is inversely related to device area, so that, given two photodiodes that are identical except for their areas, the larger photodiode should have the smaller series resistance. For photodiodes with areas less than 0.5 cm^2, lead resistance and contact resistance dominate the series resistance, and it becomes essentially constant for similar photodiodes.

Having identified all the elements of the current model, we are now ready to use it to predict some trends. In the following discussion we assume that the sum of the load resistance R_L and the photodiode series resistance R_S is small compared to the shunt resistance R_{SH} and that R_L is larger than R_S.

$$R_L + R_S \ll R_{SH}$$

$$R_L > R_S$$

Further, let's assume that the device is reverse biased and that the average photo current I_p is much larger than the average noise current I_n. These assumptions will allow us to simplify the model of Figure 10-12 to the model of Figure 10-13.

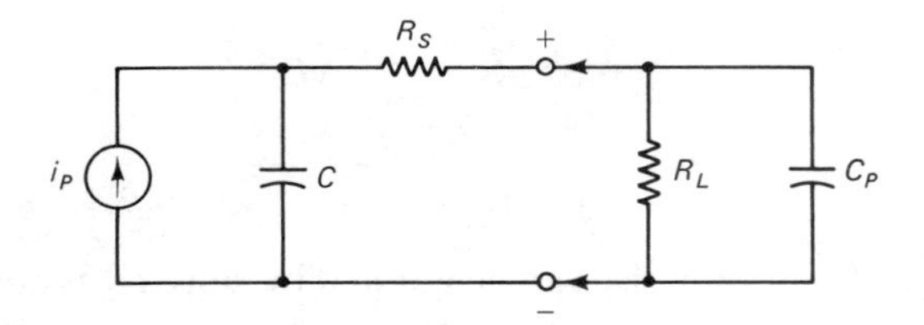

Figure 10-13 Simplified small-signal photodiode model drawn assuming that the effect of R_{SH} and I_N are negligible.

The dominant time constant of the detector system is the resistance capacitance time constant created by the junction capacitance, the photodiode series resistance, and the load resistance. Any parallel capacitance, caused by lead wires or circuit layout, that appears across the load terminals will also increase the time constant. An engineering estimate can be made of this time constant as follows (this is *not* an exact solution):

$$\tau_{RC} \equiv (R_S + R_L)(C + C_p) \tag{10-22}$$

where τ_{RC} is the time constant
R_S is the photodiode series resistance
R_L is the load resistance
C is the junction capacitance
C_p is other parallel capacitance

In addition to the time constant, there are two other time delays associated with the physics of the photodiode junction, but fortunately, these are usually small enough to be ignored. These time delays are the charge collection time delay τ_{CC} and the diffusion time τ_{diff}. The charge collection time delay is the time required for the electric field to move all the photo-excited carriers out of the depletion region. The charge collection time delay is typically $\frac{1}{2}$ ns or less. The diffusion time is the time required for the carriers caused by photons outside the depletion region to diffuse into the depletion region; it is inversely related to the magnitude of the reverse bias voltage and can be minimized with appropriate values of bias voltage. The two time delays and the time constant can be used to estimate the overall time constant using the equation

$$\tau = (\tau_{RC}^2 + \tau_{CC}^2 + \tau_{\text{diff}}^2)^{1/2} \tag{10-23}$$

Generally, the overall time constant is within a few percentage points of τ_{RC}.

When photodiodes are being used as detectors, it is desirable to have output current linearly related to the incident flux density. When the flux density reaches very small levels, the noise current can become an appreciable portion of the output current. A conservative approach to linearity would require that the signal current be at least 100 times larger than the noise current. On the linearity graphs provided by manufacturers on data sheets dashed lines are used to indicate the levels of detected current where noise current may become a problem.

The linearity of response also has a maximum current value. When the detector is reverse biased there is a saturation current value that cannot be exceeded. Increasing flux density above the value that causes the saturation current to flow will not result in any increase in circuit current. The response of the detector actually starts to become measurably nonlinear at current values that are about one-third of the saturation current. Linearity graphs that show detected current plotted against flux density are used to illustrate at what point the nonlinear response due to saturation begins. An estimate of the maximum linear current that can flow in a reverse-biased circuit is given by

$$\begin{aligned} I_{\max} &= \tfrac{1}{3} I_{\text{sat}} \\ &= \frac{1}{3} \frac{V_B}{R_S + R_L} \end{aligned} \tag{10-24}$$

where $I_{\max}$ is the maximum linear current
V_B is the reverse bias supply voltage
R_S is the photodiode's internal series resistance
R_L is the load resistance

Since quite often the value of R_S is unknown, many designers assume that it is zero, which results in an overly large estimate of maximum linear current. If the photodiode is short circuited, the current response will be linear almost to I_{sat} and equation (10-24) does not apply.

There is no simple mathematical expression that will estimate the onset of nonlinearity in the photovoltaic mode. If you are lucky, the manufacturer will provide graphs of output current versus flux density for various values of load resistance (Figure 10-14). If this information is not available, you are usually safe in assuming that linearity of response can be maintained if the voltage across the diode is held to a value less than one-sixth of the characteristic forward voltage and the

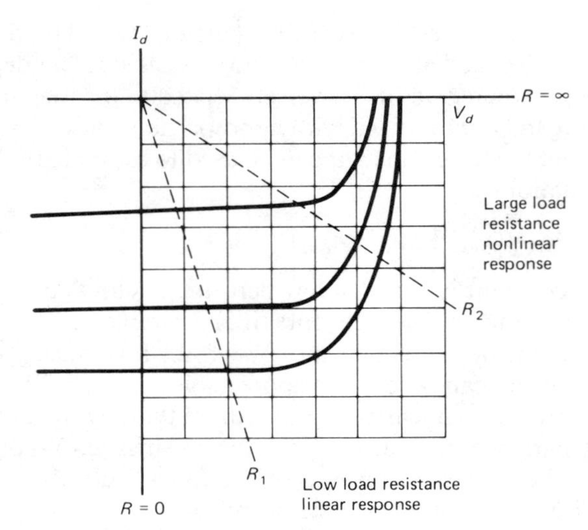

Figure 10-14 Detail of photodiode fourth-quadrant photovoltaic characteristics, illustrating the effects of large and small load resistors.

current density does not exceed 50 μA/mm^2. The current density can be estimated by dividing the dc current through the diode by the active area of the diode.

The photodiode model provides the designer with a tool to predict photodiode current output, noise performance, rise time, and linear range of operation. The size of the load resistance and the magnitude of the bias voltage are tools the designer can use to manipulate device characteristics and circuit performance. A summary of these effects for a reverse-biased photodiode is given below.

Increasing the load resistance R_L results in:

1. Increased thermal noise current
2. Slower rise times
3. Decreased linear range of operation
4. Lowered output current for very large values of R_L

Increasing bias voltage across the diode results in:

1. Little effect on output current
2. Increased shot noise current and dark current
3. Decreased capacitance due to an enlarged depletion region and therefore faster rise times
4. Increased current range of linear operation
5. Decreased series resistance R_S
6. Increased shunt resistance R_{SH}

In addition, an increase in the depth of the depletion region caused by increased bias voltage will allow longer wavelengths to be absorbed in the depletion region, which slightly enhances the response of the photodiode near the bandgap limit.

10-6 COMPARISON OF TYPICAL SPECIFICATIONS FOR PHOTODIODES

Silicon, germanium, and gallium compounds are common materials used to fabricate semiconductors. Selenium and indium compounds as well as other materials are also used. New materials are being developed and old materials improved on a continuing basis. Materials that can be used at very long and very short wavelengths are the objects of serious research.

The most immediately noticeable difference between diodes fabricated out of different materials is the spectral response. Some typical spectral response characteristics are listed in Table 10-2. As the table illustrates, each diode covers only a region of the spectrum. There is some spectral overlap between adjacent devices in the table. Measurements over a wide range of wavelengths would require the use of several different semiconductors. Of course, an alternative might be to use a thermal detector instead of a photodiode. Another noticeable difference between diodes is the responsivity at the wavelength of peak response.

As Table 10-3 illustrates, comparing photodiodes on the basis of responsivity alone can be misleading. The responsivity of the indium antimonide photodiode is

TABLE 10-2 COMMON SEMICONDUCTOR MATERIALS AND THEIR WAVELENGTHS OF PEAK RESPONSIVITY AND HALF-PEAK RESPONSIVITY

	Wavelength (nm)		
Material	Lower half-peak	λp	Upper half-peak
Selenium	400	575	650
Gallium arsenide phosphide	400	610	700
Silicon	550	840	1050
Germanium	850	1500	1700
Indium antimonide	2000	4800	5050

TABLE 10-3 TABLE COMPARING PHOTODIODE RESPONSIVITIES AND QUANTUM EFFICIENCIES AT λ_p

Type	Material	λp (nm)	R (A/W)	η
Planar	Selenium	575	0.30	0.65
Planar	Gallium arsenide phosphide	610	0.29	0.57
Schottky	Gallium arsenide phosphide	610	0.18	0.37
Planar	Silicon	840	0.50	0.74
PIN	Silicon	930	0.50	0.67
Planar	Gernanium	1500	0.70	0.58
Planar	Indium antimonide	4800	2.90	0.75

more than five times as large as the responsivity of the planar silicon diode, but the quantum efficiencies are essentially identical. In fact, except for the Schottky device the quantum efficiencies of all the devices listed are very similar. The only time that a responsivity comparison makes sense is when the operating wavelengths are the same. For example, we could make a useful comparison of the two gallium arsenide phosphide (GaAsP) devices.

Device rise time is also a major factor affecting device selection. Device rise time is a complex function of device material, construction technique, bias voltage, and surface area. All of these device characteristics acting together determine device capacitance and series resistance. These device parameters and the load resistance determine device rise time. Data sheets will list device rise times for specific sets of test conditions, but these listed values can be used as a basis for comparison only when the test conditions are similar. Typical small (area = 1 to 5 mm^2) planar diffused silicon and GaAsP devices will have rise times in the range 0.5 to 4.0 μs under similar test conditions. Comparable Schottky photodiodes will have similar rise times.

PIN devices will be much faster than planar devices of a similar active area, and rise times as small as 0.5 to 3 ns are not unusual. This is an improvement of 1000:1 over planar-type devices. Of course, the PIN bias voltage will be significantly larger than the bias voltage used with the planar devices.

When you work with large surface area devices (area > 1 cm^2) you may find that sheet capacitance rather than rise time is the listed device parameter. The data sheet will often list several devices that have differing shapes and areas but which all exhibit the same sheet capacitance. Device capacitance is found by multiplying the specified sheet capacitance by the device area. Typical values of sheet capacitance would be 23 nF/cm^2 for silicon and 39 nF/cm^2 for selenium. In general, you can expect a silicon device to be faster than a selenium device of the same area. Once the device capacitance is calculated, it can be combined with the series resistance and the load resistance to calculate rise times. These large devices can be expected to be slower than the small switching devices discussed previously. Rise times of tens or even hundreds of microseconds would not be unusual.

Photodiodes intended for use in photometric work will have a sensitivity specification measured with respect to a flux in lumens emanating from a source of constant color temperature. This type of specification will be found on the data sheets of selenium-, silicon-, and gallium-based devices, all of which have a useful level of quantum efficiency at the visible wavelengths. The data sheets of germanium- and indium-based devices would not list photometric specifications because these devices do not respond to the visible wavelengths. Incandescent lamps operated at color temperatures in the range 2800 to 2900 K are the typical specified sources. The detected current can be expected to increase linearly with increases in flux density as long as the color temperature of the lamp is held constant.

You may remember from Chapter 4 that an incandescent lamp radiates most of its energy in the infrared region, and in fact its peak emission will be at an IR wavelength. The spectral response of a silicon diode encompasses much more of this available radiation than does either a selenium or a gallium device. A silicon device will have a larger detected current and an apparently higher responsivity under these test conditions.

This does not mean, however, that the silicon device will be the most sensitive photometric detector. To make a photometric detector, an optical filter is placed between the photodiode and the source to create a system with a responsivity that corresponds to the CIE curve. Using a selenium photodiode that has a natural responsivity that is very close to that of the CIE curve, a very simple, low-loss, yellow optical filter is all that is required to create a photometric detector. The sensitivity of the finished detector system will be only a little less than that of the photodiode alone. A silicon detector, on the other hand, requires a more complex and higher loss filter to suppress its strong IR responsivity. The complete silicon-based photometric system is often less sensitive than a comparable selenium system.

Detectors that have their spectral responsivity adjusted to match the CIE curve will be relatively insensitive to increases in source color temperature. An increase in color temperature from 2800 K to 4700 K with flux density held constant will cause about a 6% increase in detected current. Over the more typical incandescent lamp range of color temperatures, 2800 to 2900 K, the change in detected current is practically negligible. Unfiltered selenium photodiodes are similarly insensitive to changes in color temperature at constant flux density because the responsivity of a selenium photodiode is so similar to the CIE curve.

Combination bench-type photometric/radiometric meters use silicon detectors because the silicon device responds to both IR and visible wavelengths. Two changeable filters are used: one a greenish photometric filter and the other a violet radiometric filter. The radiometric filter creates a responsivity curve that is flat from 400 nm to near 1000 nm. This radiometric filter attenuates the silicon detector's peak responsivity to the value of responsivity that exists at 400 nm and 1000 nm. Since the responsivity of the detector at 400 nm is only about 30% of the peak value, the filter introduces significant energy losses. In addition, the filter introduces excess loss above that required to achieve the desired filtering. Even with this filtering this radiometric detector is not useful for making measurements at wavelengths longer than 1000 nm.

When making measurements at long wavelengths, those greater than 1500 nm, it is usual not to use any filtering beyond that caused by the window material covering the photodiode. The measured values of flux density are adjusted for the known responsivity of the detectors used. The photodiodes used to make measurements at wavelengths longer than 1500 nm tend to have responses that are more sensitive to ambient temperature than are the responses of photodiodes used at shorter wavelengths. To control the magnitude of the dark current, noise currents, and responsivity of these detectors they are cooled to constant temperatures far below room temperature. Indium antimonide photodiodes, for example, are designed to be mounted in Dewar flasks, where they are cooled with liquid nitrogen to a temperature of 77 K. The cooling medium cools not only the detector but also the detector optics because a "hot" optical element would appear as an image to a long-wavelength detector.

To wrap up this discussion of specifications, the data sheet of Figure 10-15 is

PIN Silicon Photocells (Ultra-Fast Response)

(A) Type No.	Outlines (P.29, 30) / (A) Window Materials	Photosensitive Surface Size (mm)	Photosensitive Surface Effective Area (mm²)	Package (mm)	Spectral Response Range (nm)	Spectral Response Peak Wavelength (nm)	Typical Radiant Sensitivity (A/W) Peak Wavelength	Typical Radiant Sensitivity (A/W) He-Ne Laser (633nm)	Typical Radiant Sensitivity (A/W) GaAs LED (930nm)	(B) NEP typ. ($W/Hz^{1/2}$)	(C) D* typ. ($cm \cdot Hz^{1/2}/W$)
S1633 Series (Plastic Molded Types)											
S1633 ★	(20)/R	3.0×3.0	9.0	Plastic mold	400~1100	900±50	0.5	0.3	0.5	1×10^{-13} (V_R=12V)	3×10^{12} (V_R=12V)
S1633-01 ★	(20)/F				730~1120			—			
S1722 Series											
S1722 ★	(21)/K	4.1 dia.	13.2	TO-8	400~1150	900±50	0.55	0.3	0.55	1×10^{-13} (V_R=100V)	4×10^{12} (V_R=100V)
S1722-01 ★	(21)/K				320~1150						
S1722-02 ★	(21)/Q				200~1150		0.5		0.5		
S1723 Series											
S1723 ★	(12)/R	10×10	100	Ceramic 15×16.5	400~1150	900±50	0.6	0.3	0.55	2×10^{-13} (V_R=30V)	5×10^{12} (V_R=30V)
S1723-01 ★	(12)/R				320~1150						
S1723-02 ★	(11)/Q				200~1150		0.5		0.5		

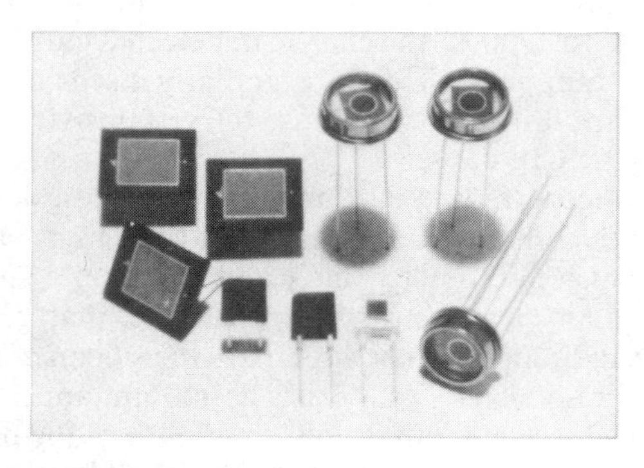

(A) ★: Newly listed in this catalog
K: Borosilicate glass window
Q: Fused silica window
R: Resin coating window
F: Visible absorbing filter window

(B) $NEP = \frac{\text{Noise Current } (A/Hz^{1/2})}{\text{Radiant Sensitivity at Peak } (A/W)}$

(C) $D^* \text{ (D-star)} = \frac{[\text{Effective Sensitive Area } (cm^2)]^{1/2}}{NEP}$

(D) t_r is the time required to transition from 10% to 90% of the peak output value. The light source is a delta function light of a laser diode (800nm) and the load resistance is 50Ω.

(E) At the −3dB point. The load resistance is 50Ω.

● **Spectral Response**

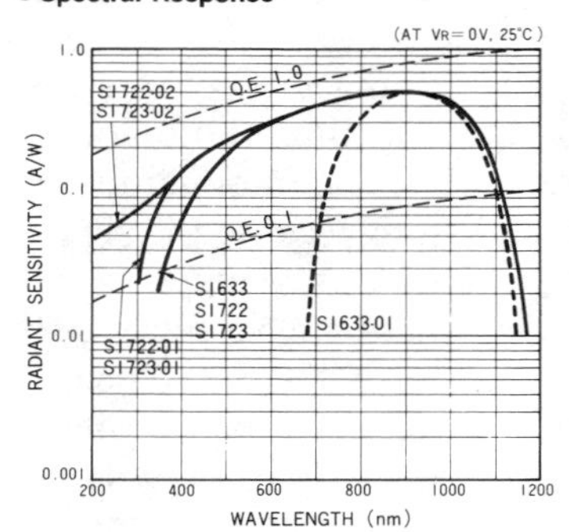

● **Temperature Characteristic of I_{sh}**

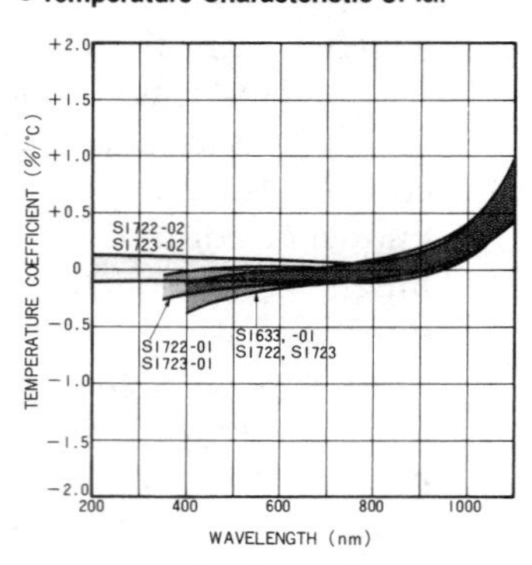

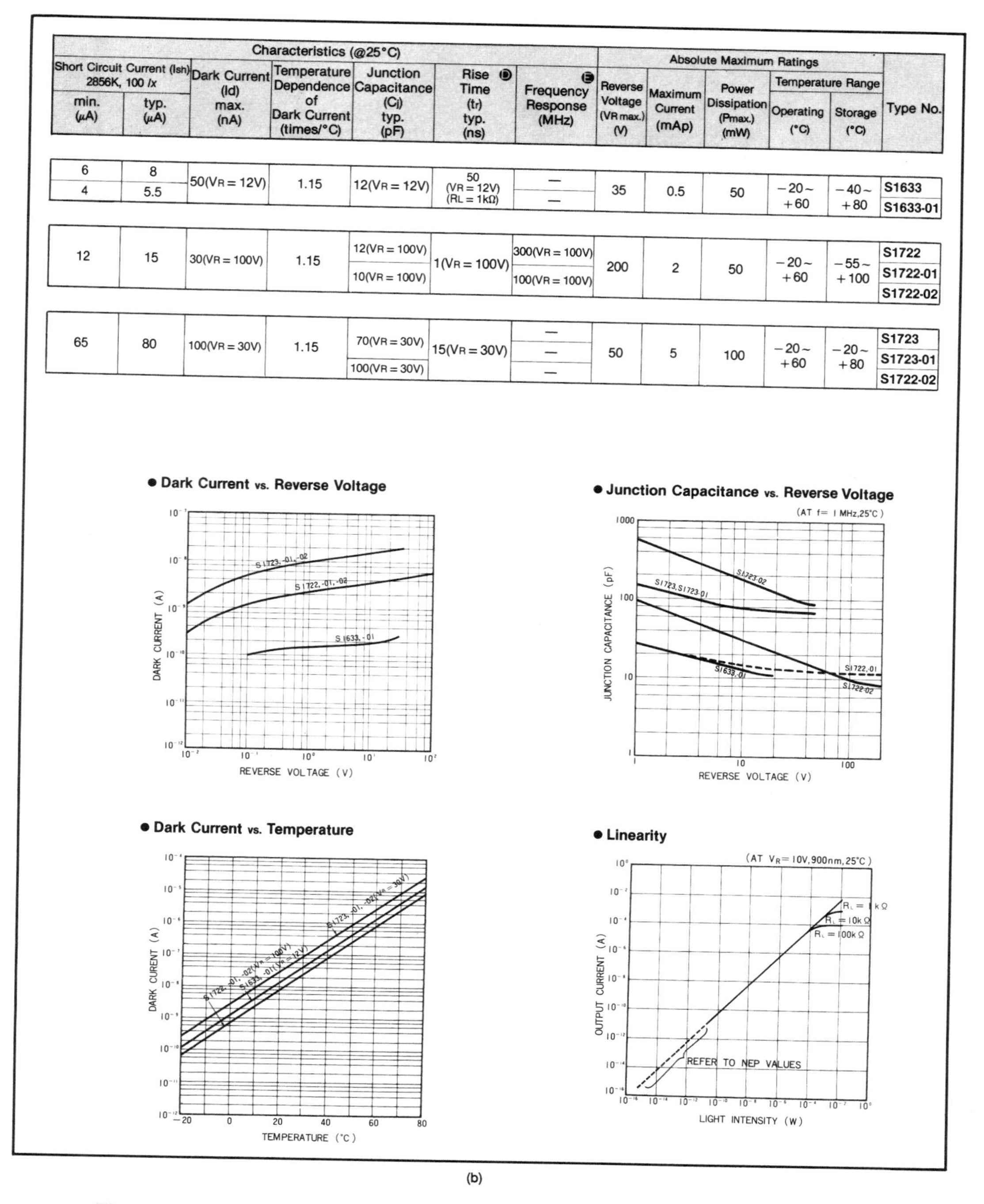

Characteristics (@25°C)							Absolute Maximum Ratings					
Short Circuit Current (Ish) 2856K, 100 lx		Dark Current (Id) max. (nA)	Temperature Dependence of Dark Current (times/°C)	Junction Capacitance (Cj) typ. (pF)	Rise Time Ⓓ (tr) typ. (ns)	Frequency Ⓔ Response (MHz)	Reverse Voltage (VR max.) (V)	Maximum Current (mAp)	Power Dissipation (Pmax.) (mW)	Temperature Range		Type No.
min. (μA)	typ. (μA)									Operating (°C)	Storage (°C)	
6 4	8 5.5	50(VR = 12V)	1.15	12(VR = 12V)	50 (VR = 12V) (RL = 1kΩ)	— —	35	0.5	50	−20~ +60	−40~ +80	S1633 S1633-01
12	15	30(VR = 100V)	1.15	12(VR = 100V) 10(VR = 100V)	1(VR = 100V)	300(VR = 100V) 100(VR = 100V)	200	2	50	−20~ +60	−55~ +100	S1722 S1722-01 S1722-02
65	80	100(VR = 30V)	1.15	70(VR = 30V) 100(VR = 30V)	15(VR = 30V)	— — —	50	5	100	−20~ +60	−20~ +80	S1723 S1723-01 S1722-02

(b)

Figure 10-15 Data sheets of PIN silicon photocells. (Courtesy Hamamatsu Corporation, Bridgewater, NJ.)

presented as being representative of a common photodiode data sheet. The particular photodiodes described are PIN silicon photodiodes. This data sheet graphically illustrates several temperature and voltage characteristics of the photodiodes and lists some additional specifications of interest.

Noise performance is characterized by both NEP and D^* specifications measured at 25°C and λp. A short-circuit mode photometric response to illumination from a lamp with a color temperature of 2856 K and a flux density of 100 lm/m^2 is specified along with the response to several monochromatic sources.

Dark-current temperature and voltage characteristics are listed and graphically illustrated. The tabulated value of dark current is a worse-case maximum value, while the graphs for the dark current illustrate average characteristics. A dark-current temperature dependence specification is listed. This specification can be used to calculate the dark current of the photodiode at various temperatures. The magnitude of the dark-current specification is a characteristic of the material used to fabricate the photodiode. Table 10-4 lists dark-current specification values for three other materials and the current multiplication factor resulting from a 10°C temperature increase.

This current versus temperature relationship can be plotted as a straight line on semilog paper as illustrated by the dark-current temperature graph on the data sheet. On this graph the current units are on the log axis and the temperature is on the rectilinear axis. If you attempt to predict values on this graph by calculation using the specified 1.15 factor, you will find that it does not work. It appears that the appropriate factor for this graph is 1.122. Shunt resistance R_{SH} is inversely related to dark current: thus, as is illustrated by the graph on the data sheet, shunt resistance decreases as temperature increases.

Reverse breakdown voltage is another important specification listed. The magnitude of reverse breakdown voltage is also material dependent. Selenium- and gallium-based photodiodes have low reverse breakdown voltages of less than 6 V. Germanium and silicon devices can have much higher reverse breakdown voltages, or tens or even hundreds of volts.

One very subtle change caused by temperature is the alteration of the detector's spectral response. Changes in temperature alter the bandgap structure of the detector, and this causes a change in spectral response. As a general rule, heating a photodiode will increase its sensitivity to longer wavelengths and shift its wave-

TABLE 10-4 DARK-CURRENT TEMPERATURE DEPENDENCE (TD) SPECIFICATIONS AND THE CURRENT MULTIPLICATION FACTOR M CAUSED BY A TEMPERATURE INCREASE ΔT OF 10°C [$M = (\text{TD})^{\Delta T}$]

Material	TD	M
Gallium arsenide phosphide	1.07	1.97
Silicon	1.12–1.15	4.05
Germanium	1.11	2.84

length of peak detection to a longer wavelength. The responsivity of the photodiode to wavelengths shorter than the wavelength of peak detection may remain unchanged or may decrease slightly. The change in detector sensitivity as a function of wavelength and temperature is specified by a temperature coefficient whose value as a function of wavelength is illustrated by a graph on the data sheet. Positive coefficient values indicate that responsivity increases as temperature increases.

One last specification related to photodiodes is fatigue. This phenomenon is most noticeable with selenium photodiodes. When a selenium photodiode is illuminated for a long time (minutes) at a fixed flux density, its current will decrease slightly, typically by about 3%. Photodiode fatigue is very similar in its effects to the light history phenomenon observed in photoresistors.

10-7 SUMMARY

Photodiodes are a class of detectors that convert incident radiant energy into electrical current. The response of these photodiodes is closest to linear when they are reverse biased or short circuited. Photodiodes have different spectral responses governed primarily by the spectral response of the semiconductor material used in their fabrication. The response time of these photodiodes can be fast, in the range of nanoseconds to microseconds. Photodiode response time is controlled by the capacitance and resistance of the diode. Silicon- and germanium-based photodiodes have spectral responses that cover a wide range of wavelengths, whereas gallium and selenium photodiodes have responses confined to the visible and near-ultraviolet region.

The maximum sensitivity of these photodiodes is limited by noise and leakage current. NEP and D^* noise specifications for these devices can be treated in the manner used in Chapter 8; leakage current can be zeroed out electronically prior to measurement. Another approach is to chop the light beam with a rotating shutter and measure only the change in current caused by the chopped beam. Photodiodes are temperature sensitive, and useful designs require the control of device temperature or compensation for temperature variations.

Compared to the other detectors discussed in this book, photodiodes tend to be more rugged, cheaper, and easier to use than either thermoelectric devices or vacuum diodes. It is possible to fabricate diode detectors out of material systems other than those discussed in this chapter. A great deal of research is being done each year, exploring new material systems for use in opto-electric devices.

All semiconductor diode detectors, however exotic, have similarly shaped spectral response curves characterized by a range of wavelengths over which the responsivity of the photodiode curve is parallel to and less than the responsivity curve of an ideal quantum detector. Photodiodes will exhibit decreasing sensitivity at wavelengths longer and shorter than those in this region of relatively constant quantum efficiency. Short- and long-wavelength responsivities are limited by absorption characteristics of the diode structure.

QUESTIONS

1. What do the terms *planar, PN, PIN, Schottky,* and *APD* describe?
2. What are the three basic photodiode circuit connections?
3. What semiconductor material exhibits the largest forward voltage?
4. On a four-quadrant plot of diode voltage and current characteristics, which quadrant illustrates reverse-bias behavior?
5. What is diode reverse breakdown voltage?
6. Which terminals of the dc power supply must be connected to the P and N materials to obtain reverse bias?
7. Is the anode terminal of a semiconductor diode P or N material?
8. Which of the four major photodiode construction techniques yields the best ultraviolet behavior?
9. Define *quantum efficiency.*
10. What factors prevent photodiode spectral response curves from following the ideal quantum detector curve?
11. Why is the photoconductive mode used for detector applications?
12. How does an avalanche photodiode work?
13. In what quadrant of the photodiode characteristics are solar diodes operated?
14. How could the quantum efficiency of a photodiode be improved?

PROBLEMS

1. What is the responsivity of an ideal quantum detector at 550 nm?
2. What is the quantum efficiency at 600 nm of a photodiode with a responsivity of 0.3 A/W at 600 nm?
3. A detector is rated as having a quantum efficiency of 0.8 between 600 and 900 nm. What is the detector's responsivity at 750 nm?
4. A reverse-biased photodiode is used in a circuit with a 10-kΩ resistor and a 10-V dc supply.
   (a) What is the saturation current?
   (b) What is the maximum electrical power that could be delivered to the photodiode by this circuit?
5. A certain photodiode has a power rating of 50 mW. The bias supply is 5 V. What is the minimum resistor size that can be used safely?
6. At 632.8 nm a photodiode has a responsivity of 0.35 A/W. The active area of the photodiode is 0.9 cm^2. The incident flux is 111 $\mu W/cm^2$. What is the detected current?
7. A photodiode has a detected current of 40 μA. The bias supply is −5 V dc. The load resistance is 62 kΩ. How much electrical power is being delivered to the photodiode?
8. A photodiode has a junction capacitance of 0.04 $\mu F/cm^2$ when reverse biased at 3 V. The material depletion voltage is 0.3 V. Estimate the junction capacitance of 5 V of reverse bias.
9. Estimate total noise current if the shot noise is 3 μA and the thermal noise is 0.5 μA.
10. Estimate the shot noise caused by a dark current of 1 μA in a system with a noise bandwidth of 10 Hz.

11. Estimate the thermal noise of a photodiode with a shunt resistance, R_{SH}, of 2 MΩ and a noise bandwidth of 10 kHz. Assume 25°C.
12. Estimate the net thermal noise current of a photodiode with a 2-MΩ shunt resistance and a 2-kΩ load resistance. The bandwidth is 10 kHz. Which resistor is making the greatest noise contribution? Assume 25°C.
13. A photodiode has a dark current of 10 μA, an average signal current of 200 μA, a shunt resistance of 2 MΩ, and a load resistance of 200 kΩ. Assume a 40-kHz bandwidth and an average temperature of 30°C for all components. Estimate the total noise current. Which noise component is largest?
14. A PIN photodiode has a junction capacitance of 25 pF and a series resistance of 350 Ω. Estimate the photodiode's rise time with a load resistor of 50 Ω and a load capacitance of 0.2 pF.
15. Two PIN photodiodes are constructed identically except for area. Date sheet values for photodiode 1 are listed. Estimate the values of C, t_r, and R_s for photodiode 2.

	Photodiode 1	Photodiode 2
Area	13.6 mm^2	6.6 mm^2
Capacitance	25pF	C
Rise time	10 ns	t_r
R_L	50 Ω	50 Ω
C_P	0.2 pF	0.2 pF
R_S	130 Ω	R_S

16. A planar silicon photodiode has a rated junction capacitance of 1300 pF and a series resistance R_S of approximately 35 Ω. Estimate its rise time with a 1000-Ω load.
17. Estimate the responsivity of an indium antimonide photodiode at 2000 nm and 5000 nm assuming 70% quantum efficiency at both wavelengths.
18. The average dark current of a S1633 silicon photodiode is measured to be 34 nA at 25°C. Estimate the dark current of this device at 10°C.

11

Radiating Diodes and Display Devices

11-1 INTRODUCTION

In Chapter 10 PN photodiodes were introduced. It was pointed out that different semiconductor material combinations would result in devices with different bandgaps of energy and therefore different spectral responses. In this chapter the properties of radiating junctions of P- and N-type semiconductor materials will be discussed. Optoelectronic light sources built from PN junctions fall into two categories: the light-emitting diode (LED) and the infrared-emitting diode (IRED). These diodes are used in both illumination and communication applications. Liquid crystal and electroluminescent lamp devices are also described in this chapter.

11-2 RADIATING JUNCTION DEVICES

When forward current flows through an LED or an IRED, photons of energy are emitted from the diode junction. These photons of energy are caused by the recombination of holes and electrons at the diode junction. The wavelengths of these emitted photons are a function of the energy-level changes that take place during recombination. These energy levels are established by the semiconductor materials used. Table 11-1 lists some common semiconductor materials and their typical radiated wavelengths. As the table indicates, most LED and IRED devices are fabricated from gallium-based materials.

A typical LED or IRED will have a relatively thick layer of N material plated on the bottom with gold. The top of the device will be a very thin P layer through which the photons are emitted. The thick N layer is usually made up of several grown layers of gallium materials doped in differing ways to enhance the desired wavelength. Typical diode construction is illustrated in Figure 11-1.

TABLE 11-1 LED MATERIALS AND WAVELENGTHS

Material	Wavelength (nm)	Comments
GaP (gallium phosphide)	520–570	Green
GaP (gallium phosphide)	630–790	Red
GaAsP (gallium arsenide phosphide)	640–700	Orange-red
GaAlAs (gallium aluminum arsenide)	650–700	Red
GaAs (gallium arsenide)	920–950	Infrared

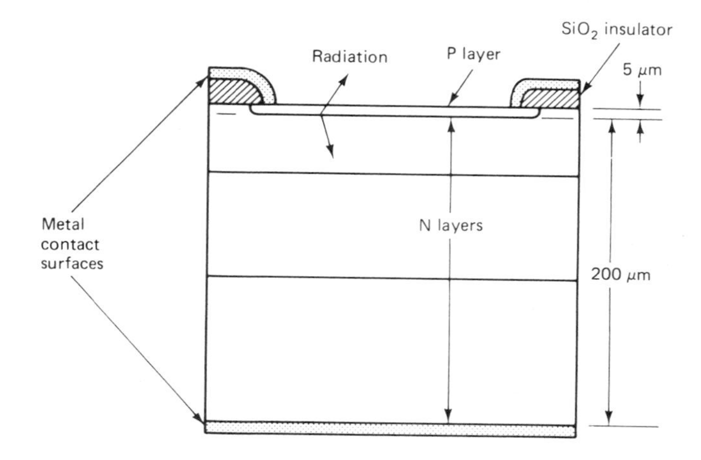

Figure 11-1 LED construction in cross section.

When the junction is forward biased, the device radiates. Gallium-based diodes have relatively high forward voltages compared to silicon and germanium diodes. The forward current versus forward voltage curve of a typical LED is noticeably less steep than that of a silicon diode. Figure 11-2 illustrates typical silicon diode and a gallium-based LED current–voltage characteristics. It is important to remember that a light-emitting diode is operated forward biased, unlike the detector diodes described in Chapter 10, which are usually operated reverse biased.

11-3 DATA SHEETS

Figure 11-3 is a typical LED data sheet. In this section various LED specifications will be explained. The first page of data lists the part numbers covered by the data sheet and a mechanical drawing of the devices. Introductory comments indicate that these are gallium arsenide phosphide devices with a red light output. The absolute maximum ratings establish temperature-related power and current failure limits. The maximum current and power specifications that follow are really temperature-related specifications. The maximum continuous forward current of 50 mA, when multiplied by the forward voltage, will yield an average power level that is close to the listed rated power of 100 mW. In general, current ratings are temperature related and a temperature derating factor of 0.2 mA/C° is listed with the current

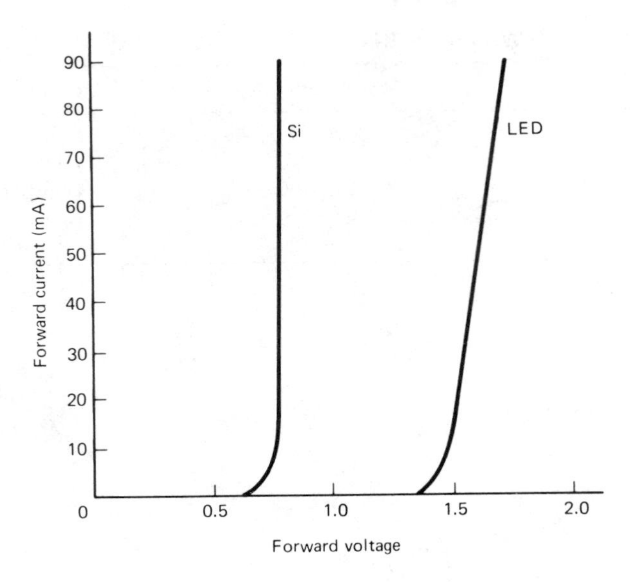

Figure 11-2 Forward characteristics of a silicon diode and an LED.

rating. This derating factor is established with respect to an ambient temperature of 50C°. The pulsed current specification of 1 A might look like a misprint until the pulse width of 1 μs and the frequency of 300 Hz is read. This means that the 1-A current is applied to the device for a 1-μs interval once every 3333 μs. This is illustrated in Figure 11-4.

This 1-A current pulse will cause some unspecified forward voltage, so for 1 μs a pulse of power will be delivered to the device. The power of this pulse averaged over the 3333-μs period will be the average power delivered to the device and will be less than or equal to the rated continuous power dissipation of 100 mW. Example 11-1 illustrates a typical pulse power calculation.

EXAMPLE 11-1

Peak Power and Average Power

Find: The peak power that a repetitive pulse can be permitted to have if the pulse width is to be 10 μs and the period is to be 100 μs. The pulse will be applied to a device with an average power rating of 200 mW.

Solution:

$$P_{\text{avg}} = P_{\text{peak}} \times \frac{\text{pulse width}}{\text{period}}$$

$$200\ \text{mW} = P_{\text{peak}} \times \frac{10\ \mu\text{s}}{100\ \mu\text{s}}$$

$$P_{\text{peak}} = 2\ \text{W}$$

T-1 3/4 (5mm) RED SOLID STATE LAMPS

HLMP-3000
HLMP-3001
HLMP-3002
HLMP-3050

TECHNICAL DATA JANUARY 1983

Features

- LOW COST, BROAD APPLICATIONS
- LONG LIFE, SOLID STATE RELIABILITY
- LOW POWER REQUIREMENTS: 20 mA @ 1.6V
- HIGH LIGHT OUTPUT:
  2.0 mcd Typical for HLMP-3000
  4.0 mcd Typical for HLMP-3001
- WIDE AND NARROW VIEWING ANGLE TYPES
- RED DIFFUSED AND NON-DIFFUSED VERSIONS

Description

The HLMP-3000 series lamps are Gallium Arsenide Phosphide light emitting diodes intended for High Volume/ Low Cost applications such as indicators for appliances, smoke detectors, automobile instrument panels and many other commercial uses.

The HLMP-3000/-3001/-3002 have red diffused lenses where as the HLMP-3050 has a red non-diffused lens. These lamps can be panel mounted using mounting clip HLMP-0103. The HLMP-3000/-3001 lamps have .025" leads and the HLMP-3002/-3050 have .018" leads.

Absolute Maximum Ratings at $T_A = 25°C$

Power Dissipation	100mW
DC Forward Current (Derate linearly from 50°C at 0.2 mA/°C)	50 mA
Peak Forward Current	1 Amp (1μsec pulse width, 300pps)
Operating and Storage Temperature Range	-55°C to +100°C
Lead Soldering Temperature	260°C for 5 sec.

Package Dimensions

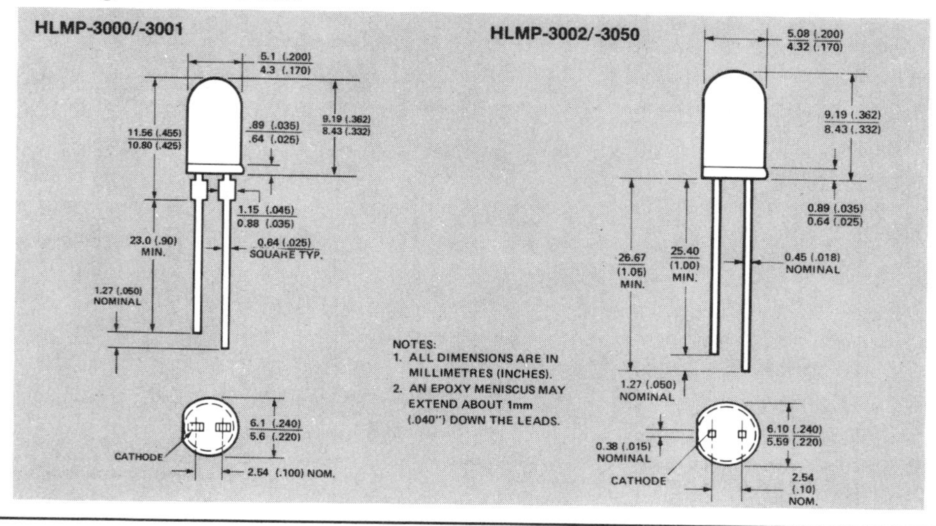

(a)

Figure 11-3 LED data sheet: (a) maximum ratings; (b) electrical characteristics. (Courtesy Hewlett-Packard Company, Palo Alto, CA.)

Electrical Characteristics at $T_A = 25°C$

Symbol	Parameters	HLMP-3000			HLMP-3001			HLMP-3002			HLMP-3050			Units	Test Conditions
		Min.	Typ.	Max.	Min.	Typ.	Max.	Min.	Typ.	Max.	Min.	Typ.	Max.		
I_V	Luminous Intensity	1.0	2.0		2.0	4.0		0.8	3.0		1.0	2.5		mcd	$I_F = 20mA$
λ_{PEAK}	Wavelength		655			655			655			655		nm	Measurement at Peak
τ_s	Speed of Response		10			10			10			10		ns	
C	Capacitance		100			100			100			100		pF	$V_F = 0$, f = 1MHz
V_F	Forward Voltage	1.4	1.6	2.0	1.4	1.6	2.0	1.4	1.6	2.0	1.4	1.6	2.0	V	$I_F = 20mA$
V_{BR}	Reverse Breakdown Voltage	3	10		3	10		3	10		3	10		V	$I_R = 100\mu A$
θ_{JC}	Thermal Resistance		100			100			100			100		°C/W	Junction to Cathode Lead
$2\theta½$	Included Angle Between Half Luminous Intensity Points, Both Axes		90			90			90			24		Deg.	$I_F = 20$ mA

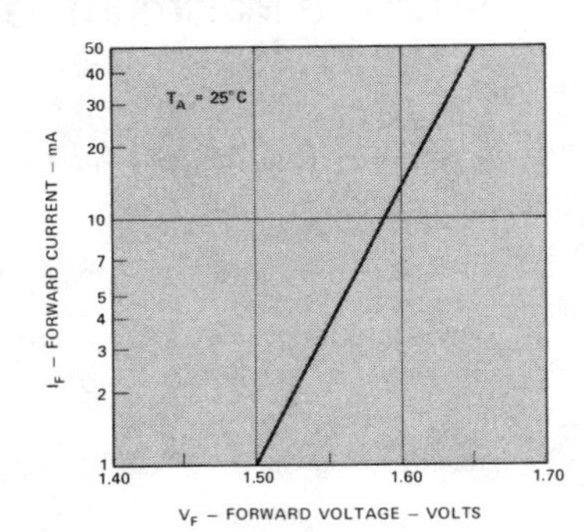

Figure 1. Forward Current Versus Forward Voltage Characteristic For HLMP-3000/-3001/-3050/-3002.

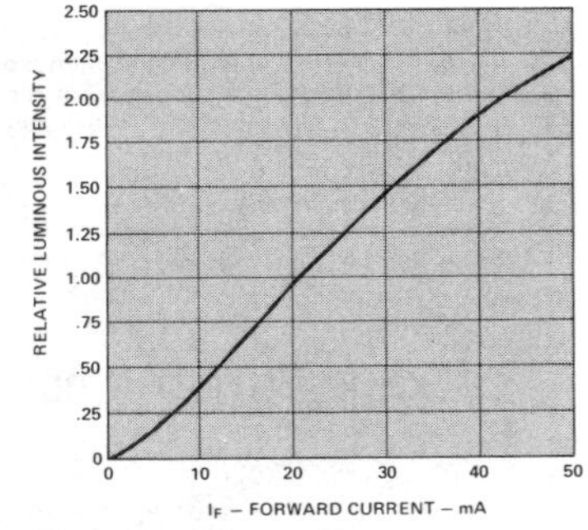

Figure 2. Relative Luminous Intensity Versus Forward Current For HLMP-3000/-3001/-3050/-3002.

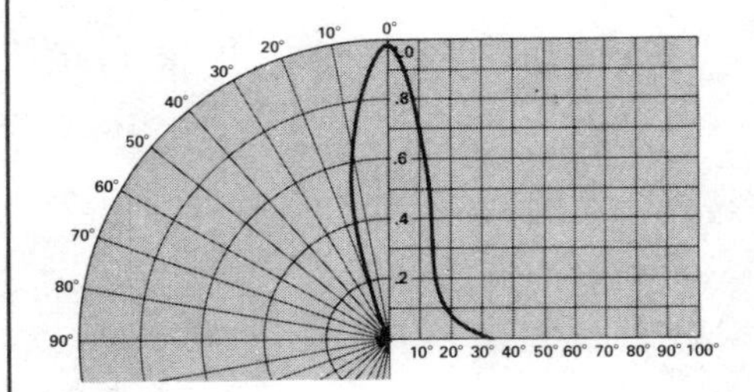

Figure 3. Relative Luminous Intensity Versus Angular Displacement. T-1 3/4 Lamp. HLMP-3050.

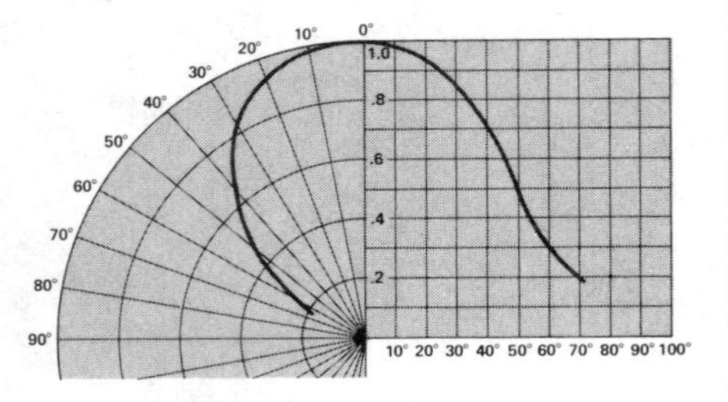

Figure 4. Relative Luminous Intensity Versus Angular Displacement For HLMP-3000/-3001/-3002.

(b)

Figure 11-3 (continued)

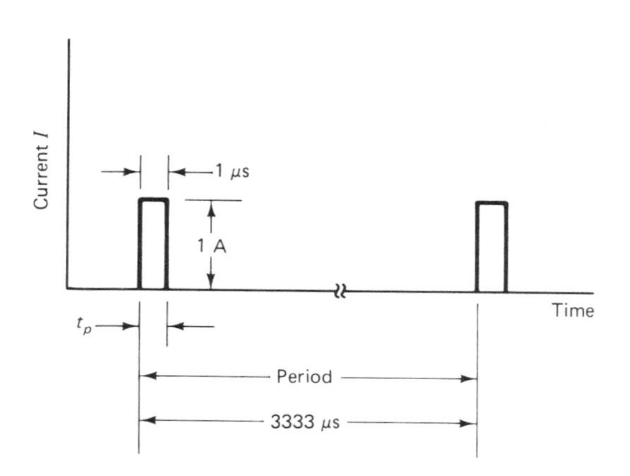

Figure 11-4 Current pulse with an amplitude of 1 A and a pulse width of 1 μs at a period of 3333 μs.

Care must be taken not to misapply the power-averaging technique illustrated in Example 11-1. An important assumption in the calculation is that the device will not reach a destructive temperature level during the "on" pulse period. The rate at which a device heats is related to its thermal time constant, which is a function of mass, surface area, radiation, and conduction characteristics. If the pulse width of the applied electrical power pulse is significantly shorter than the thermal time constant of the package, the procedure of Example 11-1 can be safely used. As the pulse width of the applied electrical power pulse approaches the thermal time constant of the package, you run the risk of overheating the device during the power pulse period. Unfortunately, thermal time constant data are not usually specified, so if long pulse widths are desired, some experimentation may be required.

A small standard metal semiconductor package number TO-5 has a typical thermal time constant of about 1 minute. Many LED packages are not only smaller than the TO-5 package, but also fabricated out of plastic. It is reasonable to assume that the thermal time constant of an LED will be much less than 1 minute. To put us back on a numerical footing, in Example 11-1 a pulse width of 10 μs repeated once every 100 μs yielded a permitted peak power of 2 W. Blindly applying the same equation, a pulse width of 1000 s repeated once every 10,000 s would yield the same permitted peak power. But 1000 s is slightly more than 16 minutes. A power pulse of 2 W applied to the device for 16 minutes would undoubtedly destroy it. For typical LED packages, pulse frequencies of 1 kHz or greater are required for safe operation at elevated peak powers.

The table of electrical characteristics of Figure 11-3 lists parameters more typical of actual operating conditions. The first specification describes the axial luminous intensity and illustrates some of the differences among the four different part number devices listed on this data sheet. The next specification indicates that maximum output intensity occurs at 655 nm.

The "speed of response" specification describes how fast the emission of these devices can be turned on and off. This specification defines the 10 to 90% time interval, that is, the rise-time value. Diode capacitance is listed as a convenience for the circuit designer who has to develop the circuitry to turn this device on and off.

A range of forward voltages is listed for a 20-mA forward current. In addition, a graph of forward current versus forward voltage is provided. Note that the vertical current axis of this graph is a log axis. The listed reverse breakdown voltage is 3 to 10 V, the implication being that a reverse voltage much greater than 3 V may cause breakdown. Three volts is quite a low reading for a diode reverse breakdown voltage, but with gallium-based semiconductors values much greater than 30 V are unusual and values of 5 V or less are common.

The thermal resistance of this LED is defined in terms of the metal cathode lead wire, not the plastic case. Effective heat sinking of this type of device can be done only at the cathode lead, not at the surface of the plastic case. Inside the device, the semiconductor chip is mounted on the cathode lead, and the thermal resistance listed on the data sheet is from the junction (the chip) to the cathode lead. Although there is no information about the thermal resistance from the lead to the ambient, which is determined primarily by the socket or circuit board to which the lead is connected, a typical soldered circuit board connection would have a thermal resistance of 300 to 400° C/W between the lead and the ambient.

The last specification listed in the table is the included angle between the half luminous intensity points. This data point is supplemented by combined polar and rectilinear intensity versus angle graphs. A luminous intensity versus forward current graph is provided, which illustrates the relatively linear relationship that exists between radiated luminous intensity and forward current.

11-4 PHOTOMETRY MEASUREMENTS OF AN LED

Visible light-emitting diodes are often used in display applications. In a display application the human eye will be the final detector. For this reason LEDs must be characterized by photometric units of measurement. The preferred specification is luminous intensity, specified in lumens per steradian or candela; this provides a figure of merit for device comparison.

Photometric measurement is a difficult task. Uncertainties of 4 to 5% on the measurement of standard lamps relative to the international standards are usual, and it is reasonable to assume that engineering laboratory measurements of photometric quantities will have a greater uncertainty. Thus it is important to select photometric measurement techniques that are easy to set up and interpret. Luminous intensity measurements meet these criteria.

As a practical matter, the photometric quantity actually measured by a detector is luminous flux density measured in lumens per unit area. The area involved is the area of the detector and the flux measured is the flux falling on the detector. In the far field, luminous flux density is related to luminous intensity by the square of the distance between the source and the detector:

$$I_v = E_v d^2 \tag{11-1}$$

where I_v is the luminous intensity in lumens/steradian
E_v is the flux density in lumens/unit area
d is the distance between the source and the detector

Repeatable and accurate results are obtained when the distance between the source and the detectors is at least 10 times as large as the diameter of either the source or the detector. Precision is enhanced if the source and detector diameters are of the same order of magnitude. Luminous intensity measurements are made with respect to a given axial line from the source, so care must be taken to eliminate reflected rays. This can be accomplished by making the measurement in a baffled box or in a large darkroom with no reflecting surfaces near the source or the detector. These are the same considerations discussed in previous chapters with respect to radiometric measurements.

Photometric calibration of the detector used for these measurements presents a particularly difficult problem. Most photometric detectors are calibrated to yield a representative photometric measurement for energy distributed over the entire visible wavelength range. The accuracy of the detector's simulation of the CIE curve at any specific wavelength could be in error by 5 to 10%. This type of error is quite common at the longer wavelengths above 650 nm and the shorter wavelengths below 475 nm. Care should be taken to calibrate the photometric detector at the wavelengths that the LED will emit.

The radiating surface of the LED semiconductor is a very small Lambertian emitter, and the distance to the observer is extremely large by comparison with the surface diameter. Conceptually, a better model of the LED radiating surface would be a point source radiating into a hemisphere. The focusing action of the package holding the diode will determine the actual beam pattern. Typical beam patterns have included angles of 20 to 90° between half-intensity points (Figure 11-5).

Optical bench measurements or integrating sphere measurements can be used to characterize the photometric output of these devices. Figure 11-6 illustrates a simple optical bench system for measuring the photometric intensity of an LED. This system has four functional components:

1. A light-tight enclosure coated on the inside with nonreflecting material.
2. A movable baffle to prevent any weak reflections from the enclosure sides from reaching the detector. This baffle may also have an adjustable aper-

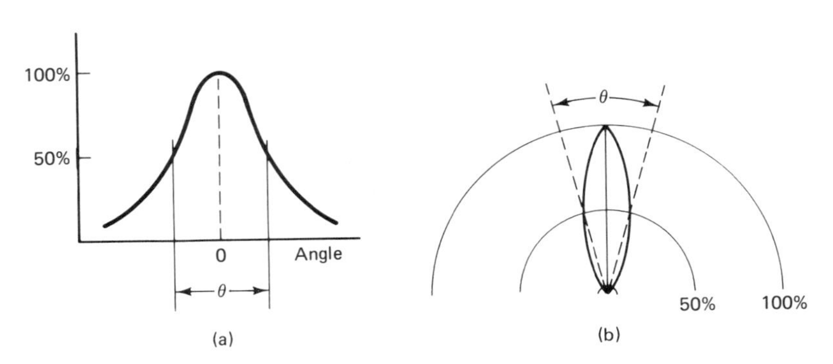

Figure 11-5 Spatial radiation patterns illustrating angle of half-intensity, $\theta^{1/2}$: (a) rectilinear graph; (b) polar graph.

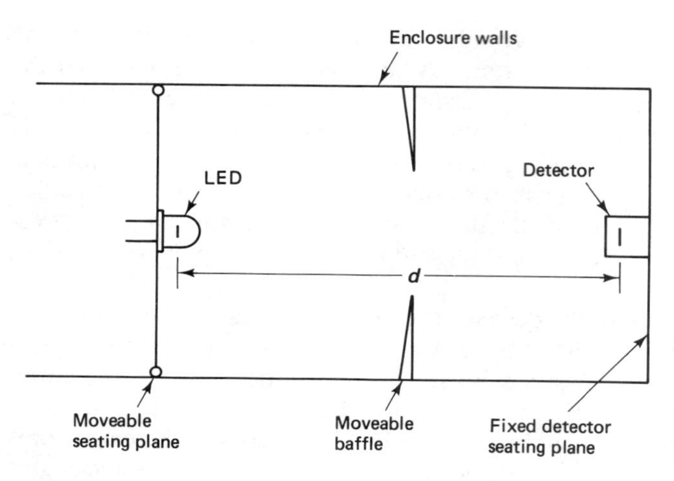

Figure 11-6 Cross section of enclosure used for the measurement of intensity.

ture. If the enclosure is large enough and adequately coated with absorbing materials, this baffle may not be needed.

3. A movable seating plane to position and hold the LED under test.
4. A detector calibrated for the wavelengths of interest. The detector will be connected to an appropriate indicating instrument.

Not shown in the drawing, but assumed, is an appropriate power supply for the LED.

If the distance d is known and the flux density varies according to the inverse-square law, the intensity can be calculated based upon d and the luminous flux measured by the detector.

$$I_v = \frac{F_v d^2}{a} \tag{11-2}$$

where I_v is the far-field luminous intensity
F_v is the luminous flux, in lumens
a is the detector's active area
d is the distance from the radiating surface of the LED to the detecting surface of the detector.

As a practical matter, d cannot be known exactly. The position of the radiating surface of the LED with respect to the mechanical mounting surface of the LED package is at best known only approximately. Similarly, the location of the detecting surface in the detector head may be ambiguous. A solution to this problem is to make two flux density measurements at two different distances. In this method, the measured distances between the detector seating plane and the LED seating plane are used as the distance data points.

Referring to Figure 11-7, we can derive an equation for I_v, the luminous intensity, based on two measurements of flux at two different distances. Let

F_1 be the flux measured at d_1
F_2 be the flux measured at d_2

a be the detector area
d_1, d_2, d_1', d_2', x_1, and x_2 be defined in Figure 11-7
$x = x_1 + x_2$

Then

$$I_v = \frac{F_1(d_1')^2}{a}$$

$$I_v = \frac{F_2(d_2')^2}{a}$$

Substituting for d_1' and d_2' yields

$$I_v = \frac{F_1(d_1 - x)^2}{a}$$

$$I_v = \frac{F_2(d_2 - x)^2}{a}$$

$$d_1 - x = \left(\frac{aI_v}{F_1}\right)^{1/2}$$

$$d_2 - x = \left(\frac{aI_v}{F_2}\right)^{1/2}$$

$$d_2 - d_1 = \left(\frac{aI_v}{F_2}\right)^{1/2} - \left(\frac{aI_v}{F_1}\right)^{1/2}$$

$$I_v = \frac{(d_2 - d_1)^2}{[(a/F_2)^{1/2} - (a/F_1)^{1/2}]^2} \qquad (11\text{-}3)$$

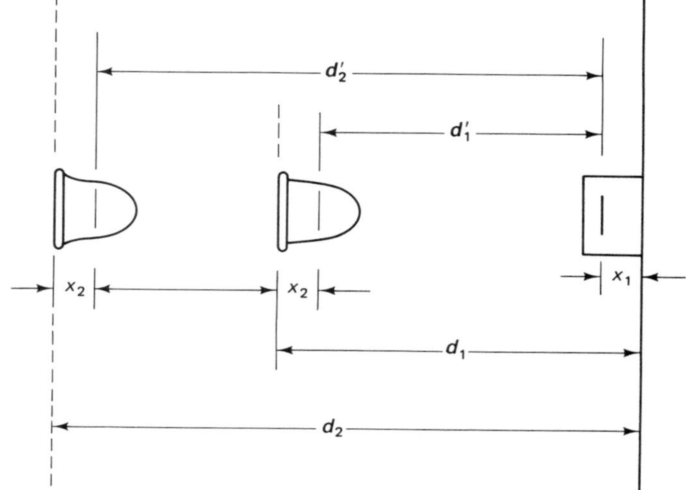

Figure 11-7 Two-distance method of intensity measurement.

Although equation (11-3) is not simple, it is solvable, and more important, it is expressed in terms of known and readily measured quantities. Of course, there is more than one way to solve the measurement problem. A simpler and very adequate solution is to make the d terms much, much larger than the x terms. Then the x terms may be ignored and d can be assumed to be equal to d'. As stated earlier, the measurement of F_v may be in error by as much as 5 to 10%. If the d terms were 100 times larger than the x terms, the error caused by assuming that d is equal to d' would be about 2%.

Alternatively, the detector area and the distance between the detector and the LED can be chosen so that the detector just intercepts the entire cone beam radiated from the LED. As was pointed out in previous chapters, the flux density at the center of the beam is much higher than it is at the edge of the beam. By intercepting the entire beam an average flux density is measured and an average intensity is calculated.

An integrating sphere can also be used to characterize the output of an LED in terms of average intensity (Figure 11-8). The LED is positioned inside the sphere and a baffle is located between the sphere and the detector. The baffle has the same reflecting coating as the sphere surface and blocks direct radiation paths between the LED and the detector. The detector "sees" only fully integrated radiation. Fully integrated radiation is radiation that is spread uniformly over the sphere surface. The amount of power the detector "sees" is related to the detector and sphere size, the reflectance of the sphere coating, and the percentage of the sphere surface occupied by other ports. These various factors are compensated for by two constants of the measurement system: the throughput of the sphere, t, which compensates for reflectance and port areas, and the detector constant, K, which compensates for detector area and location.

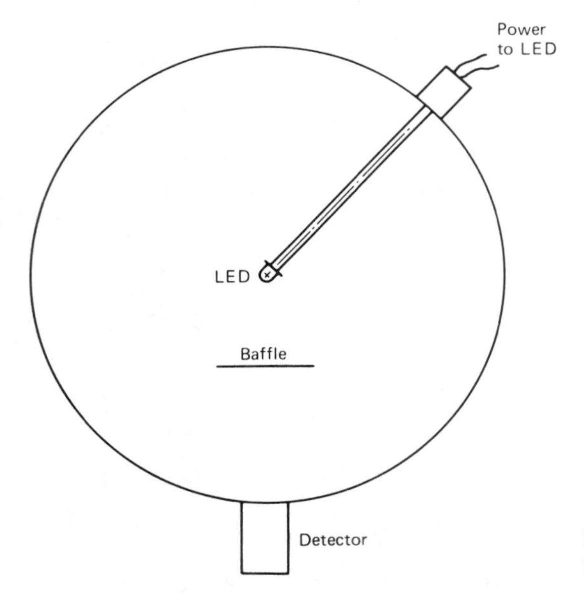

Figure 11-8 Integrating sphere measurement of LED output.

$$P_d = tKP_r \tag{11-4}$$

where P_d is the detected power or detected luminous flux
t is the sphere throughput
K is the detector constant
P_r is the total radiated power (flux) from the LED

In an established and calibrated setup the t and the K values are fixed and the total radiated flux can be calculated directly using the detected flux value and the constants of the measurement system. Occasionally, the total luminous flux value is divided by 4π and the LED is specified in terms of mean spherical candlepower. This specification was discussed in Chapter 4, where it was applied to incandescent lamps.

11-5 USING INFRARED EMITTERS

Infrared emitters are used in communication and industrial control systems. The designer of these systems needs a method of estimating the minimum flux density that will reach the detectors. Typical IRED specifications list only the radiated intensity. This IRED technique of specification does not lend itself to estimation of the flux density experienced by the detector.

Infrared devices developed for communication systems have more detailed specifications. The devices described by Figure 11-9 are intended for use with fiber optic cables. These thin glass fibers have a central core area that carries the radiation. The intensity information needed for this application will be related to these very small core diameters. Note that these data are listed in terms of power for two common core diameters. This is consistent with the needs of the system, where the basic question is how much power can be introduced into the fiber.

In addition to power level, the fiber optic system designer needs precise information about wavelengths and temperature variations. Starting at the top of the specifications list, note that at a fixed temperature, individual devices of the same part number may have peak wavelengths anywhere in a 100-nm range. The next specification, spectral width, describes the range of wavelengths over which an individual device will radiate. This specification is illustrated by the emission spectrum graph. The next two specifications describe the changes in the radiated spectrum as a function of temperature. These specifications indicate that the peak wavelength will increase at a rate of 0.5 nm/C° and that the radiated spectrum will widen at a rate of 0.2 nm/C°. You may recall that photodiodes also exhibit a spectral shift favoring the longer wavelengths when heated. Radiated output power is also radiated to device temperature as illustrated by the graphs of Figure 11-9. With a fixed forward current the radiated output power will decrease as the device heats up.

Some infrared devices are intended for use in industrial control systems, in which they will radiate across a short distance to a detector. A convex lens on the device top forms the beam pattern, and a detector aligned on axis with one of these devices and intercepting a large enough angle will intercept essentially all of the

LASER DIODE INC.

LDT-362, LDT-362E
LDT-60005, LDT-60005E
LDT-60001

1300nm EDGE EMITTING LED SERIES

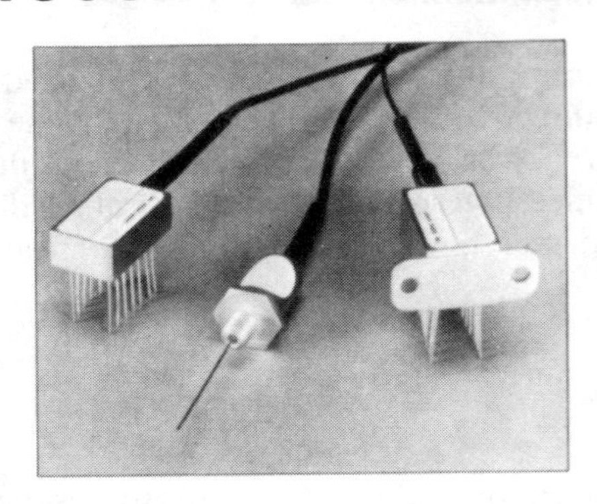

FEATURES

- High Peak Power
- Fast Rise Time
- Single and Multi mode Fiber Optic Pigtail Options
- Hermetic and Non-Hermetic Packages
- Thermo-electric Cooler Option
- Military Qualified Package Available
- High Reliability and Coupling Stability

DESCRIPTION

Laser Diode, Inc.'s (LDI) InGaAsP, edge emitting, light emitting diode (LED) has a peak transmission wavelength of 1300 nanometers (nm). The diode is offered in three package styles: a 14 pin dual inline package (DIP) with a flange, a low profile 14 pin DIP and an LDL-9F package. The DIP packages are hermetically sealed. These packages are offered both with single and multi mode fiber optic pigtails. A thermo-electric cooler and thermistor may be included in the flange package to provide temperature stability. All of the package styles contain a similar diode with the high quality electro-optical specifications stated below.

ELECTRO-OPTICAL CHARACTERISTICS OF THE DIODE (AT 25°C)

Parameters	Symbol	Min.	Typ.	Max.	Units
Wavelength*	λ	1270	1300	1330	nm
Spectral Width	Δλ		70	90	nm
Spectrum vs. Temperature Coefficient			0.5	0.65	nm/°C
Spectral Width vs. Temperature Coefficient			0.5		nm/°C
Optical Rise Time	T_R		4.0		nsec
Optical Fall Time	T_F		4.0		nsec
Output Power	P_O				
• Into 50 micron core, 0.2 N.A. fiber at 150mA					
Option 1		40	60		μW
Option 2		80	100		μW
Option 3		150	200		μW
• Into 9 micron core, single mode fiber at 150mA					
Option 1		4			μW
Option 2		8			μW
Option 3		50			μW
Avg. Power Decrease with Increase in Temperature			-1.5		%/°C
Avg. Power Increase with Decrease in Temperature			+5.0		%/°C
Absolute Maximum Ratings					
• Forward Current	Imax			200	mA
• Forward Voltage	V_F			2.0	V
• Soldering Time @ 260°C				10	sec

* Wavelength may be specified with a ± 20nm tolerance for a central wavelength between 1240 and 1330 nm

LASER DIODE, INC.

(a)

Figure 11-9 Edge-emitting communications-grade LED data sheets. (Courtesy: Laser Diode Inc., New Brunswick, NJ.)

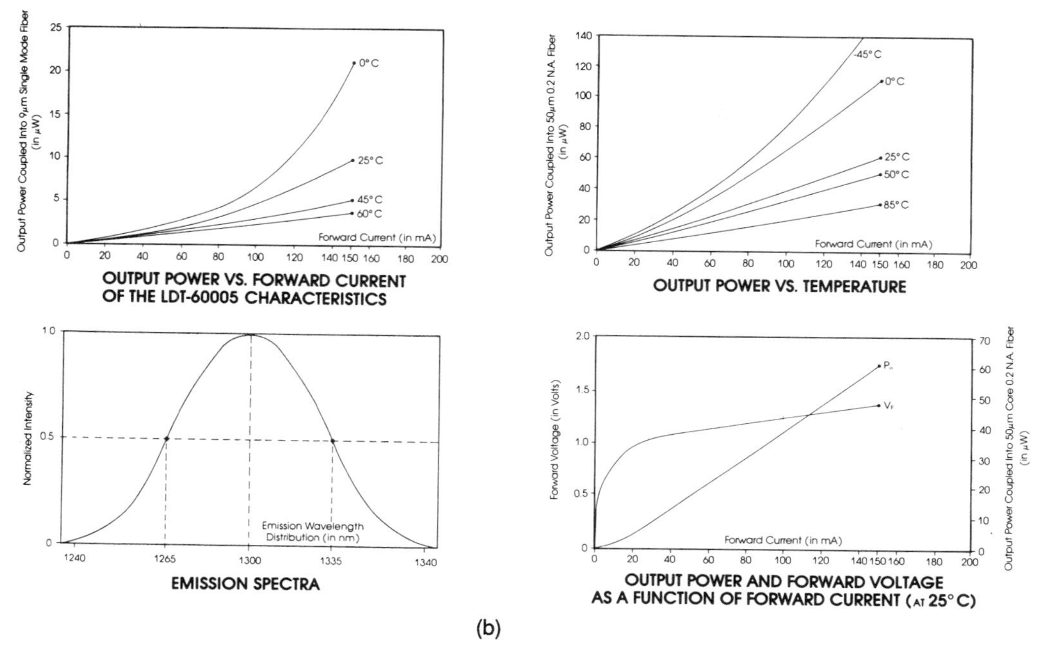

Figure 11-9 *(continued)*

power radiated from the device. If the detector-to-IRED distance is kept constant and the detector is made smaller, it will intercept a smaller included angle. The average measured power, however, will be higher, while the total measured power will be less. Over short distances the average intercepted flux density depends on the size of the included angle of measurement.

Figure 11-10 is a plot of relative intensity versus intercepted solid angle for a source with a half-intensity beamwidth of 20°, which corresponds to a solid angle of 0.10 sr. As this graph illustrates, when the intercepted solid angle gets smaller, the intensity increases. This occurs because the intensity is not uniform across the beam front. A circular beam has a Gaussian intensity distribution, with the maximum intensity at the beam center. The intensity decreases exponentially with radial distance away from the center.

Before attempting performance calculations, the operating conditions and characteristics of the infrared emitter and its location with respect to the detector must be known. The physical size of the detector and its operating characteristics must also be known. When doing calculations using data sheet specifications it is assumed that the detector and the source are aligned so that maximum intensity falls on the detector. Source and detector operating temperatures should be specified so that temperature derating factors can be used.

Simple IRED and detector pairs are often used in an industrial setting, where the interruption of the beam provides some information about the process. For example, when a material bin is full, the beam is interrupted, and when the material

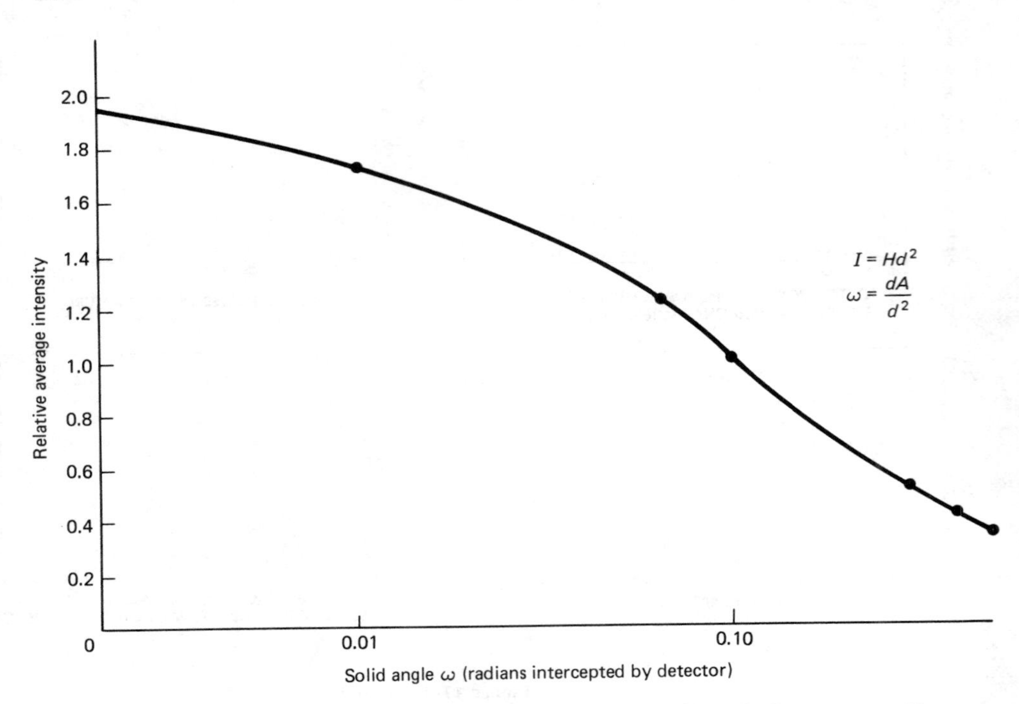

Figure 11-10 Relative average intensity versus intercepted solid angle for a source with a half-intensity angle corresponding to 0.10 sr.

levels falls below a predetermined value, the beam is reestablished and an alarm signal results.

In this and the preceding section we have discussed some of the photometric and radiometric vocabulary associated with LED and IRED devices. In general, an understanding of flux density, intensity, and the solid angle are required to analyze and describe the radiation characteristics of these devices.

11-6 LIQUID CRYSTALS AND ELECTROLUMINESCENT LAMPS

Liquid-crystal displays and electroluminescent lamps are often used in conjunction with each other. A *liquid-crystal display* is a device that creates a visual image by controlling the transmission of light through a polarization process. Electroluminescent lamps are often used as the blacklight source for this application. In this section we discuss the operating principles of these two devices.

If you use a pocket calculator, chances are it has a liquid-crystal display. The alphanumeric information from the calculator is displayed as black characters against a gray background. The actual display module is made up of a number of individual liquid-crystal elements. By applying the appropriate electrical signal, these individual elements can be made to appear gray or black, and by selecting the appropriate arrangement of elements the alphanumeric characters can be created.

When these elements have voltage applied to them, they are said to be biased. In this section we discuss the physical process that causes these elements to appear as gray or black.

The liquid-crystal material is an organic material which at room temperature has a milky appearance and is in a liquid state. At low temperatures the material will become a crystalline solid. At elevated temperatures the liquid begins to clear. This liquid-crystal material is sandwiched between two plates, at least one of which is transparent and both of which can support electrical conduction.

When polarized light energy passes through an element biased with a voltage less than a critical value, V_c, its polarization is rotated 90° [Figure 11-11(a)]. When the voltage applied across the element exceeds a saturation level, V_{sat}, the polarized light passes through the element without any change in polarization. For voltage values between these extremes the liquid-crystal element will cause the polarization of the light to be rotated by angles between 0 and 90°. Two techniques have been developed that allow this phenomenon to be used in a display. One technique employs a diffuse light source mounted behind the display and the second uses a mirror; both methods use two polarizing filters. Figure 11-12 illustrates a back-lighted system; with a voltage less than V_c applied across the liquid-crystal element a bright spot is obtained [Figure 11-12(a)]. When a voltage greater than V_{sat} is used, a dark spot is obtained [Figure 11-12(b)]. The degree to which the liquid-crystal element changes from being a 90° polarizer to being a zero rotation path is a function of the magnitude of the bias voltage applied and the device temperature.

As illustrated in Figure 11-13, there is a critical voltage that must be exceeded before the liquid-crystal element begins to alter transmission through the system as it changes state from a 90° polarizer to a zero rotation path. For the system of Figure 11-12, transmittance through the system will be at a maximum when the crystal causes the greatest rotation. The change in the polarization state from 90° begins at the voltage value V_c, and zero degree polarization occurs when the applied voltage equals V_{sat}. Maximum contrast between the transmit and no-transmit states could

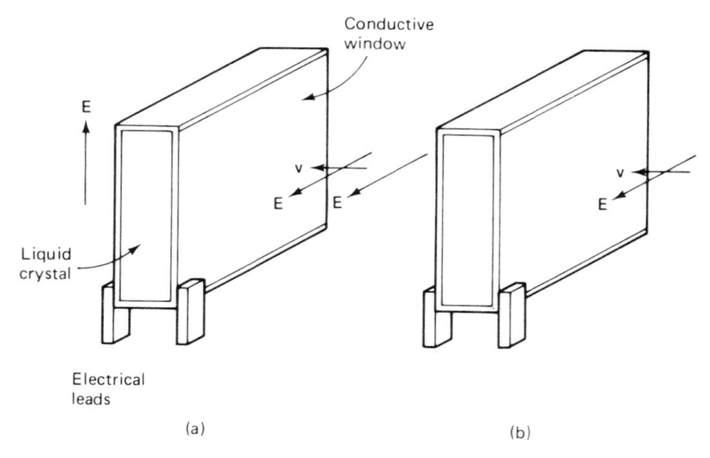

Figure 11-11 Liquid-crystal element: (a) voltage $< V_c$; (b) voltage $> V_{sat}$.

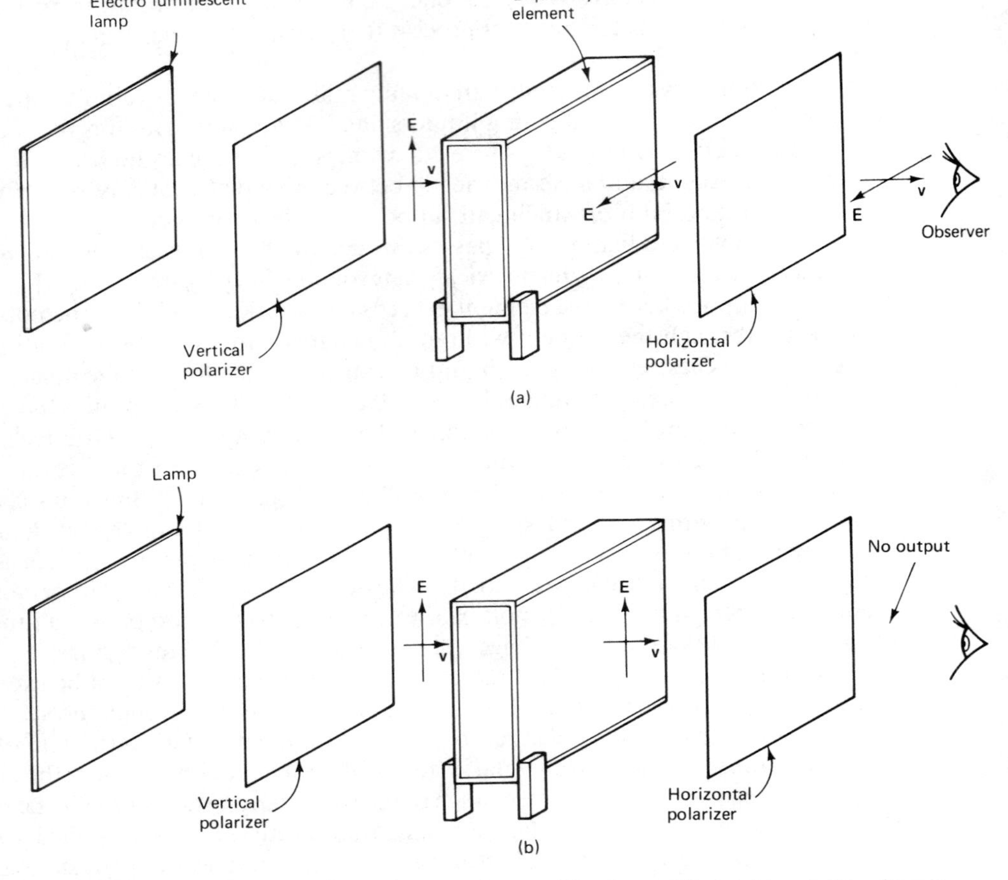

Figure 11-12 Backlighted liquid-crystal element: (a) voltage < V_c, bright spot; (b) voltage > V_{sat}, dark spot.

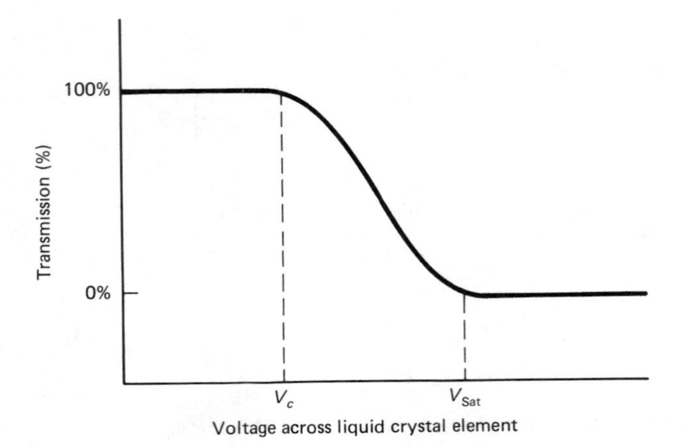

Figure 11-13 Transmission through the liquid-crystal system of Figure 11-12 as a function of bias voltage.

be achieved if the bias voltage were always below V_c at one extreme and always above V_{sat} at the other. Unfortunately, both voltage values vary with temperature, and in practice some change in relative brightness may occur as device temperature changes. The required value of saturation voltage can be as small as 3 V and is usually no larger than 20 V.

These devices are biased with a time-varying voltage waveform such as a sine wave or a square wave. The saturation voltage is expressed in terms of the rms value of this voltage. Frequencies of a few kilohertz have been used, but most commercial devices are designed to operate at frequencies of 30, 60, or 100 Hz. Typical commercial units fail if more than a few millivolts of a constant or dc voltage is present in the bias voltage. The average value of the bias voltage waveform must be close to zero.

For electrical circuit analysis, the liquid crystal can be modeled as a small capacitor in parallel with a large resistor. The ac bias voltage causes resistive and capacitive currents. The capacitive current component is typically 50 times larger than the resistive current. The electronic bias circuits must be designed to drive a load that appears to be a capacitor.

In commercial devices the individual liquid-crystal elements are called *segments* or *dots*. The area of each element determines how much resistance and capacitance it exhibits. As the area of an element is increased, its resistance decreases and its capacitance increases. Table 11-2 lists typical sheet resistance and capacitance values for liquid-crystal devices. These sheet resistance and capacitance values can be used to estimate the current required to drive a liquid-crystal display element.

TABLE 11-2 SHEET RESISTANCE AND CAPACITANCE VALUE FOR LIQUID-CRYSTAL DISPLAYS

Sheet capacitance, S_c	3400 pF/in^2
Sheet resistance, S_r	44 M$\Omega \cdot$ in^2

EXAMPLE 11-2

Estimation of the Current Required to Drive a Liquid-Crystal Display Element

Given: **(1)** Element area = 0.032 in^2

(2) Voltage = 5 V_{rms}

(3) Frequency = 60 Hz

Find: The element current I.

Solution

1. Estimate the resistance of the element.

$$R = \frac{S_r}{a}$$

$$= \frac{44 \text{ M}\Omega \cdot \text{in}^2}{0.032 \text{ in}^2}$$

$$= 1375 \text{ M}\Omega$$

2. Estimate the capacitance of the element.

$$C = S_c a$$

$$= \frac{3400 \text{ pF}}{\text{in}^2} \times 0.032 \text{ in}^2$$

$$= 108.8 \text{ pF}$$

3. Estimate the resistive and capacitance current.

$$I_R = \frac{E}{R}$$

$$= \frac{5 \text{ V}}{1375 \text{ M}\Omega}$$

$$= 3.6 \text{ nA}$$

$$I_C = E \times 2\pi fC$$

$$= 5 \text{ V} \times 2\pi \times 60 \text{ Hz} \times 108.8 \text{ pF}$$

$$= 0.21\ \mu\text{A or } 210 \text{ nA}$$

4. *Conclusion:* The resistive bias current required is negligible compared to the capacitive bias current required. Essentially, this device requires 210 nA of capacitive bias current.

The values calculated in Example 11-2 are representative of those specified for a single segment of an alphanumeric display module. Such displays usually have seven segments per character and at least four characters, for a total of 28 segments. The capacitance exhibited and the current drawn by a complete module would be 28 times as large as those calculated in this example.

Before ending the discussion of liquid crystals a brief description of the internal processes that cause polarization is in order. The polarization of the radiation is caused by the interaction of the radiation and the liquid-crystal molecules. For purposes of our discussion the molecules will be modeled as thin bars represented by two-headed arrows. When the molecular arrows and the electric field of the radiation lie in the same plane, the electric field of the radiation will tend to align itself with the molecules [Figure 11-14(a)]. If the molecular arrows are parallel to the velocity vector of the radiation, there will be no discernible interaction between the electric field of the radiation and the molecules [Figure 11-14(b)].

The liquid-crystal display element is constructed so that with zero voltage applied, successive layers of molecules have the arrangement illustrated in Figure 11-15. These molecules of the unbiased liquid crystal lie in the same plane as the electric field of the incident radiation, so the electric field will interact with them.

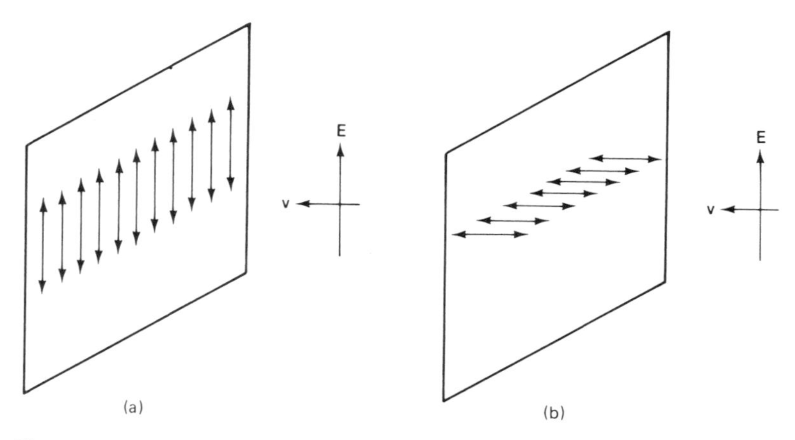

Figure 11-14 Two possible molecular alignments of the first surface of a liquid-crystal display element relative to the incident electric field: (a) interaction will occur; (b) no interaction will occur.

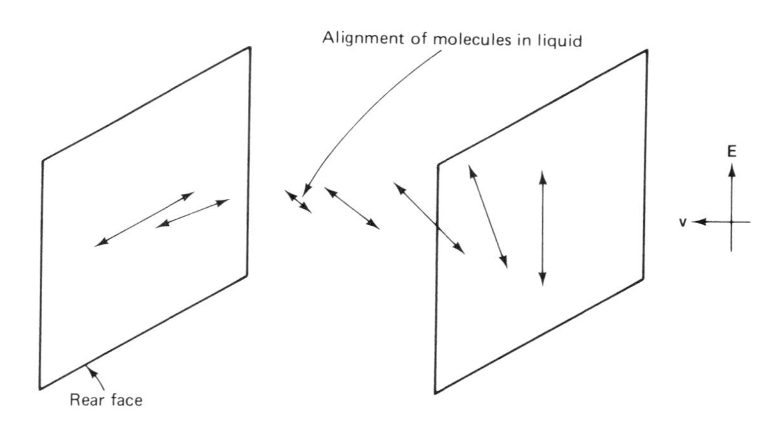

Figure 11-15 Molecular alignment in the liquid between the faces of a liquid-crystal element when the bias voltage is less than V_c.

The relative position of the molecules rotate 90° as the cell is traversed. The electric field of the radiation will have its polarization rotated by 90° as it maintains alignment with the molecules while traveling through the unbiased element.

When a bias voltage greater than V_c is applied across the cell, the molecules begin to line up along the electrical lines of force caused by the applied voltage. When the element is biased with a voltage greater than V_{sat}, the molecules assume the orientation illustrated in Figure 11-14(b). When the molecules are oriented in this manner, the radiation will cross the cell without a change in polarization.

The liquid-crystal element is very popular as a display for calculators because of its low current requirements. In calculator applications the display system uses ambient light and reflection from the back of the element to obtain the contrasting gray and black states (Figure 11-16). When the voltage across the element is less

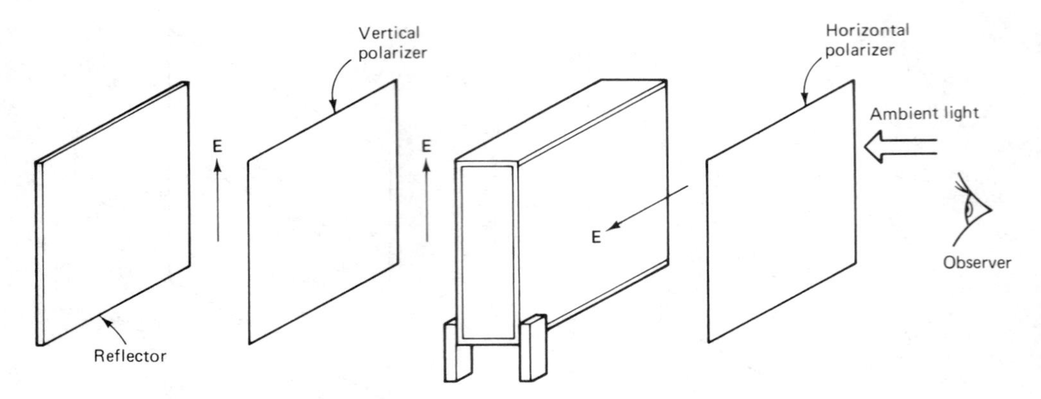

Figure 11-16 Calculator-style, reflector-backed, liquid-crystal display element when the bias voltage is less than V_c.

than the critical voltage V_c, the incident light passes through the cell and the two polarizers and is reflected off the mirror back to the observer. This results in the gray state. When the saturation voltage is applied across the element, incident light that is horizontally polarized by the front polarizer is "stopped" by the rear vertical polarizer, so no reflection takes place. This results in the dark or black state.

The time required by the liquid crystal to make these changes of state can be quite long. State-changing times of 100 ms would not be unusual for large segments. This is not a serious problem with alphanumeric displays. Development work is being done to improve the switching time of these devices to obtain speeds compatible with computer graphic displays. These graphic displays use very small segments each individually driven by a switching transistor, are often rear illuminated, and are being employed in military equipment and laptop computers as a substitute for the much larger cathode ray tube. At a recent trade show a color TV was shown that used a liquid display.

Electroluminescent lamps are used in applications where their flat shape, low operating temperature, and diffuse radiation are an advantage. Some applications would be in safety lighting, instrument illumination, and the backlighting of liquid-crystal displays. These lamps have three features that make them particularly suited to this last application:

1. *Size.* Electroluminescent lamps are a few tenths of an inch thick and available in a variety of rectangular and circular shapes. They can be made in shapes that custom fit a display.
2. *Operating temperature.* These lamps operate at very close to the ambient temperature. They are not a source of heat like an incandescent lamp.
3. *Uniformity of brightness.* These lamps are very uniform diffuse sources of light. They are very close to being ideal Lambertian sources.

These devices consist of a phosphor dielectric sandwiched between two conductive electrodes. One of the conductive electrodes will be a transparent polymer, through which the light will be emitted. The other layer is usually opaque and coated with a thin metal layer. The phosphor dielectric filling has very fine phosphor powder suspended in a nonconducting transparent binding material. The phosphor powder is uniformly distributed throughout this binder, with the individual particles being isolated from each other.

When an alternating current is applied across these devices the phosphor material is excited by the electric fields into radiation. The exact mechanism of this excitation is still being researched. To the external electric circuit, electroluminescent lamps look like a parallel capacitive and resistive load; thus the current drawn by these lamps will increase as frequency is increased. Commercially available lamps are rated for operation at 115 V ac 60 Hz and 115 V ac 400 Hz, and will be about three times brighter at 400 Hz than they will be at 60 Hz. It is possible to operate these lamps at reduced voltages, which will cause reduced brightness. The radiation emitted decreases in magnitude very rapidly with decreasing voltage and reaches zero at voltages in the range 40 to 60 V ac.

11-7 SUMMARY

The devices discussed in this chapter fall into two broad wavelength-based groupings: visible and infrared. The infrared devices are diode junctions that emit infrared radiation when the junction is forward biased. These devices are used in communications systems and in industrial control sensing systems.

Visible-light radiators called light-emitting diodes (LEDs) are also diode junctions that emit radiation when forward biased. Light-emitting diodes are used primarily in indicator and display applications as a low-power, low-heat substitute for incandescent lamps. Occasionally, LEDs will be used in very short distance (10 m or less) communication links and in industrial control sensing systems.

Electroluminescent lamps are low-intensity low-temperature Lambertian sources that radiate when connected to an ac voltage source. Liquid-crystal displays do not radiate at all. These devices alter the polarization of the available light and are used in conjunction with polarizers to create light and dark spots. They operate from low-amplitude ac voltage. Liquid-crystal displays are often used in conjunction with electroluminescent lamps.

QUESTIONS

1. What effect does an increase in forward current have on the relative intensity from an LED?
2. What effect does an increase in temperature have on the spectral distribution of an LED?
3. At what wavelengths do infrared diodes operate?
4. Why does intensity decrease as the area of the beam intersected increases?
5. What happens to total intercepted power as the area of the beam intersected increases?

6. What is the typical forward voltage range of an LED?
7. Describe what happens to light propagating through a liquid-crystal element when the voltage across the element is greater than the saturation voltage.
8. What happens to the light output of an electroluminescent lamp when the frequency of the applied voltage is increased?
9. Why is the top P layer of an LED so thin?

PROBLEMS

1. An LED is going to be pulsed with 2 A of current. The pulse width is 0.5 μs and the frequency is 5 kHz. What is the average current delivered to the LED?
2. Refer to the derating specification on Figure 11-3, which is specified in terms of ambient temperature. If the ambient temperature is 60°C, estimate the maximum safe current for the LED.
3. Refer to Figure 11-3. Using the rated typical value of luminous intensity, estimate the luminous intensity of an HLMP-3000, whose forward current is 30 mA.
4. Refer to Figure 11-9. How much power is being delivered to the LED when the forward current is 100 mA?
5. A photometric detector with a circular active area of 1 cm^2 is used to measure the radiated output of an LED. The detector system yields constant readings of flux for distances between the detector and the LED that are less than 1.6 cm. Measured values of flux decrease for distances between the devices that are greater than 1.6 cm. The constant measured flux is 1 mlm. Estimate the average luminous intensity of the LED and the total spatial angle of radiation.
6. A detector with an active area of 0.1 cm^2 is used to measure the flux density of an LED. The following data are obtained. Estimate the luminous intensity of the LED.

Distance (cm)	Flux Density (μlm/cm^2)
5	121
10	30

7. An integrating sphere has a throughput of 0.2 and a K value of 0.9. The detector output device yields a reading of 0.2 min. What is the total radiated flux from the LED?
8. An infrared emitter and a silicon detector are going to be used as an intrusion alarm detection system across an entranceway. A reasonable estimate of radiated intensity is 1.3 mW/sr. The doorway is 36 in. wide. The detector has an area of 1 cm^2 and a responsivity of 500 μA/mW. Estimate the detector current when the beam is unbroken.
9. Refer to Table 11-2. A graphic liquid-crystal display module is rectangular in shape with a dot format of 600 × 200. Each dot is 0.8 mm × 0.3 mm. With a 10-V_{rms} 60-Hz power supply, estimate the current drawn by each dot and the maximum current that could be drawn by the display module.
10. The brightness of an electroluminescent lamp increases at a rate approximately proportional to the square root of the frequency of the applied voltage. The brightness of a lamp is 12 cd/m^2 at 60 Hz; estimate its brightness at 400 Hz.

12

Transistors, Phototransistors, and Opto-Isolators

12-1 INTRODUCTION

Transistors are semiconductor devices fabricated with three alternating P and N layers. These devices are available individually and they are also fabricated into integrated circuits where they perform switching and amplifying actions. The phototransistor is a quantum detector. As was the case with a photodiode, the current flow through a phototransistor is proportional to the incident flux density. Phototransistors or photodiodes are used together with a LED or IRED to create an opto-isolator.

12-2 TYPES OF TRANSISTORS

With three layers there are two possible combinations of PN materials; transistors may have a PNP layer configuration or a NPN layer configuration. The configuration used determines the transistor's electrical characteristics. Each of the configurations has a different schematic symbol so that the type used may be indicated on a schematic. Figure 12-1 illustrates both layer configurations and their schematic symbols.

As Figure 12-1 illustrates, each lead has a letter designation, which is the first letter of the terminal name: C, collector; B, base; and E, emitter. On the emitter is drawn an arrow that indicates the direction of conventional current flow. In normal operation, current will flow out of the emitter of the NPN transistor, or into the emitter of the PNP transistor.

The fixed dc voltage levels applied to the transistor terminals are called the *bias voltages*. If the bias voltages are applied so that the transistor is conducting

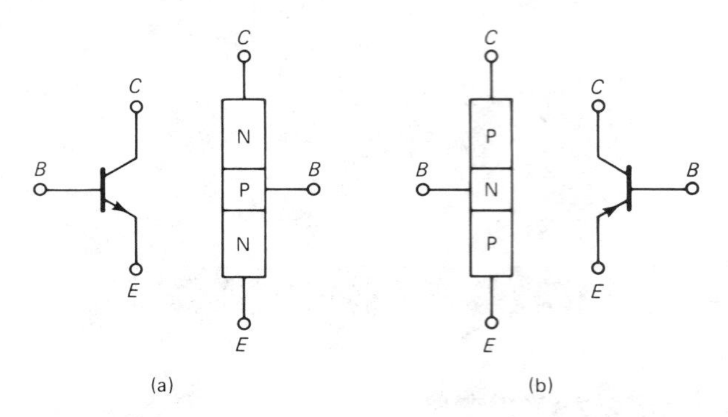

Figure 12-1 (a) NPN and (b) PNP transistor configurations and symbols.

current, the transistor is said to be "ON." If the bias voltages are applied so that the transistor is not conducting current, the transistor is said to be "OFF." The relative polarity of the base-to-emitter voltage determines whether the device is on or off.

When an NPN transistor is biased ON, current is flowing out of the emitter terminal. This current flowing out of the emitter terminal will be the sum of the currents flowing into the collector and base terminals. For conventional current to flow into the collector and base and out of the emitter terminal, the collector and base terminals must be more positive than the emitter terminal. In amplifier configurations the collector will be more positive than the base. Figure 12-2 illustrates an NPN transistor biased ON.

When a PNP transistor is biased ON, current is flowing into the emitter terminal. This current flowing into the emitter terminal will be the sum of the currents flowing out of the collector and base terminals. For conventional current to flow out of the collector and base terminals and into the emitter terminal, the collector and base terminals must be less positive than the emitter. In amplifier configurations the collector will be more negative than the base. Figure 12-3 illustrates a PNP transistor biased ON.

In both the NPN and the PNP cases the emitter current is the sum of the base and collector currents.

$$I_E = I_C + I_B \tag{12-1}$$

Due to the physical nature of a transistor, the collector current is proportional to the base current. The ratio of collector current to base current is called H_{FE} and for most transistors is on the order of 100.

$$H_{FE} = \frac{I_C}{I_B} \simeq 100 \tag{12-2}$$

In many simple analysis problems it is safe to assume that I_C is equal to I_E because I_B is so small by comparison to either. When a transistor is biased ON, the magnitude of the base-to-emitter voltage will be the same as the magnitude of the forward voltage typical of a diode of the same material. Table 12-1 lists the common base–emitter voltages that will be encountered.

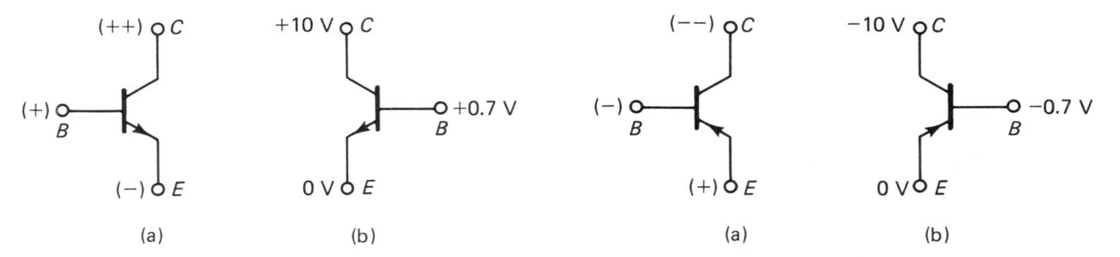

Figure 12-2 Biasing an NPN transistor ON: (a) general; (b) example.

Figure 12-3 Biasing a PNP transistor ON: (a) general; (b) example.

TABLE 12-1 BASE–EMITTER VOLTAGES FOR ON TRANSISTORS

Material	Base–Emitter Voltage (V)
Silicon	0.5–1.0
Germanium	0.3–0.4
Gallium	1.2–2.0

12-3 TRANSISTOR AS A SWITCH

When a mechanical switch is closed, current flows through the switch contacts. Therefore, when a mechanical switch is open, zero current flows through the switch contacts. When a transistor is ON, current flows through the path between its collector and emitter terminals. When a transistor is OFF, no significant current flows through it. A transistor does not provide the same absolute isolation between opposite sides of the circuit that is provided by the mechanical switch. The collector and emitter are always in electrical contact, and there will always be small temperature-related "leakage" currents flowing through the transistor.

Transistors are used as switches in computers and sensing applications. A transistor switch is turned on or off by the potential and current at the base terminal. If the base-to-emitter junction is forward biased, base current will flow and the transistor switch will be on. If the base-to-emitter diode is reverse biased, no base current flows and the transistor switch will be off.

The bias voltages will be applied to the transistor switch so that the transistor is either normally on or normally off. An input signal voltage, applied across the base-to-emitter junction, can then be used to cause the transistor switch to change state from on to off or off to on. Figure 12-4(a) illustrates a normally ON circuit, with the collector and base connected to a bias source V_{CC} of the polarity required to forward bias the base-to-emitter junction. The normally OFF circuits of Figure 12-4(b) have the collector connected to a source of the correct relative potential for conduction, but the base and emitter are connected to the same potential. This means that the base-to-emitter junction has zero volts across it and therefore the transistor is OFF.

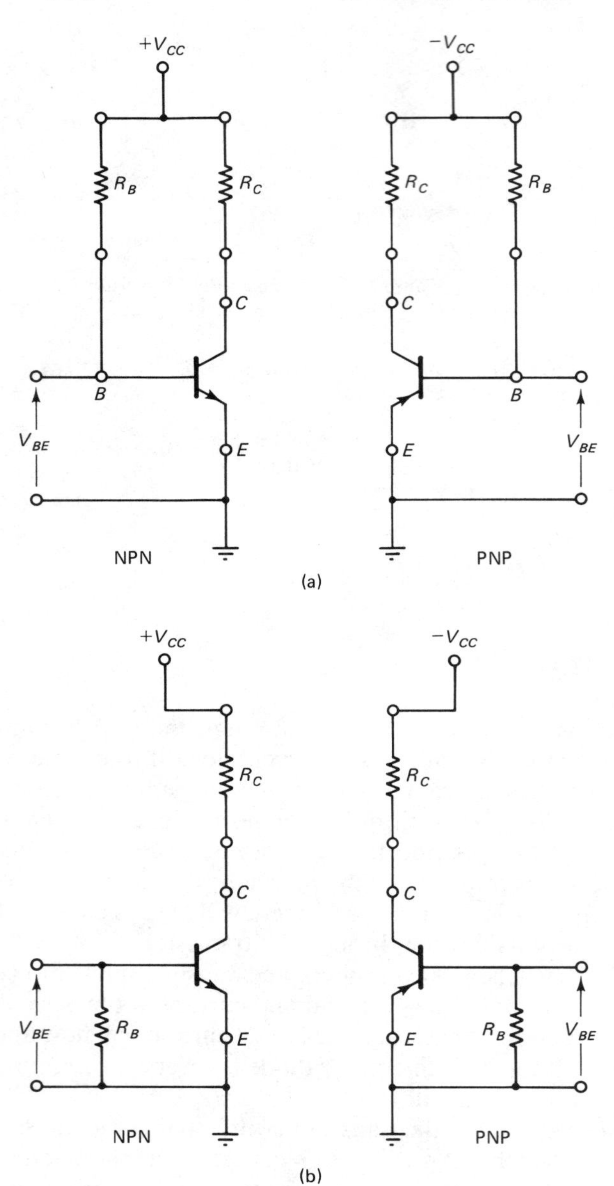

Figure 12-4 Transistor switching circuits: (a) normally ON circuits; (b) normally OFF circuits.

Figure 12-5(a) illustrates the voltage pulses required to shut off the normally on transistors, and Figure 12-5(b) illustrates the voltage pulses required to turn on the normally off transistors. The resistors in series with the input signal voltage pulses limit the current flow into the base–emitter junction and the voltage across the junction. Examples 12-1 and 12-2 present simple analyses of normally ON and normally OFF circuits, respectively.

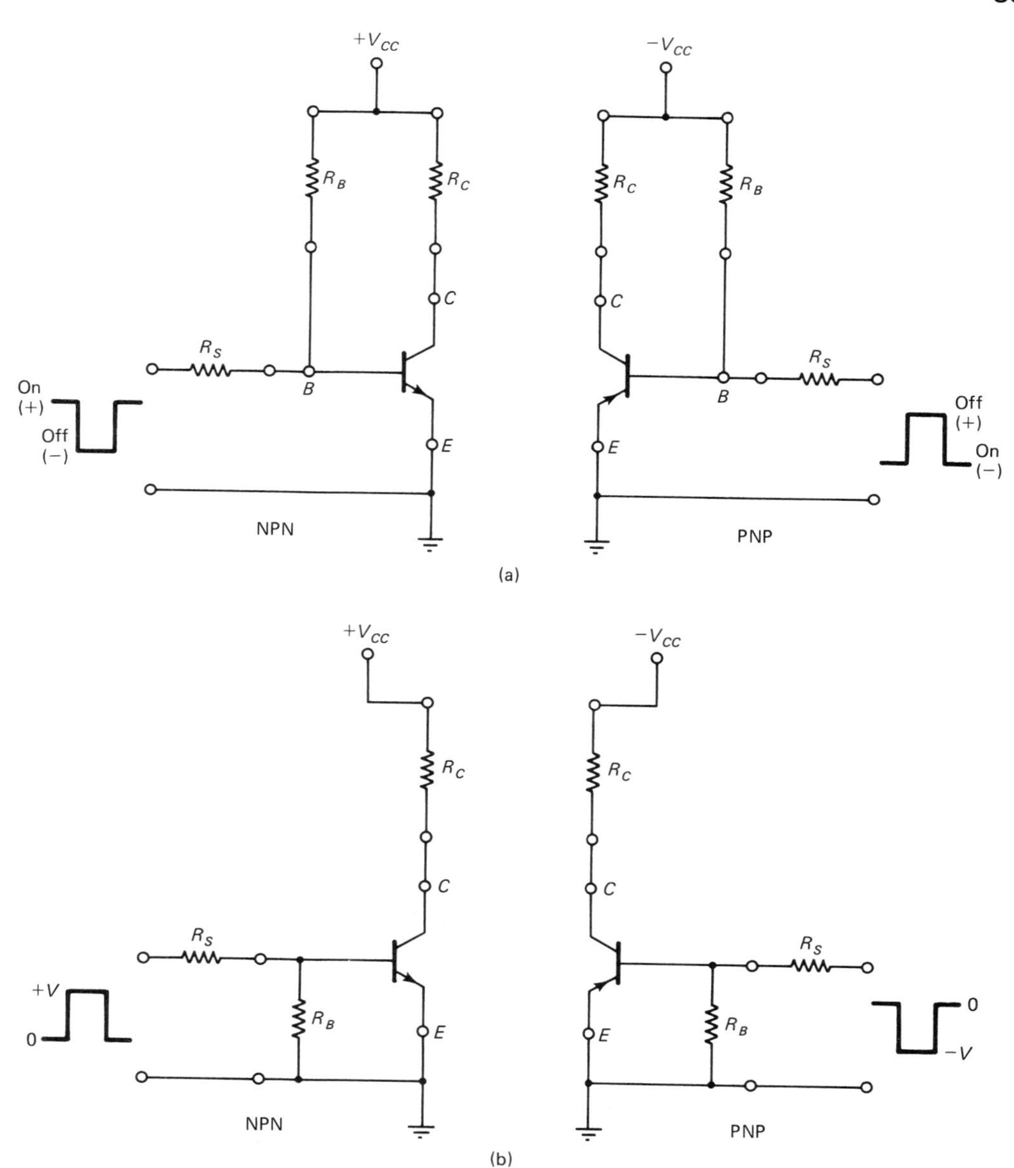

Figure 12-5 Pulses required to actuate transistor switches: (a) turning OFF normally ON switches; (b) turning ON normally OFF switches.

EXAMPLE 12-1

Analysis of a PNP Normally ON Circuit

Given: That the device is silicon, so that $V_{BE} \cong -0.7$ V, and that $H_{FE} = 100$.

Find: The bias voltages and currents for the circuit illustrated in Figure 12-6.

Solution:

$$V_{R_B} = V_{CC} - V_{BE}$$
$$= -10\text{ V} - (-0.7\text{ V}) = -9.3\text{ V}$$
$$I_B = \frac{V_{RB}}{R_B} = \frac{-9.3\text{ V}}{1\text{ M}\Omega} = -9.3\ \mu\text{A}$$
$$\frac{I_C}{I_B} = H_{FE}$$
$$I_C = I_B \times H_{FE} = -9.3\ \mu\text{A} \times 100 = -930\ \mu\text{A}$$
$$V_{CE} = V_{CC} - I_C R_C$$
$$= -10\text{ V} - (-930\ \mu\text{A}) \times 10\text{ k}\Omega$$
$$= -0.7\text{ V}$$

EXAMPLE 12-2

Analysis of an NPN Normally OFF Circuit

Given: That the device is silicon, so that $|V_{BE}| \simeq 0.7$ V, and that $H_{FE} = 100$.

Find: The circuit voltages and currents for the circuit illustrated in Figure 12-7:
(a) When the input voltage pulse is at zero volts.
(b) When the input voltage pulse is at +5 V.

Solution: **(a)** When the input generator is at zero volts, the base voltage and current will both be zero. Therefore,

$$I_C = H_{FE} \times I_B = 0$$

and

$$V_{CE} = 5\text{ V} - I_C R_C = 5\text{ V}$$

(b) When the input voltage pulse goes to +5 V, current flows through resistors R_S and R_B and into the base; the transistor will then be turned ON. For purposes of analysis the +5-V pulse will be represented on the schematic as a constant +5-V voltage supply. Analyzing this circuit, which is illustrated in Figure 12-8,

$$I_{R_B} = \frac{V_{BE}}{R_B}$$

$$= \frac{0.7 \text{ V}}{22 \text{ k}\Omega}$$

$$= 31.8 \ \mu\text{A}$$

$$I_{R_S} = \frac{5 \text{ V} - 0.7 \text{ V}}{100 \text{ k}\Omega}$$

$$= 43 \ \mu\text{A}$$

$$I_{R_S} = I_{R_B} + I_B$$

$$I_B = I_{R_S} - I_{R_B}$$

$$= 11.2 \ \mu\text{A}$$

$$I_C = H_{FE} \times I_B = 100 \times 11.2 \ \mu\text{A}$$

$$= 1.1 \text{ mA}$$

$$V_{CE} = 5 \text{ V} - I_C R_C = 5 \text{ V} - (1.1 \text{ mA}) \times 3.6 \text{ k}\Omega$$

$$= 1.04 \text{ V}$$

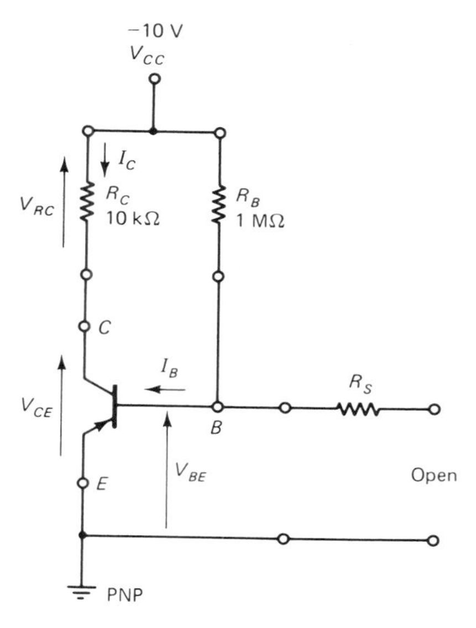

Figure 12-6 Normally ON PNP circuit for analysis in Example 12-1.

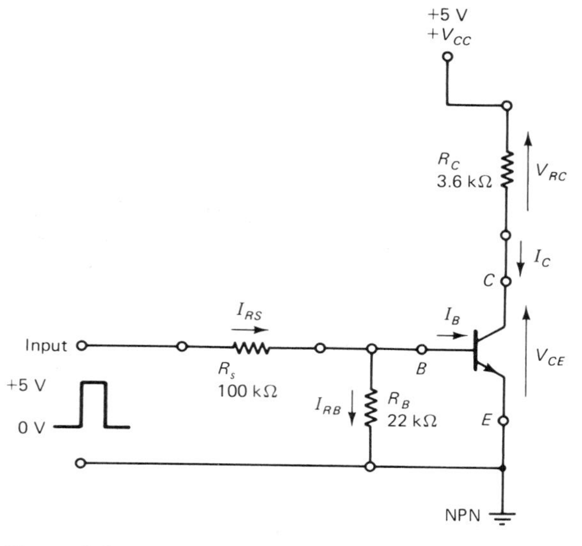

Figure 12-7 Normally OFF NPN circuit analyzed in Example 12-2.

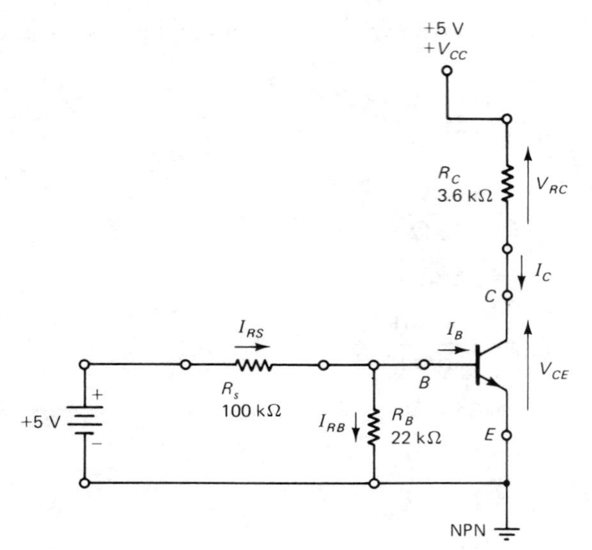

Figure 12-8 Normally OFF NPN circuit turned ON with the +5-V level of the input pulse, Example 12-2.

Examples 12-1 and 12-2 illustrate the analysis of simple transistor switching circuits. Example 12-2 illustrates the analysis of a normally OFF circuit in the normally OFF state and when it is turned ON.

Switching circuits are designed so that when the transistor is ON, the voltage V_{CE} is at a small value close to V_{BE} called the *saturation voltage,* V_{sat}. When the transistor is in this saturated state the collector current is at its maximum possible value. This maximum possible collector current is called the *saturation current,* and its magnitude is determined by the supply voltage V_{CC}, the collector resistor R_C, and the value of V_{sat}.

$$I_{sat} = I_{C\max} = \frac{V_{CC} - V_{sat}}{R_C}$$

An engineering assumption often made is that the transistor's saturation voltage is zero, in which case the saturation current is obtained simply by dividing the supply voltage by the collector resistor value. Table 12-2 summarizes the voltage and current characteristics of the switching circuit's ON and OFF states.

TABLE 12-2 VOLTAGE AND CURRENT CHARACTERISTICS OF SWITCHING CIRCUITS

State	I_C	V_{CE}
ON	I_{sat}	0 to V_{sat}
OFF	0	V_{CC}

Contrary to what might be thought initially, a switching transistor is OFF when a large voltage is measured across its collector-to-emitter terminals, and ON when the collector-to-emitter voltage approaches zero.

12-4 TRANSISTORS AS LINEAR AMPLIFIERS

Transistors are used not only as switches but also as linear amplifiers either individually (discrete) or in integrated circuits. Students of electronics spend entire courses learning to analyze and design transistor amplifiers. In this short section the amplification process will be discussed and the voltage levels that might be encountered, pointed out.

The bias current and voltage configuration of a transistor amplifier is different from that of a transistor switch. Transistor switches are designed to operate at one of two extreme conditions: either off (zero current) or saturated. Linear amplifiers are fixed biased with a collector current about halfway between zero and I_{sat}. The collector-to-emitter voltage across the transistor in a linear amplifier can be anywhere from several volts to several tens of volts, and it is often set at value close to one-half V_{CC}. The base-to-emitter bias voltage of a linear amplifier will be near the low end of the typical base-to-emitter voltages. Figure 12-9 illustrates a typical linear amplifier, and Example 12-3 analyzes the amplifier.

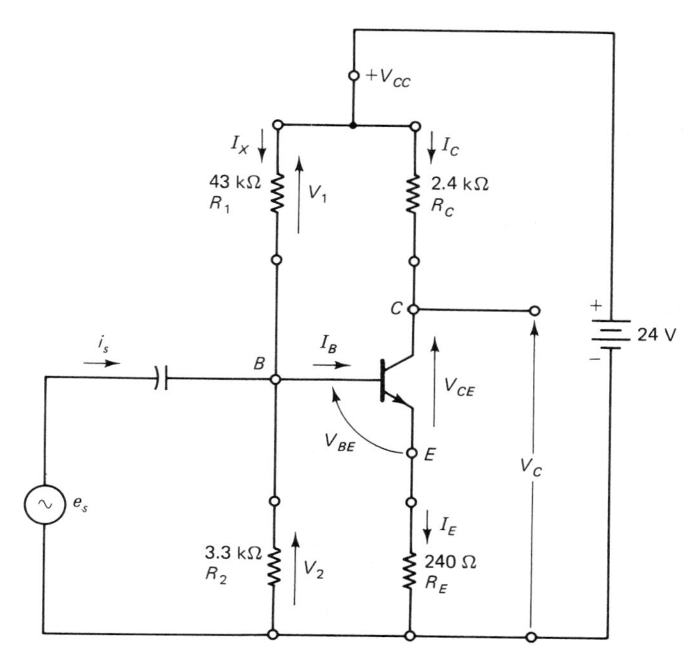

Figure 12-9 Linear amplifier analyzed in Example 12-3.

EXAMPLE 12-3

Linear Amplifier Biasing and Operation

Given: That I_x is much greater than I_B. Treat the series resistors R_1 and R_2 as an unloaded voltage divider.

Find: The bias voltages and currents for the circuit illustrated in Figure 12-9.

Solution:

$$I_x \simeq \frac{24\text{ V}}{43\text{ k}\Omega + 3.3\text{ k}\Omega} = 518\ \mu\text{A}$$

$$V_2 = I_x R_2$$

$$= 518\ \mu\text{A} \times 3.3\text{ k}\Omega$$

$$= 1.71\text{ V}$$

If $V_2 = 1.71$ V

and $V_{BE} \simeq 0.6$ V, then

$$V_E \simeq V_2 - V_{BE}$$

$$= 1.11\text{ V}$$

$$I_E = \frac{V_E}{R_E} = \frac{1.11\text{ V}}{240\ \Omega} = 4.63\text{ mA}$$

Assume that $I_E \simeq I_C$. Then

$$I_B = \frac{I_E}{H_{FE}} = \frac{4.63\ mA}{100} = 46.3\ \mu\text{A}$$

Check the assumption that $I_B < I_x$.

$$46.3\ \mu\text{A} < 518\ \mu\text{A} \qquad \text{(so the assumptions are correct)}$$

Assume that $I_E = I_C$. Then

$$V_C = V_{CC} - I_C R_C$$

$$= 24\text{ V} - (4.63\text{ mA})(2.4\text{ k}\Omega)$$

$$= 24\text{ V} - 11.11\text{ V}$$

$$= 12.89\text{ V} \qquad \text{(this voltage is the collector bias voltage)}$$

$$V_{CE} = V_C - V_E$$

$$= 12.89\text{ V} - 1.11\text{ V}$$

$$= 11.78\text{ V}$$

To demonstrate the amplification process, assume that the source e_s causes a +0.5-V rise in the base voltage V_2. This 0.5-V change is the input signal voltage. This new value of base voltage will cause all the voltages and currents in the circuit to assume new values, as illustrated below.

$$V_2 = 1.71 + 0.5 \text{ V}$$
$$= 2.21 \text{ V}$$
$$V_E = V_2 - V_{BE}$$
$$= 1.61 \text{ V}$$
$$I_E = \frac{V_E}{R_E}$$
$$= \frac{1.61 \text{ V}}{240 \ \Omega} = 6.71 \text{ mA}$$

Assume that $I_C \simeq I_E$. Then

$$V_C = V_{CC} - I_C R_C$$
$$= 24 \text{ V} - (6.71 \text{ mA})(2.4 \text{ k}\Omega)$$
$$= 24 \text{ V} - 16.1 \text{ V}$$
$$= 7.9 \text{ V}$$

Determine the change in the collector voltage caused by the change in base voltage.

$$\Delta V_C = V_C \text{ (original)} - V_C \text{ (new)}$$
$$= 12.89 \text{ V} - 7.9 \text{ V}$$
$$= 4.99 \text{ V} \qquad \text{(this 4.99-V change is the output signal voltage)}$$

$$|\text{voltage gain}| = \frac{\text{change in collector voltage}}{\text{change in base voltage}}$$

$$\frac{\Delta V_C}{\Delta V_2} = \frac{4.99 \text{ V}}{0.5 \text{ V}} \simeq 10$$

Since output voltage V_C decreased as input voltage V_2 was increased, the gain is said to be negative. Something that always seems to cause trouble in understanding linear transistor amplifiers is the difference between the static bias voltages and bias currents and the changing signal voltages and signal currents. Bias voltages and currents are the fixed values of voltages and currents present when dc power is applied to the circuit before the signal generator e_s is connected, and are caused by the fixed V_{CC} direct-current source and the various circuit components. In the circuit of Example 12-3 the bias value of the voltage V_C is 12.89 V and the bias value of the emitter current is 4.63 mA. Signal voltages and currents are caused by the signal generator e_s and are equal to the amount of *change* in the bias values. Signals are related to the amounts of *change*—that is the key concept. The circuit analyzed in Example 12-3 had its collector voltage change from 12.89 V to 7.9 V; the collector signal voltage was *the amount of voltage change,* in this case 4.99 V.

If the signal generator e_s had an output voltage that varied sinusoidally, the base voltage V_2 would vary sinusoidally about the bias voltage level of 1.71 V and the collector voltage would vary sinusoidally about its bias level of 12.89 V. As illustrated above, the magnitude of the collector voltage will decrease as the base voltage increases. The magnitude of the collector voltage will increase when the base voltage decreases. This process causes a voltage sine wave at the collector that is 180° out of phase with the voltage sine wave at the base. For this example the amplitude of the voltage sine wave at the collector will be 10 times larger than the amplitude of the voltage sine wave at the base. These circuits are called *linear amplifiers* because there is a linear proportionality between the amplitude of the input signal voltage and the amplitude of the output signal voltage. This proportionality is independent of the amplitude of input signal and it is established by the design of the amplifier.

These circuits will behave as linear amplifiers as long as the current through the transistor does not come close to zero or I_{sat}. If the change in collector current caused by the signal source causes the collector current to approach zero and I_{sat}, the voltage waveform observed at the collector will exhibit noticeable flattening of the positive and negative peaks. This process is called *distortion* and the amplifier is said to be *overdriven.* Waveform distortion can be cured by either reducing the amplitude of the input signal or by reducing the gain of the amplifier.

12-5 PHOTOTRANSISTORS

A *phototransistor* is a transistor whose base current is caused by the incident radiation and whose collector-to-emitter current is therefore dependent on the incident radiation. The collector–base junction of the phototransistor acts like a photodiode and converts photons to charge carriers, causing a photon-produced base current I_p. This base photon current causes a collector current in the same manner as the electrical base current I_B in a conventional transistor.

$$I_C = H_{FE} \times I_p \tag{12-3}$$

Sometimes an electrical contact is also brought out from the phototransistor's base region, in which case an electrical base current I_B is also present. The collector current will then be a function of both I_B and I_p.

$$I_C = H_{FE}(I_B + I_p) \tag{12-4}$$

Phototransistors can be used as linear amplifiers; however, they are more commonly used as switches. Switching speeds are usually 10 μs or longer, which makes them useful as detectors in slow systems such as machinery controllers or bulk processes. Silicon phototransistors are really too slow to be useful in most communication applications.

These devices come in a variety of packages. Three typical ones are:

1. Single phototransistor per package with a simple lens or window.
2. Photo-Darlingtons, which are a special connection of a phototransistor and a conventional transistor in one package.
3. Photon-coupled isolators, which are packages containing an IRED and a detector such as a phototransistor, a photo-Darlington, or a photodiode.

The attractive feature of the phototransistor as compared to a photodiode is the gain HFE. In this sense the phototransistor and the avalanche photodiode both have an advantage over a simple photodiode. The collector-to-emitter current of a phototransistor will be greater than that of a planar diffused photodiode with the same active area by the HFE factor. A comparable avalanche photodiode may have more "gain" than a phototransistor. All three devices start with a quantum-related photo current. The phototransistor and the avalanche photodiode use different processes to "multiply" the number of photon-generated charge carriers, thus increasing the amount of current flowing.

12-6 SPECIFICATIONS OF PHOTOTRANSISTORS

Typical data sheets will list the maximum ratings first. The maximum ratings list those voltages, currents, power levels and temperatures that will cause device failure. The current, power, and temperature ratings are interrelated as discussed in previous chapters in sections on thermal management.

The voltage ratings will have some odd-appearing subscripts and require some explaining. Table 12-3 lists the voltage ratings and their explanations.
The voltage ratings may have the letters BR appearing before the subscript. An example would be $V_{(BR)CEO}$. The letters BR designate the reverse breakdown voltages.

The optical characteristics section of a phototransistor data sheet will list the current response of the phototransistor. The phototransistor will develop a collector current often called (I_L) in response to an incident flux density. A common source of flux is a lamp with a rated color temperature near 2870 K. Sometimes the response of the phototransistor to flux from a monochromatic source or a specified LED or IRED may be listed. There will also be a dark-current specification. The

TABLE 12-3

Rating	Meaning
V_{CEO}	Voltage measured from emitter to collector with the base terminal open or the base-to-emitter junction darkened.
V_{CBO}	Voltage measured from base to collector with emitter terminal open.
V_{EBO}	Voltage measured from base to emitter with the collector terminal open. Notice that this is a reverse bias voltage.

dark current, also called I_{CEO}, is the collector-to-emitter current that will flow when the phototransistor is not illuminated. The current response of these devices to increases in flux density is not very linear, so a single specified value of current response is of limited utility. To get around this problem, response curves are often provided. Curves of spectral response and angular response are also usually found on data sheets. The spectral response curve of a phototransistor will be very similar to the spectral response of a photodiode of the same material.

Figure 12-10 illustrates the construction and schematic of an opto-isolator. These devices are usually mounted in a case that allows easy connection to a printed circuit board. Often there are two transistors mounted in the case and connected together in the manner illustrated. The first transistor is the phototransistor. Its emitter current is the base current of the conventional second transistor. This type of connection, called a Darlington connection or Darlington pair, will yield a much higher total collector current for a given amount of flux density than will a phototransistor alone. Unfortunately, Darlington connections have slower response times than single transistors.

Instead of a Darlington pair, an opto-isolator may have a single phototransistor or a photodiode as the detecting element. The source is usually a gallium arsenide IRED which is mounted in the same package, as shown. A voltage pulse applied across the IRED causes a photon pulse that is coupled to the detecting element. The space between the IRED and the detecting element provides electrical isolation between the input voltage pulse and the output detected current. This type of isolation can be very desirable in biomedical and industrial control applications.

The isolation characteristics of these devices are usually expressed in three ways: resistance, capacitance, and breakdown voltage. These measurements are made between the IRED and the detector side of the device. The breakdown voltage represents the voltage difference that can exist between opposite sides of the device before it fails. Depending on your point of view, this type of device can be considered an optical signal coupler or an electrical circuit isolator.

The thermal problems of this device are a little more complicated than those we have looked at previously. An opto-isolator has two heat sources: the IRED and the detector. In addition to simple self-heating in response to their individual power dissipations, these two semiconductors will heat each other. Heat energy will

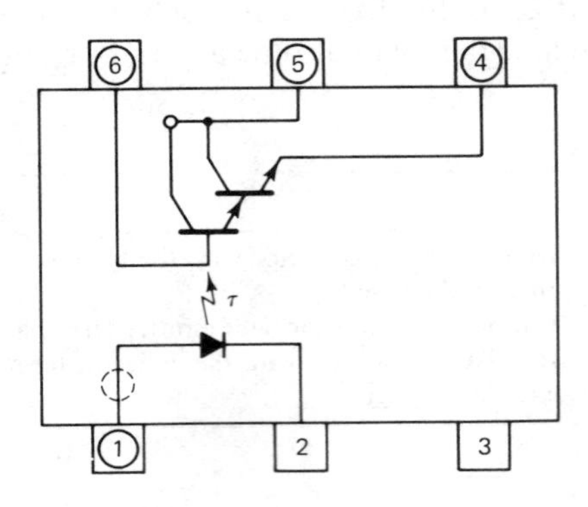

Figure 12-10 Optoisolator package with schematic overlay showing IRED- and Darlington-connected phototransistors.

flow from the hotter semiconductor to the cooler semiconductor. The designer's job is to keep both semiconductors below the manufacturer's recommended maximum temperature. The following equation is used for these calculations:

$$\Delta T = \theta(P_H + KP_C) \qquad (12\text{-}5)$$

where ΔT is the difference between the ambient temperature and the rated maximum temperature
θ is the junction-to-ambient thermal resistance
P_H is the largest of the two power dissipations, the hottest semiconductor
K is an empirical thermal coupling coefficient
P_C is the smaller of the two power dissipations, the cooler semiconductor

---

EXAMPLE 12-4

Calculation of Device Dissipation in an Opto-Isolator

Given: **(1)** Maximum temperature = 100°C
(2) Thermal resistance = 500°C/W
(3) Thermal coupling coefficient = 0.2

Find: P_i, the maximum permitted device dissipation when the dissipation of both devices is equal and the ambient temperature is 75°C.

Solution:

$$\Delta T = 100°\text{C} - 75°\text{C} = 25°\text{C}$$

$$25°\text{C} = (500°\text{C/W})\,(P_i + 0.2P_i)$$

$$P_i = \frac{25°\text{C}}{(1.2)(500°\text{C/W})}$$

$$= 42 \text{ mW}$$

In most cases the two devices will not dissipate equal amounts of power. Before power thermal management calculations are performed in these cases, we need to know which semiconductor will be hotter. The power value, P_i, computed in Example 12-4 is required to determine this, and Example 12-5 illustrates the method that would be used.

EXAMPLE 12-5

Permitted Device Dissipations for an Opto-Isolator

Given: A coupler with the following specifications:

(1) Maximum temperature = 100°C
(2) Thermal coupling coefficient = 0.2

(3) Thermal resistance to ambient = 500°C/W

Find: The maximum permitted phototransistor dissipation if the IRED dissipates 80 mW and the ambient is 35°C.

Solution: **1.** Estimate P_i.

$$P_i = \frac{100°C - 35°C}{500°C/W\,(1.2)}$$

$$= \frac{65°C}{500°C/W\,(1.2)}$$

$$= 108 \text{ mW}$$

Thus the IRED will be the cooler device because its dissipation is less than P_i. If the IRED power had been greater than P_i, it would be the hotter of the two devices.

2. Calculate the maximum permitted phototransistor dissipation.

$35°C = 500°C/W\,[P_T + (0.2)(0.08 \text{ W})]$

P_T will be the phototransistor dissipation.

$P_T \leq 114$ mW

Finally, a method of estimating the average dissipation for each semiconductor device is required. For the IRED, if current and voltage are constant,

$$P = I_d V_d \tag{12-6}$$

where P is the dissipated power
I_d is the IRED current
V_d is the IRED voltage caused by the IRED current

If the IRED were driven by a voltage pulse, the peak power caused by the pulse would be determined and then the power would be averaged using the techniques discussed in Chapter 11. The transistor dissipation is given by the product of the collector-to-emitter voltage and the collector current.

$$P = V_{CE} I_c \tag{12-7}$$

If the phototransistor is switched on and off, the switched power will be averaged, as illustrated in Example 12-6.

EXAMPLE 12-6

Estimation of Photo-Darlington Dissipation in a Circuit

Given: **(1)** $R_C = 100\ \Omega$

(2) $V_{CC} = 10$ V
(3) $I_C = 0$ to 50 mA peak
(4) $t_p = 100$ μs pulse width
(5) $f = 1$ kHz

Find: The dissipation of the Darlington transistor pair. Refer to Figure 12-11.

Solution: **1.** Note that when $I_C = 0$, $P_d = 0$.

2. When $I_C = 50$ mA,

$$V_{CE} = V_{CC} - I_C R_C$$
$$= 10\text{ V} - 50\text{ mA} \times 100\ \Omega$$
$$= 5\text{ V}$$

3. Calculate the peak power.

$$P = I_c V_{CE}$$
$$= 50\text{ mA} \times 5\text{ V}$$
$$= 250\text{ mW}$$

4. Calculate the average power.

$$P_{AVG} = 250\text{ mW} \times 100\ \mu\text{s} \times \text{kHz}$$
$$= 25\text{ mW}$$

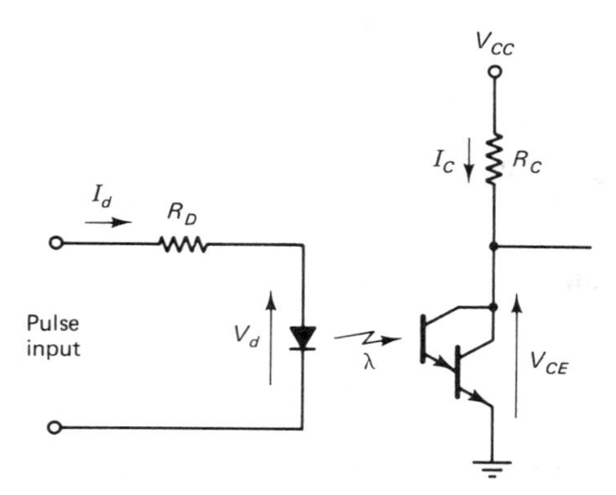

Figure 12-11 Opto-isolator components connected to an external circuit.

12-7 SUMMARY

Phototransistors are three-layer semiconductor devices that act as photodetectors. These devices will yield greater current per unit flux density than a photodiode of equivalent active area. Phototransistors are not particularly linear detectors, and they tend to be slower than photodiodes. Phototransistors can be used as switches and amplifiers, and are often used as detectors in opto-isolators.

PROBLEMS

1. A silicon transistor with an H_{FE} of 100 is used as a switch in a circuit with a collector resistor of 1 kΩ and a V_{CC} of 10 V. What is the saturation current?
2. Estimate the base current for the transistor in Problem 1 when the transistor is saturated.
3. Estimate the bias collector current, I_C, and the bias collector-to-emitter voltage for the circuit in Figure 12-9 if V_{CC} is decreased to 20 V.
4. Refer to Figure 12-11. I_C is specified as being equal to 60 mA when I_d is 10 mA. Let $R_D = 750\ \Omega$ and assume that $V_d \simeq 1.3$ V when the IRED is on. Estimate the collector current I_C when a 3-V voltage pulse is applied to the circuit input. Assume that transistor saturation is not a problem.
5. An opto-isolator has a rated maximum temperature of 100°C, a rated maximum thermal resistance of 150°C/W, and a rated thermal coupling coefficient of 0.25. Estimate the permitted transistor dissipation when the diode dissipation is 15 mW and the ambient is 40°C.
6. An opto-isolator is used in a circuit such as the one illustrated in Figure 12-11, where $R_D = 1000\ \Omega$. The voltage pulse applied to input has an amplitude of 5 V. A typical diode drop is 1.3 V. Estimate the IRED peak current and average power dissipation if the applied pulse has a 50% duty cycle.
7. For the circuit of Figure 12-11, estimate the average transistor dissipation for the following conditions:

$$V_{CC} = 10\ \text{V} \qquad R_C = 1000\ \Omega$$

$$\frac{I_C}{I_d} = 8 \qquad \text{Duty cycle 50\%}$$

$$I_d = 4\ \text{mA peak}$$

8. An opto-isolator has an IRED with an average dissipation of 15 mW and a transistor with an average dissipation of 55 mW. If the maximum rated operating temperature is 100°C, the thermal resistance is 300°C/W and the thermal coupling coefficient is 0.3. Estimate the maximum safe ambient temperature.
9. The isolation resistance of an opto-isolator is rated as being $1 \times 10^{10}\ \Omega$. If a surge voltage of 1000 V occurs on the IRED side of the package, estimate the surge current through the package. Assume that the detector side of the package is connected directly to ground.
10. With a 1000-Ω collector resistor the bandwidth of an opto-isolator is found to be 3 kHz. When the collector resistor is decreased to 500 Ω, the bandwidth increases to 6 kHz, and when the collector resistor is increased to 2000 Ω, the bandwidth decreases to 1.5 kHz. Estimate what the bandwidth will be if the collector resistor is decreased to 100 Ω. This is similar to the gain–bandwidth characteristics of the pyroelectric detector's amplifier discussed in Chapter 8.

13

Semiconductor Lasers

13-1 INTRODUCTION

In this chapter the semiconductor laser diode is discussed. Before starting this chapter you might want to review the general principles of lasers discussed in Chapter 6. In this chapter we introduce the fundamental operating concepts relating to laser diodes. Laser diodes have very pronounced responses to changes in current and temperature, and these interactions are discussed in detail. A section is devoted to examining the data sheet descriptions of two infrared laser diodes and a high-power laser array. In that section temperature and current effects are illustrated. After this discussion of laser diode specifications and the effects of temperature on them, thermal and electrical control circuits are discussed. Special attention is paid to the operating characteristics of the semiconductor thermoelectric cooler. Finally, two simple laser diode applications are discussed.

13-2 FUNDAMENTAL IDEAS

Like the various lasers discussed in Chapter 6, the semiconductor laser system has three fundamental parts:

1. *Photon source.* Hole–electron recombinations in the semiconductor provide the photons.
2. *Feedback.* Photons are fed back into the recombination region by reflection, causing stimulated emission.
3. *Energy source.* In this case the source is a supply of electric current called injection current.

In semiconductor lasers the energy transitions in semiconductors take place between the valence and conduction bands. A common graphic method used to illustrate this process is shown in Figure 13-1. Line AA represents permitted conduction-band energies, and line BB represents permitted valence-band energies. The horizontal axis represents electron momentum. When the energy minimum of the conduction band is at the same value of momentum as the energy maximum of the valence band, the material is said to be a *direct bandgap material.* In a direct bandgap material, no change in momentum is required of an electron moving between the valence-band maximum and the conduction-band minimum. This facilitates the exchange of energy between the electrons and the feedback photons, which have energy but no momentum. Practical laser diodes are fabricated out of direct bandgap materials such as gallium arsenide and indium arsenide.

Indirect bandgap materials are materials in which the minimum energy of the conduction band is not at the same momentum as the maximum energy of the valence band. For the absorption or emission of photons to take place, the momentum difference must be obtained from the lattice vibrations of the material. Thus the exchange of energy between photons and electrons is an inefficient two-step process that is not suitable as a laser feedback mechanism. In electronics terms we could think of this situation as having too little loop gain to sustain oscillations. Indirect bandgap materials such as silicon, germanium, and gallium phosphide are not used to fabricate lasers.

Lines E_1 and E_2 illustrate two of many possible transitions in a direct bandgap material. When absorption takes place, an electron moves from line BB up to line AA. When radiation takes place, an electron moves down from line AA to line BB. These transitions up or down can take place only if there is a vacancy (a hole) for the electron to occupy. By appropriate doping and the applications of an electric field, the situation illustrated in Figure 13-1(b) can be created. In this case the low-

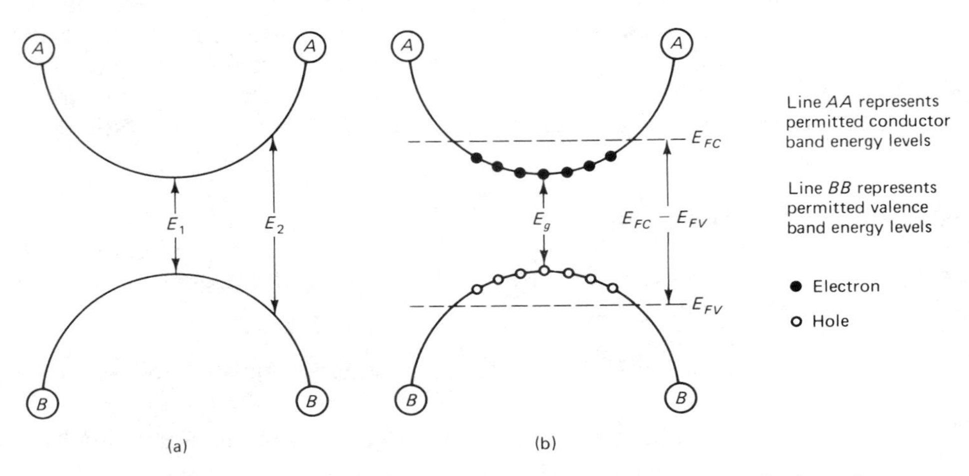

Figure 13-1 Semiconductor energy levels (vertical) and momentum (horizontal) in a laser diode: (a) two possible transitions E_1 and E_2; (b) gap energy E_g and Fermi energy boundaries E_{fc} and E_{fv}.

est energy levels of the conduction band are occupied by electrons and the highest energy levels of the valence band are occupied by holes. A population inversion has been created.

The energy level E_{fc} represents the highest energy level at which conduction band electrons will exist, and the energy level E_{fv} represents the lowest energy level at which holes will exist. These two boundary levels are called the *Fermi energies.* When a photon with an energy greater than the gap energy E_g and less than the difference between the Fermi energies passes through the PN junction, it can cause an electron to fall from the conduction band to the valence band. The energy released by the electron as it changes state will be the same as the energy of the stimulating photon, and thus we achieve stimulated emission.

Electrons will move down from the conduction band both spontaneously and as a result of stimulated emission. The external power supply acts to replace the electrons in the conduction band. At low current levels the spontaneous process dominates and the junction will radiate in a manner similar to an IRED. As the current is increased, more electrons are injected into the recombination region and the number of photons produced increases. The reflection mechanism feeds most of these photons back into the junction. The injection current electrons are then required to replace electrons that changed state due to stimulated emission rather than spontaneous emission. The observed result of this process is that the junction begins to radiate over a very small range of wavelengths, as in a laser.

The current level at which lasing action begins is called the *threshold current.* The magnitude of the threshold current is related to the materials used to fabricate the laser diode, the doping levels, the geometry of the junction, and the device temperature. As you will see, this last parameter, temperature, is a very important consideration when working with laser diodes.

The design of an injection laser diode must allow for both the injection or current into the junction and the reflection of photons into the junction. When reading articles and data sheets you will encounter the terms *homojunction, heterojunction, double heterojunction* (DH), and *buried heterostructure* (BH) to describe different types and arrangements of semiconductor layers.

Conceptually, a homojunction laser would look like the device illustrated in Figure 13-2. Here a single type of semiconductor material is doped P and N and a simple PN junction is formed. The designation *homojunction* refers to the fact that a single type of semiconductor material is used. Forward current flow is established across the junction from top to bottom. The ends of the device are cleaved smooth, forming mirrors that reflect photons back into the junction, providing the feedback required for laser action. While elegant in its simplicity, this device has one major drawback; at room temperature the current densities associated with laser action are so large that the device becomes a smoke-emitting diode rather than a laser diode. Homojunction laser diodes are, at the moment, laboratory animals that operate only when cooled with liquid nitrogen.

Practical laser diodes have single or double heterostructures. A heterostructure implies that more than one type of basic semiconductor material is used in fabricating the device. In a single heterostructure one side of the junction would be one type of semiconductor material doped to be P material, and the other side of the

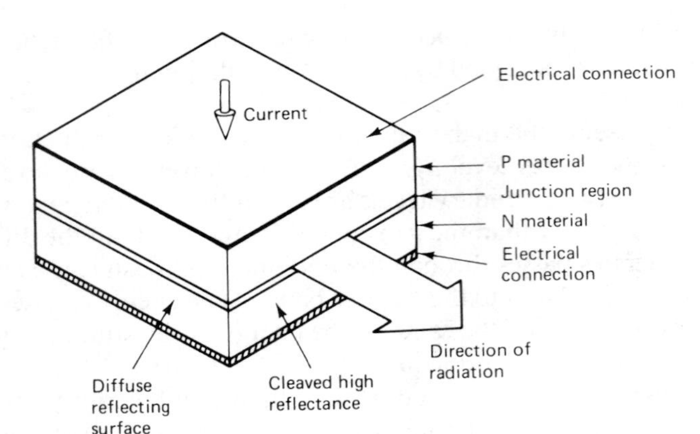

Figure 13-2 Homojunction PN laser diode.

junction would be a different semiconductor material doped to be N material. In a double heterostructure the material type is altered again after the junction is crossed.

Conceptualizations of single heterojunction, double heterojunction, and buried heterojunction layering systems are illustrated in Figure 13-3. The various layers can have thicknesses as small as 30Å in materials which have molecules that are 2 or 3 Å thick. It is unlikely that any layer would be much thicker than 1 μm. In earlier chapters it was emphasized that the spectral characteristics of quantum detectors and radiating diodes are established by the gap energies of the materials used to fabricate the device. This is also true for laser diodes, but the heterojunction

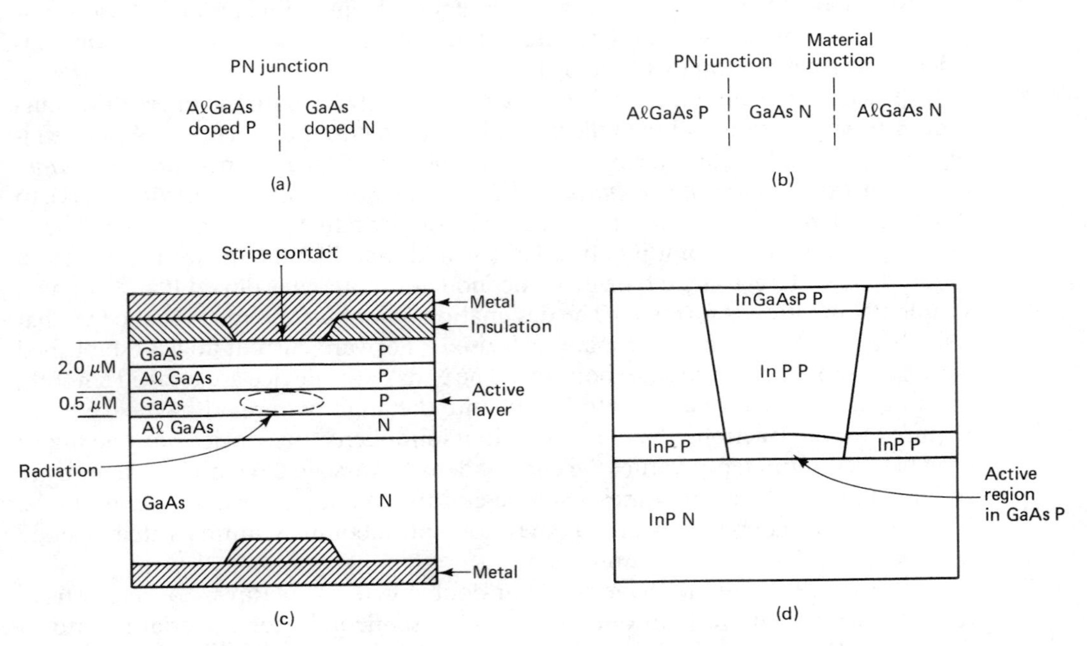

Figure 13-3 Heterojunction laser diode types: (a) heterojunction; (b) double-heterojunction; (c) strip contact double-heterojunction; (d) buried channel laser diode.

structure allows the engineer to fine-tune resultant bandgaps and thus the spectral characteristics of the laser. Individually, the different materials have different bandgap energies. Because the layers are so thin and in such close physical proximity, their electrical characteristics interact and establish well-defined and limited ranges of gap energies for the completed structure. The layers may extend across the entire width of the semiconductor chip, as in the homojunction device. It is possible, however, to create a small junction channel along the central axis of the device, thus creating a buried heterojunction, as illustrated in Figure 13.3(d).

In most practical semiconductor lasers, the radiation is emitted from small spots centered on the smooth cleaved ends. The implication of this is that the recombination region containing the lasing action is by some method confined to a channel along the central axis of the semiconductor device. The vertical boundaries of the lasing channel are determined by the semiconductor's layering structure. The active layer containing the channel will have a higher index of refraction than that of the surrounding layers. The optical energy is confined to the active layer in the vertical direction by reflection off the interfaces between the active layer and the adjacent lower index layers. The horizontal boundaries, the sides of the channel, may be established by changing the index of refraction of the active layer along the horizontal axis or by controlling the current flow path so that current flows only through the desired lasing region. Some devices use both approaches. The stripe contact laser of Figure 13-3(c) confines the spot to a channel in the center of the active region by confining the current flow path through the laser structure to the channel region.

The shape of the emitting spot may be rectangular or elliptical, and due to diffraction, the resulting beam will have an elliptical shape. These laser diodes are sometimes called *edge-emitting diodes,* where the reflecting side of the diode semiconductor chip is considered the edge. The emission pattern of a typical edge-emitting laser diode is illustrated in Figure 13-4.

Various types of optical systems are used to alter the expanding beam's shape and sometimes to collimate the radiation. These optical systems are often incorpo-

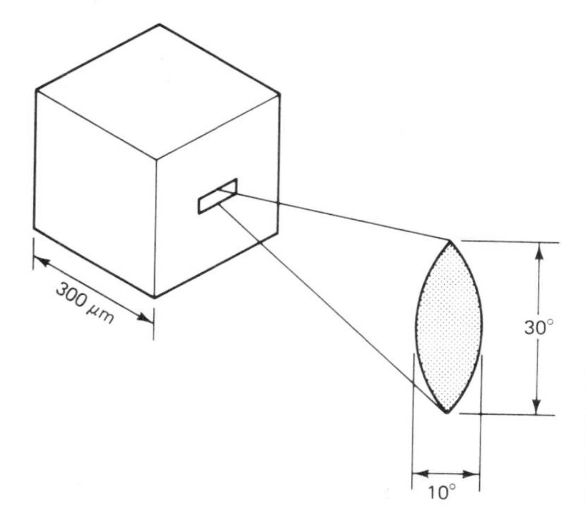

Figure 13-4 Typical elliptical beam pattern emitted from the edge of a laser diode. (*Note:* The largest angular diffraction is caused by the smallest dimension of the emitting surface.)

rated directly into the laser diode package. Some packages are designed for use with fiber optic cables. These packages will have a piece of optical fiber mounted in a manner that makes best use of the laser radiation. The user of this kind of diode package must splice a fiber onto the fiber "pigtail" coming out of the package.

In addition to feedback by reflection from the ends of the channel, a second kind of feedback, called *distributed feedback,* can be used. In a distributed feedback laser, the interface between two of the layers is etched so that it has a regular periodic pattern of undulations. This pattern distributes feedback into the lasing channel all along the channel's length, hence the name "distributed feedback." At the present time, this type of laser is being researched for use in analog and television video transmission systems and it is not widely available. Practical commercial lasers depend on the reflection from the cleaved faces of the semiconductor chip for feedback. Semiconductor materials have an index of refraction of 3.5 to 3.6, yielding a reflection of about 30% at each end of the laser channel when radiating into air.

The basic semiconductor laser is a single heterostructure or double heterostructure device. Current flow paths and layering structures are arranged to create a recombination region, a channel, along the central axis of the device. The cleaved ends of the semiconductor chip form mirrors that feed photons back into the recombination region. When the current flow is large enough, the stimulated emission caused by the feedback from reflections becomes the dominant photon-creating effect in the device's channel, and laser action takes place.

13-3 ELECTRICAL, OPTICAL, AND THERMAL INTERACTIONS

Injection laser diodes are "diodes" and have a voltage versus current characteristic very similar to that of an LED or IRED. The most common injection laser diode materials are GaAs and GaAlAs. GaAs and GaAlAs devices generally operate at wavelengths in the range 750 to 900 nm. Devices containing indium will operate at longer wavelengths in the range 1180 to 1580 nm, and there are laser diodes that will operate at visible wavelengths. A single laser diode has a turn-on voltage on the order of 1.5 to 2.5 V. Under pulsed conditions, short-term forward voltages of 5 to 25 V may be applied across a device. Some laser packages have multiple junctions connected electrically in series or in symmetrically arranged series parallel combinations. When properly cooled and pulsed, these multijunction packages can tolerate forward terminal voltages of several hundred volts.

Figure 13-5 illustrates curves of forward voltage and radiated power for a continuous-wave GaAlAs laser as a function of drive current, with junction temperature held constant. Notice that radiated power increases only very slightly, from 0 mA to about 18 mA. In this range of current values, the device is operating similarly to an IRED. Above 18 mA, radiated power begins to rise very dramatically with increases in input current. The diode is operating as a laser for currents larger than 18 mA. At this temperature and for this particular device, the threshold current is 18 mA.

The relationship between the change in radiated power and the change in drive current illustrated in Figure 13-5 appears to be relatively linear at this device tem-

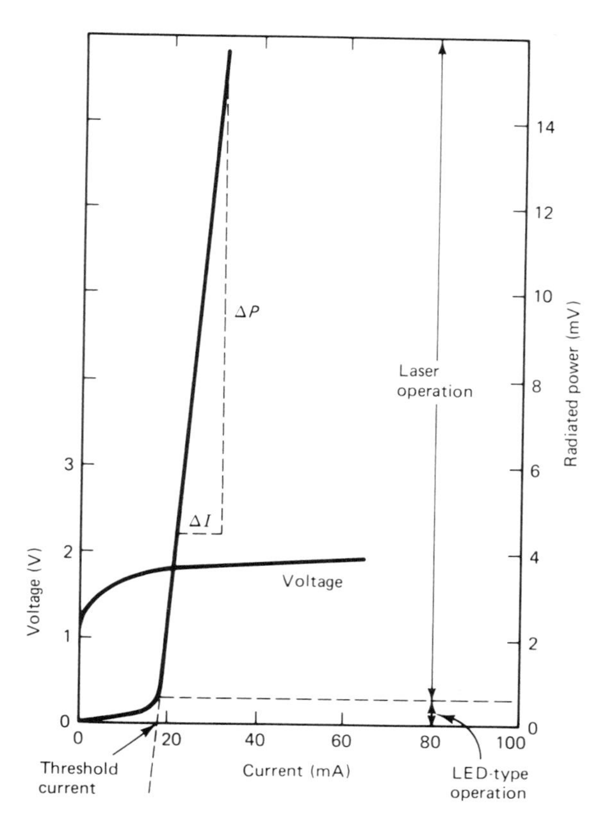

Figure 13-5 Laser radiated power and diode forward voltage versus laser diode forward current for a continuous-wave GaA1As laser diode.

perature for currents between 20 and 30 mA. The slope of this curve is called the *differential efficiency.* The linear appearance of the slope is deceiving, however; if the device were biased on 25 mA and driven with a sine wave of 5 mA peak to peak, you would find that the harmonic distortion would be at least several percent. In general, lasers are less linear transducers of current to radiation than are LEDs and are more suited to digital than to analog applications. It is hoped that distributed feedback lasers will overcome this linearity drawback.

Some older laser diode types exhibit a kink in the radiation versus current characteristics (Figure 13-6). This kink is caused by the laser changing TEM modes and/or by material defects in the channel region of the laser diode. The curve above the kink often exhibits a decreasing slope. This decreasing slope characteristic is called *power saturation.* This type of nonlinearity is undesirable even for digital applications.

By far the biggest problem in the current radiation relationships is the temperature sensitivity of the threshold current:

$$I_{th} = I_0 \exp\left(\frac{T - T_0}{K}\right) \tag{13-1}$$

where I_{th} is the required threshold current at temperature T

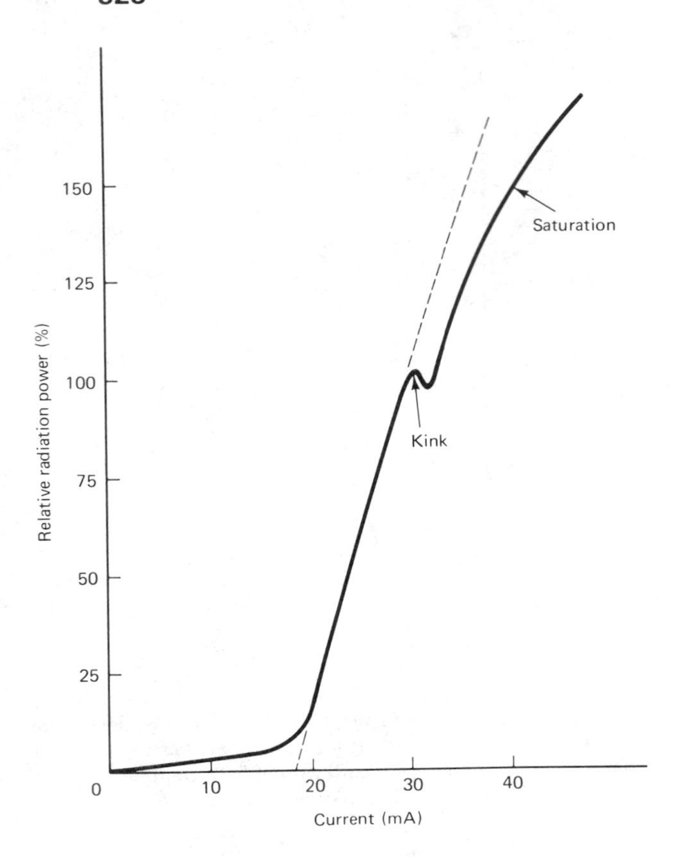

Figure 13-6 Laser diode relative radiated power versus diode forward current; curve illustrates kink and saturation.

I_0 is the required threshold current at temperature T_0
K is an empirical constant for the device with units of temperature

Equation (13-1) defines the relationship between temperature and threshold current, which is illustrated in Figure 13-7.

To understand the importance of this problem, a brief discussion of device application is required. Laser diodes are most often operated as pulsed devices. To obtain fast switching times and reliable radiated amplitudes, the diode is biased with a current at or above the threshold level and the signal current pulse is the additional increment of current required to achieve a specific level of radiated power. For the device described in Figure 13-7, the threshold current would be 25 mA at 25°C and the incremental current pulse required to achieve a peak radiated output of 6 mW would have an amplitude of 11 mA. Now imagine that for some reason the device heats up to a temperature of 80°C but the bias current remains constant at 25 mA. When the incremental current pulse occurs, the total current through the device increases to 36 mA, but this is below the required threshold current of 40 mA for a temperature of 80°C, so no laser radiation occurs. If the bias current had been increased to the required 40-mA threshold level, the 11-mA incremental current

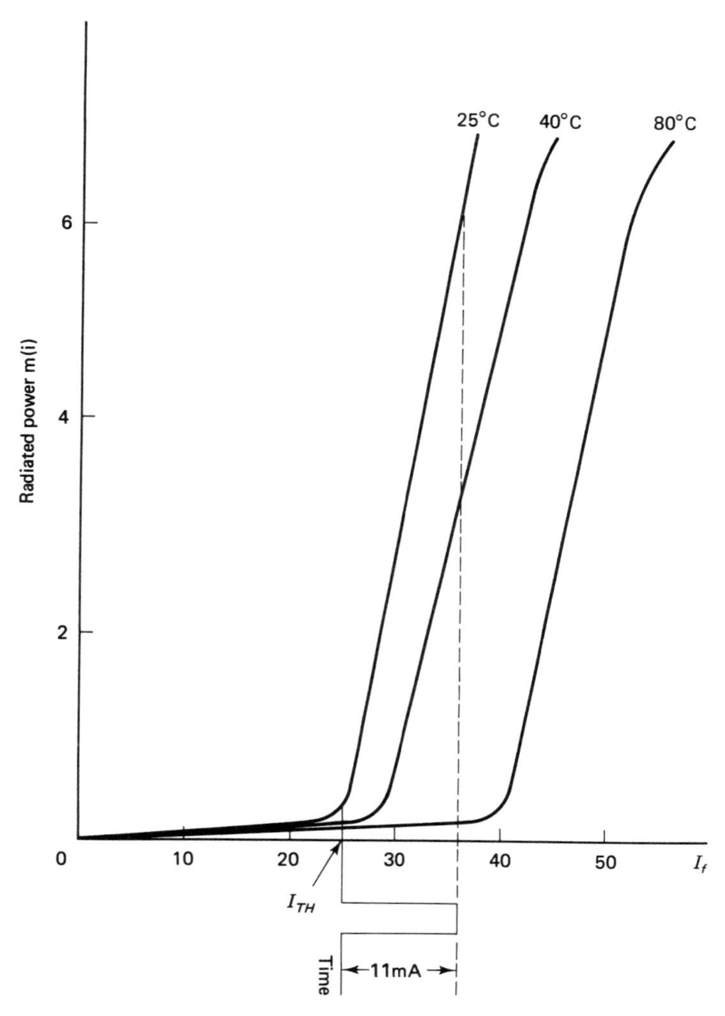

Figure 13-7 Illustration of changing values of threshold current as a function of device temperature.

pulse used previously would have caused a radiated pulse of 6 mW, just as it did at temperature of 25°C.

Reliable operation of a laser diode requires either that the temperature of the device be controlled, or that bias current be increased to the required threshold current for each device temperature. The major drawback of adjusting only bias current is that as bias current increases, the device gets hotter, requiring more bias current to achieve the same radiated output power. This type of design often ends up at a very high device temperature before it stabilizes at fixed temperature, bias current, and radiation levels.

Thus far, two relationships between radiated power and current have been established.

1. There is a minimum value of current, called the threshold current, I_{th}, which must be exceeded before lasing action will occur. The magnitude of this threshold current is related to the device temperature as described by equation (13-1).
2. There is an approximately linear relationship between radiated power and changes in device current above threshold, called the *differential efficiency*. This relationship is illustrated in Figure 13-5.

The spectral purity of the device's radiated output and the value of the dominant radiated wavelength are also dependent on the current through the device and the device's temperature. Two wavelength phenomena are noted as the magnitude of the current is changed:

1. At low current values, near threshold, the diode's radiated output will consist of a distribution of spectral lines like that illustrated in Figure 13-8(a).
2. As current increases the number of spectral lines decreases and the center of the distribution shifts to a longer wavelength.

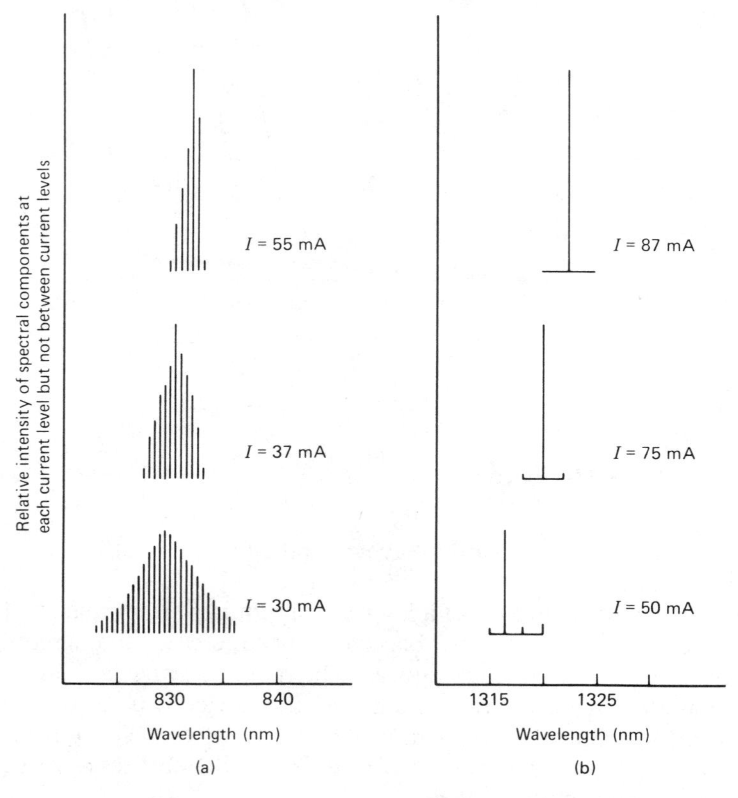

Figure 13-8 Variation of radiated spectra as a function of diode drive current: (a) gallium-based multimode DH laser diode; (b) indium-based single-mode buried channel laser diode.

These phenomena are illustrated for a common DH laser and a very fancy and expensive *single-mode laser* in Figure 13-8.

The magnitude of this central wavelength shift will vary significantly from laser to laser. Values anywhere between 0.1 and 20 nm are possible depending on device characteristics and the current magnitudes. Typical shifts of 2.5 nm occur between threshold and I_{max}. The spectral components observed at low current levels are a result of the interaction of the longitudinal modes of the laser cavity and the broad range of wavelengths permitted by the energy-band structure of the semiconductor. As injection current and radiated power increase, the feedback characteristics of the laser cause the power to be concentrated into fewer and fewer longitudinal modes, resulting in fewer spectral components. There is usually a point of diminishing returns with respect to spectral purity beyond which increases in drive current cause heating and shifts in wavelengths without any significant improvement in spectral purity.

When the laser is pulsed, the shift in operating wavelength and therefore in frequency as a function of increasing current is referred to as *chirping*. As the incremental current pulse increases in amplitude it causes the laser diode radiation to scan rapidly from the relatively broad spectral distribution at the shorter wavelengths to the relatively spectrally pure longer wavelength. This complex phenomenon degrades the shape of the radiated pulse, as illustrated in Figure 13-9. The radiated pulse can exhibit noticeable ringing, particularly if the bias current is set too low or if the diode is cut off prior to being pulsed. In general, the amplitude of the ringing will decrease as bias current is increased. Device fabrication techniques that lead to single-mode lasers also tend to minimize this ringing effect.

We should not overlook the fact that the semiconductor laser is a diode and that the same problems will be encountered when turning on a laser diode as are encountered when turning on any other diode. Two predominant problems encountered when turning on diode junctions with current and voltage pulses are voltage overshoot and current delay times at turn-on and turn-off. When a diode that is off is turned on by a large current pulse, the voltage across the diode can exhibit a pronounced peak near the beginning of the pulse; some diodes even exhibit a damped oscillation; this is illustrated in Figure 13-10. Such a voltage response can

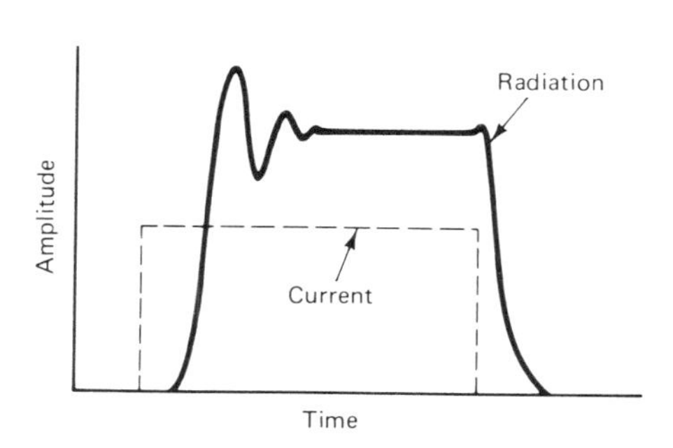

Figure 13-9 Radiation pulse from a "cutoff" laser diode in response to a current pulse. (Note the radiated pulse delay and ringing.)

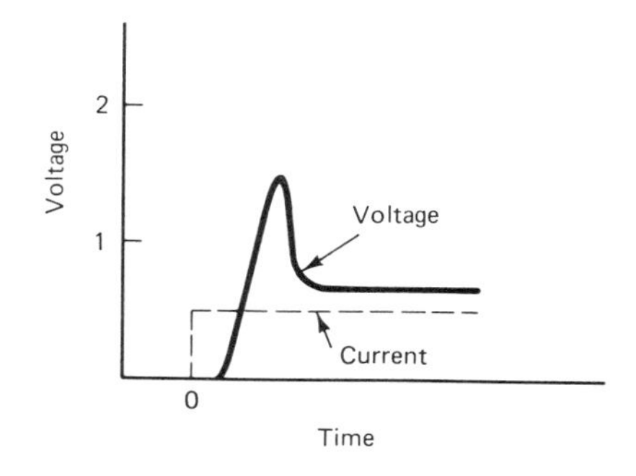

Figure 13-10 Voltage across an initially OFF diode junction in response to a current pulse from a current source.

make it difficult for the electronic driver circuit to maintain a constant current.

If a turned-off diode is turned on by a voltage pulse, a time delay occurs before current starts to flow, and diode current will continue to flow for some time after the voltage pulse shuts off. This is illustrated in Figure 13-11. Additional considerations for the laser diode are the turn-on and turn-off delays of the radiation with respect to the current flowing through the diode.

All of these delays are dependent on the physical displacement of electrical charge that must take place in the diode before current flow and radiation can occur. If the diode bias current is sufficient to bias the diode above the knee voltage on the diode forward voltage and current curve, these delays can be minimized. Similarly, the diode voltage overshoot or ringing can be diminished by biasing the diode ON.

Figure 13-5, discussed previously, illustrates both diode voltage and radiated output as a function of current. At the temperature where these data were taken, the lasing threshold current is about 18 mA, and this current falls on the diode forward voltage curve above the nonlinear portion of that curve. The good news is that practical values of threshold current will significantly decrease the various delay and overshoot effects, particularly if the drive circuit is a good current source. At very high switching speeds, however, small residual delays will become significant, and they can be attenuated only by increasing bias current above threshold, or by changing devices, or both.

The speed of response of a given laser can be "fine tuned" by selecting bias currents above the minimum required threshold current. In a recent experiment using sine-wave modulation, it was found that by increasing the bias current from $2I_{\text{th}}$ to $6I_{\text{th}}$, the sine-wave cutoff frequency of the diode increased by a factor of 1.7.

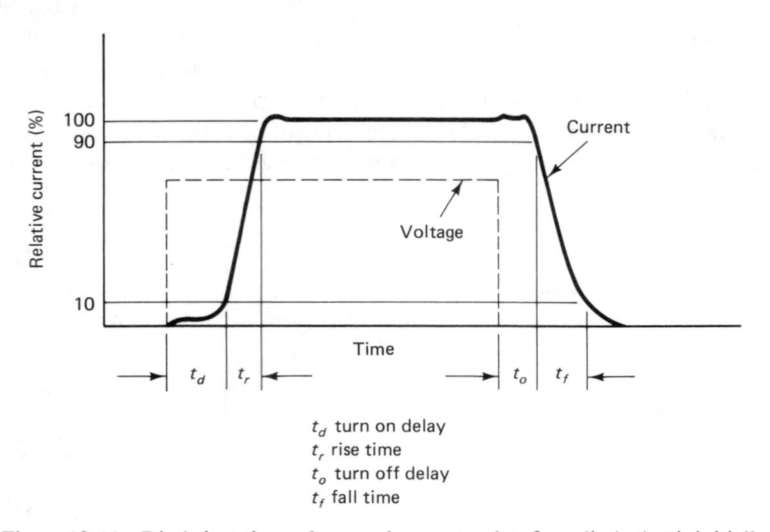

Figure 13-11 Diode junction voltage and current pulses for a diode that is initially OFF driven from a voltage source.

For this experiment, the magnitude of the sine-wave cutoff frequency factor was approximately proportional to the square root of the ratio of the bias currents.

As a sort of "mini-summation" we can conclude that laser diodes must be biased at the threshold current in order to obtain a constant radiated pulse amplitude, and that the device performance can be optimized for some particular switching speed by adjusting the bias current to a level above the threshold current.

There are several specific noise problems associated with the use of laser diodes. Two of the most common are discussed here. One is associated with the spectral characteristics of the diode arising from its longitudinal modes and is called *modal noise.* The other noise problem is caused by radiation being reflected back into the laser diode channel from outside the diode and is called *reflection noise.* Modal noise problems are caused by the spectral characteristics of the laser diode. A measured spectrum is a picture of the average spectral output of the laser taken over a very long time period compared to the rate at which changes can take place in the diode. There are two different ways in which the displayed spectrum can be interpreted:

1. The laser is a multimode radiator; it can be thought of as radiating simultaneously all the spectral components displayed at the relative amplitudes displayed.
2. The laser is a single-mode device; it radiates only one wavelength at any given instant. The individual spectral components observed represent a summation of all the different single wavelengths the device radiated during the period of observation. The amplitudes of the different spectral components represent their frequency of occurrence during the observation period. Large-amplitude spectral components are components that occurred frequently and low-amplitude components occurred infrequently.

The single-mode wavelength changes described under (2) above are called *mode hops.* Mode hopping can cause amplitude variations at the optical system's output due to the different attenuations and propagation velocities that each individual wavelength will undergo. Mode hopping is a random phenomenon that can cause random variations in radiated amplitude, which produce amplitude noise.

The process taking place in the multimode laser, described under (1) above, is conceptually similar to that of the single-mode laser. In the multimode situation, multiple spectral components are always present, but the amount of power assigned to any given component can vary from instant to instant. The spectrum analyzer shows the time-averaged amplitudes of the components. As the spectral power distribution varies from one instant to the next, the optical system's output will also vary.

Reflection noise is the result of an entirely different set of circumstances. The output wavelengths and power of the semiconductor laser are determined by the interaction of the laser's reflection feedback mechanism and its internal photon-creating characteristics. If a reflection surface is established outside the laser that can feed significant energy back into the active region, a whole new set of performance criteria is established. This new external reflector can cause large shifts in power and phase, and may alter the wavelength of operation. Given the very small dimensions of the laser diode emitting surface, this feedback alignment is difficult to achieve and is often intermittent. It may occur for brief instants when the external reflecting surface is momentarily shifted into alignment by vibration or temperature-related mechanical movements. If potential reflections can be held to values of less than one-thousandth of the radiated power, this source of noise can be eliminated.

Laser diodes are new devices and their reliability is still a matter for study and conjecture. The following list covers most of the known factors that adversely affect a laser's useful life:

1. Handling during assembly and testing
2. Short-term externally generated electrical or magnetic pulses
3. Current levels
4. Temperature
5. Peak radiation levels
6. Inherent device degradation

Factors 1 and 2 are really two different descriptions of the same failure mechanism. The physical dimensions of these devices are very small; thus if a large voltage spike appears across the device or a large current pulse flows through it, the device can be severely damaged. The short-term static discharge inherent in the assembly, testing, and applications environment can all cause device failure or reduce device lifetime; therefore, persons, tools, and instruments should be grounded to prevent static discharges. The electrical drive circuits must be shielded and filtered against the electromagnetic pulses that occur in the applications environment.

The current or, more precisely, the current density through the diode directly affects laser diode lifetime. The useful life of a laser diode can be reduced by as much as a factor of 4 when the device current density is doubled. The relationship between temperature and useful life is less precisely known, but clearly, hot devices fail faster. This is discussed further in the next section.

High radiation levels apparently degrade the reflecting properties of the laser's reflecting surfaces. The surface actually appears to erode away. As the degradation proceeds, output power decreases and spectral content changes.

When the diodes are fabricated there are small defects in the materials of the laser channel, the various semiconductor layers, the reflecting surfaces, and the electrical contacts made to the device. With the passage of time, the stresses imposed by the temperature cycling of the package and radiation absorption in the material cause these defects to increase in size and effect. Therefore, we can conclude by saying that the useful life of these devices depends on device current, device temperature, and the underlying diode structure.

13-4 LASER DATA SHEETS

In this section we examine laser diode data sheets and illustrate the manner in which manufacturers describe and specify some of the phenomena that we discussed in Section 13-3. The data sheets of Figure 13-12(a) tabulate data for a typical high-power (40-mW) infrared laser diode. The first data table in Figure 13-2(a) provides a tabulation of the important characteristics of this diode. In the heading we find that it is a high-power (40-mW) device operating at a nominal 830-nm wavelength. The device exhibits a single TEM mode and is intended for use in digital disk systems.

The absolute maximum output power is rated as 40 mW with the case held at 25°C (see the little parenthetical note at the top right-hand side of each table). The reverse voltage lists two values: a 2-V level, which is the specified maximum safe reverse voltage for the laser diode, and 30-V level for the "monitor." This package contains both the laser diode and a PIN detector diode. The detector diode is called the monitor diode or the rear facet diode. The term *rear facet* refers to the fact that the detector diode is mounted inside the package behind the laser diode; the *front facet* of the laser radiates through the package window. The 30-V monitor voltage represents the maximum reverse bias voltage that can be applied to this PIN diode detector. Three temperature ratings are given; the −10 to +50°C operating case temperature is the most important from a systems standpoint.

The second table in Figure 13-12(a) lists typical diode operational specifications, all of which are made with a case temperature of 25°C. The threshold current and operation current specifications are instructive because they illustrate the wide variance that can occur for laser diodes of the same type at the same temperature. The forward laser diode voltage value can also vary measurably from device to device. The next specification shows that the wavelength specification might be more clearly specified as 830 ± 15 nm. Monitor current refers to the detector diode current that will occur at a fixed laser diode radiation level of 30 mW and with a detector reverse bias voltage of 15 V. The 100:1 difference between the acceptable minimum and maximum detector current values indicates that circuits using this detector current will have to provide some adjustment capability for the wide variance that can exist. The next six specifications describe the angular and position characteristics of the radiated beam. The last specification listed in the table is the differential efficiency.

The angular data indicate that the radiated beam is about 2.8 times as wide in the perpendicular direction as it is in the parallel or horizontal direction. This is the classic elliptical radiation pattern. The very thin top-to-bottom dimension of the radiating layer causes the greatest diffraction of the beam, in this case typically 27°.

The third table in Figure 13-12(a) lists the detector's sensitivity, dark current, and capacitance with a 15-V reverse-bias voltage. These photodiode specifications were all discussed in Chapter 10. The graph numbered 90-1 in Figure 13-12(b) illustrates the forward current and voltage characteristics at three different temperatures. The break in the forward voltage versus current curve occurs at a value of about 1.5 V, which is typical for GaAlAs diodes. At threshold current levels of 40 to

LT015MD/MF

Features

- High power (maximum optical power output: 40 mW)
- Wavelength: 830nm
- Single transverse mode

Applications

- Optical disk memories
- Medical apparatus
- Optical floppy disks
- Optical memory cards
- Information processing equipment

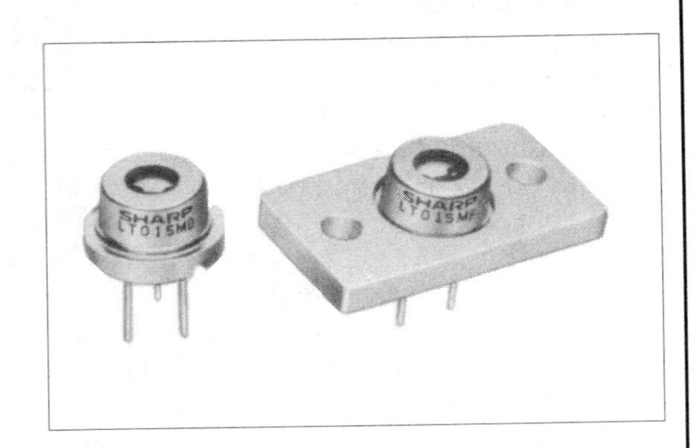

Absolute Maximum Ratings

(Tc=25°C)

Parameter		Symbol	Ratings	Units
Optical power output		Po	40	mW
Reverse voltage	Laser	V_R	2	V
	PIN		30	
Operating temperature*1		Topr	−10~+50	°C
Storage temperature*1		Tstg	−40~+85	°C
Soldering temperature*2		Tsol	260 (less than 5 seconds)	°C

*1 Case temperature *2 At point 1.6 mm from lead base

Electro-optical Characteristics*1

(Tc=25°C)

Parameter			Symbol	Condition	Ratings MIN	TYP	MAX	Units
Threshold current			Ith	—	—	60	80	mA
Operating current			Iop	Po=30mW	—	95	130	mA
Operating voltage			Vop	Po=30mW	—	1.75	2.2	V
Wavelength*2			λp	Po=30mW	815	830	845	nm
Monitor current			Im	Po=30mW V_R=15V	75	250	750	μA
Radiation characteristics	Angle*3	Parallel to junction	θ//	Po=30mW	8	9.5	14	deg
		Perpendicular to junction	θ⊥	Po=30mW	20	27	38	deg
	Ripple			Po=30mW	—	—	±20	%
Emission point accuracy	Angle		Δφ//	Po=30mW	—	—	±2	deg
			Δφ⊥	Po=30mW	—	—	±3	deg
	Position*4		Δx, Δy, Δz	—	—	—	±80	μm
Differential efficiency			η	$\frac{20\text{mW}}{I_F(30\text{mW})-I_F(10\text{mW})}$	0.5	0.8	1.1	mW/mA

*1 Initial value
*2 Single transverse mode
*3 Angle at 50% peak intensity (full width at half-maximum)
*4 Not specified for LT015MF

Electrical Characteristics of Photodiode

(Tc=25°C)

Parameter	Symbol	Condition	Ratings MIN	TYP	MAX	Units
Sensitivity	S	V_R=15V	—	8.3	—	μA/mW
Dark current	I_D	V_R=15V	—	—	150	nA
Terminal capacitance	Ct	V_R=15V	—	8	20	pF

(a)

Figure 13-12 High-power 40-mW laser diode data sheets: (a) specifications; (b) operation characteristics. (Courtesy Sharp Electronics Corporation, Mahwah, NJ.)

LT015 Series Characteristics Diagrams

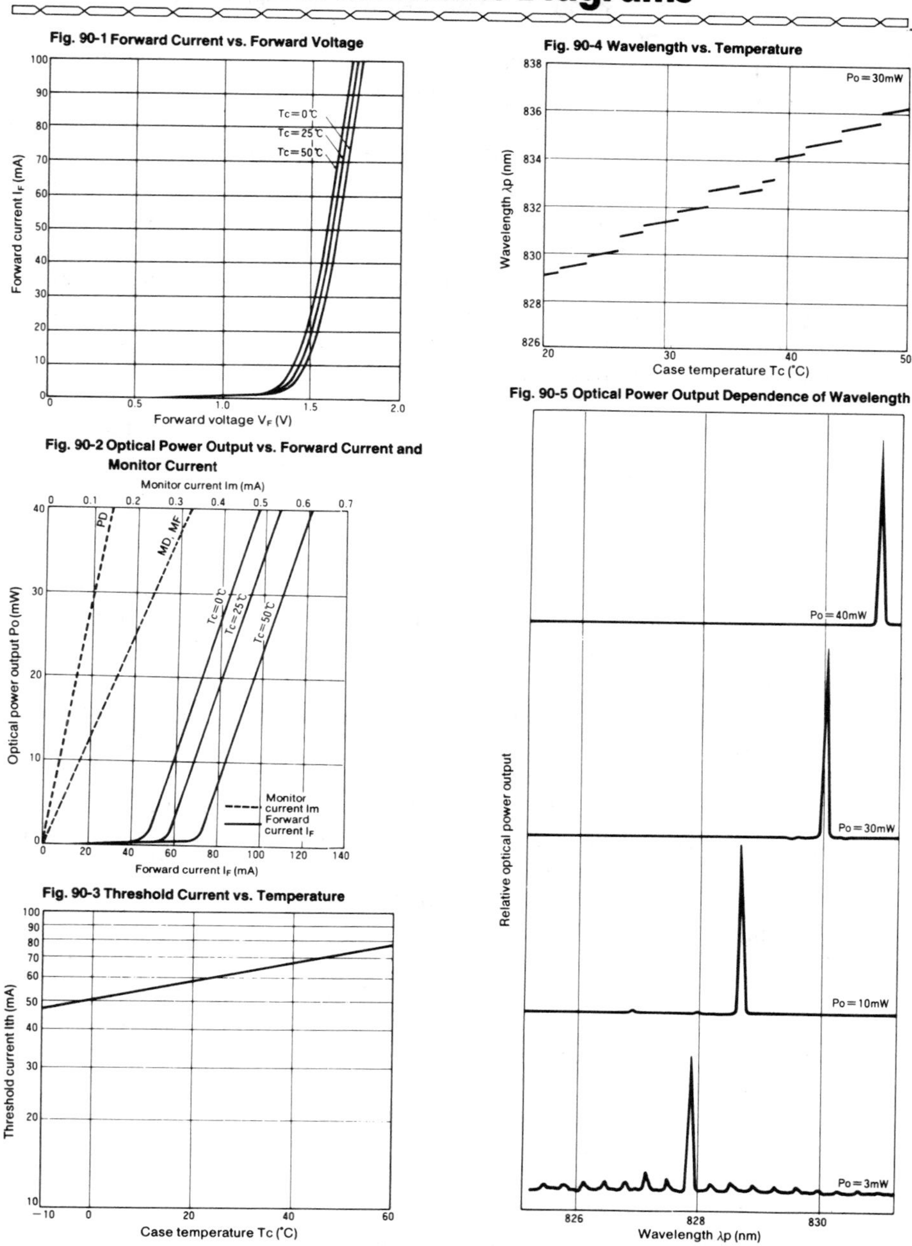

Note: All data on this page is typical only, and is not intended as a specification. The shapes of these curves can be used as a general reference, but the actual characteristics will vary from device to device.

(b)

Figure 13-12 *(continued)*

50 mA the forward voltage drop would be closer to the 1.75 V specified on the data table. There is a small decrease in forward voltage evidenced as temperature increases; this phenomenon occurs with all diodes.

The graph numbered 90-2 in Figure 13-12(b) illustrates both radiated power versus laser current (the solid dark lines) and detector diode current. The detector current (top axis of the graph), which is a function of radiated power, is described by the dashed line. The three solid graph lines describing laser diode operation clearly illustrate the increase in required laser diode threshold current as a function of case temperature. The linear current response of the detector diode to increasing radiation is also illustrated.

The graph numbered 90-4 in Figure 13-12(b) illustrates the change in wavelength as a function of case temperature. The graph numbered 90-5 illustrates the change in wavelength as a function of radiated power. Two important pieces of information are conveyed by these graphs:

1. As a laser diode's temperature increases, so does its radiated wavelength.
2. Over a typical operating temperature range of 20 to 50°C, the wavelength change for this diode will be about 5 nm.

The graph numbered 49-5 in Figure 13-13 illustrates the spectral characteristic of a low-power (5-mW) laser diode evaluated at four power levels. The multimode characteristic of the radiation is clearly evident at low radiated power levels, which would correspond to low laser diode current levels. For current levels capable of causing 5-mW levels of radiation, the device approaches a single longitudinal mode of operation. The high-power (40-mW) device discussed previously also operates in a single mode at all but the lowest (3-mW) radiated power level. At this low current and radiation level, the spectral graph [90-5 in Figure 13-12(b)] shows small sidebands similar in amplitude to those found on the low-power laser data sheets at the same radiated power level. Both of these data sheets illustrate the characteristic increase in wavelength as the current levels, and therefore the radiated power levels, increase. For these lasers the wavelength shifts are about 2 to 3 nm for the changes in radiated power level illustrated.

Figure 13-12(a) illustrates two package options. These two packages are very similar in size and shape to discrete semiconductor packages. One package option is a cylindrical device package mounted on a small rectangular heat sink. The heat sink is 20 mm wide and is slightly longer than $\frac{3}{4}$ in. The cylindrical device has a diameter of 9 mm, which is about $\frac{1}{3}$ in. The semiconductor chips inside the package have a frontal width of about 300 μm or about 12 thousandths of an inch.

These data sheets do not contain any reliability specifications. To illustrate reliability trends, the graph of Figure 13-14 was created by averaging reliability data from several manufacturers. The graph illustrates the relationship that exists between useful life and laser case temperature. Each line represents a different level of accumulated failures for this "average" laser diode. Accumulated failure levels can be interpreted as follows. Imagine that your company produced 100,000 laser diodes whose typical case temperature was 40°C under continuous use. In the first 192 days after product release, 500 angry customers will have called. By $2\frac{1}{2}$ years

LT020 Series Characteristics Diagrams

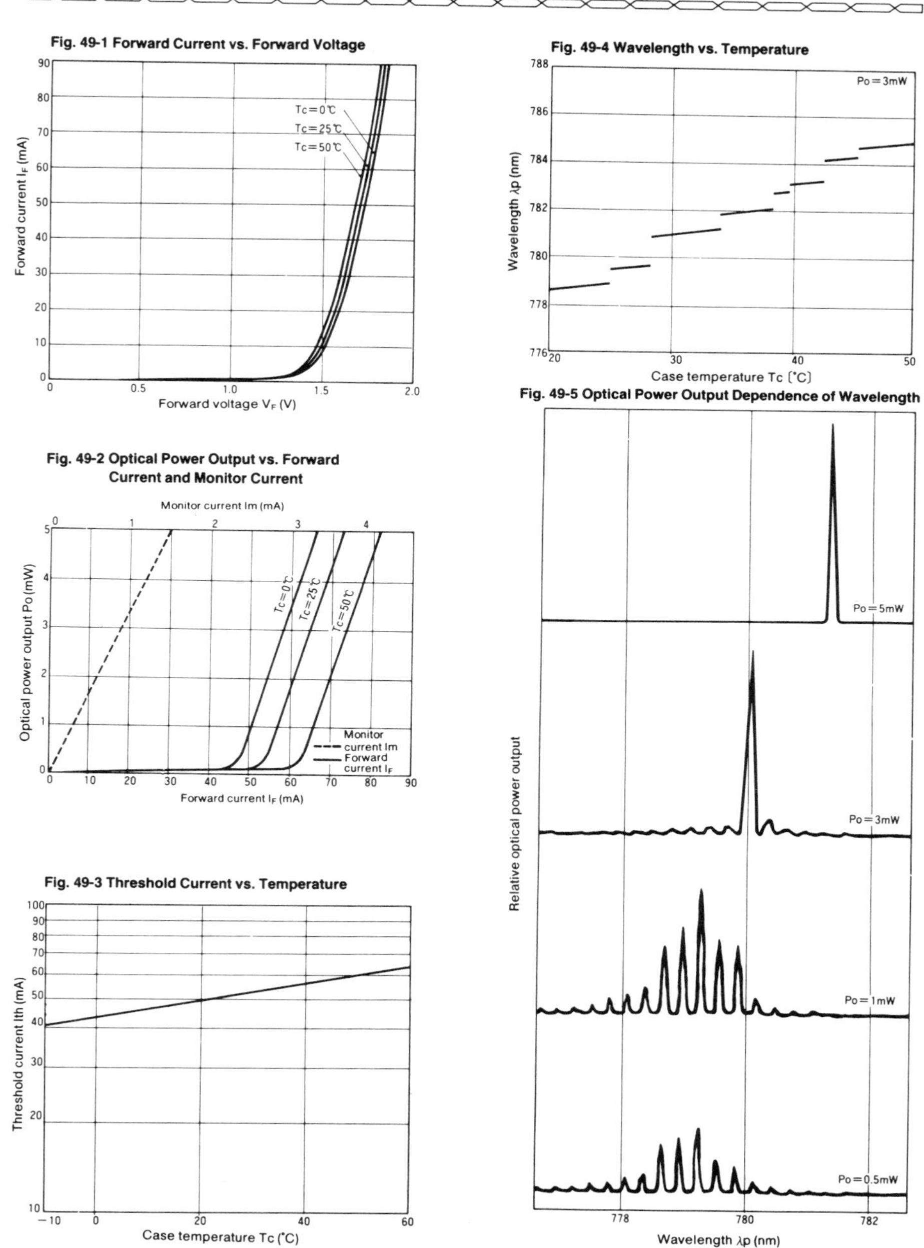

Note: All data on this page is typical only, and is not intended as a specification. The shapes of these curves can be used as a general reference, but the actual characteristics will vary from device to device.

Figure 13-13 Data sheet illustrating low-power (5-mW) laser diode operating characteristics. (Courtesy Sharp Electronics Corporation, Mahwah, NJ.)

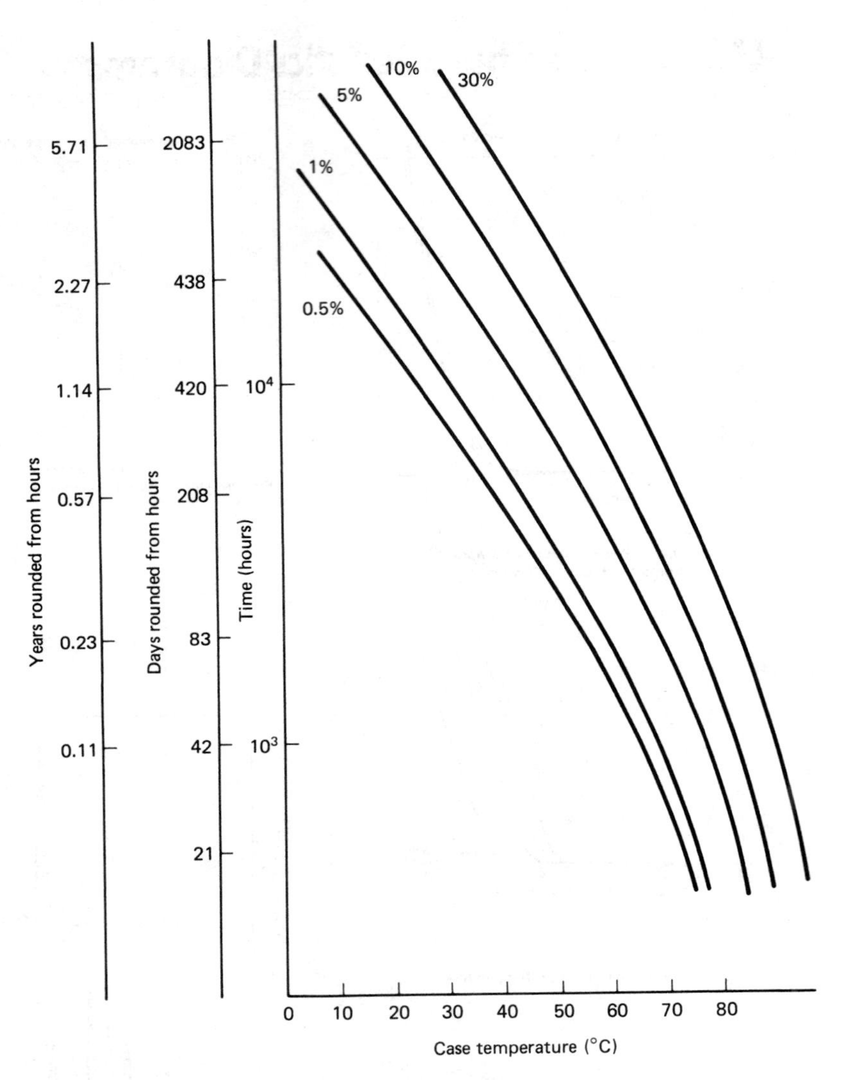

Figure 13-14 Average temperature failure data.

after product release, 10,000 angry customers will have called. As the graph illustrates, whatever level of accumulated failure you find acceptable, the time required to reach it is reduced by increasing the case temperature. As the technology matures, the useful life of devices at any given temperature should improve, but the general shape of the curves will be very similar to these. As the diode's average temperature is increased, its useful life is decreased.

Currently, most laser diodes are being used by the telecommunications industry or in digital disk applications. In these applications the ideal source would operate at a fixed power level and frequency and be coherent. There are other applications, however, where the net radiated power level is more important than coherence or spectral purity. An example would be using a semiconductor laser diode as a pump for a crystal laser structure. The devices described on the data sheets of Figure 13-15 are high-power laser diode arrays. These devices, containing multiple laser diodes in a single package, are capable of delivering short-term peak power levels in excess of 1000 W.

These devices packages are operated at much higher voltage, current, and power levels than we normally associate with semiconductor lasers. The lowest-power-level package listed on Figure 13-15(b) is the LD-210, which is specified as radiating 60 W when driven by a pulse with a peak power of 1125 W. This power level is achievable only when a very low duty factor of 0.02% is used. Converting peak power to average powers yields an average radiated power of 12 mW and an average electrical power of 225 mW. Duty factor and peak power level can be traded off against each other and radiated power levels between 12 mW and 60 W can be obtained. A nontrivial specification is the current pulse width of 200 ns given on Figure 13-15(a). Although we do not have thermal time constant data for these devices, it is reasonable to assume that a 200-ns pulse is short enough to prevent device overheating during the pulse period.

It is important to understand that the percent duty factor specification of 0.02% can be employed usefully in calculations of peak and average power only when the power pulse width is narrow compared to the thermal time constant of the laser diode. For these diodes, that value of pulse width is apparently 200 ns. A pulse with a duration of 20-μs and a frequency of 10 Hz would also have a duty factor of 0.02%. But a 20-μ pulse would be 100 times longer than the specified 200-ns pulse. A maximum rated power the 20-μs pulse would probably overheat the laser diode during the pulse period.

Laser diode arrays achieve high power levels by spreading the dissipation over a number of devices and by keeping the pulse width narrow and the duty factor small. The various devices described in Figure 13-15 have between 8 and 120 diodes in a package. The peak power radiated by any one diode is between 6 and 11 W and conversion efficiencies are between 3 and 5.5%. Temperature rise caused by self-heating is a problem and these arrays are mounted in packages that are readily attachable to a heat-sinking structure or which are themselves heat sinks.

Diode arrays can also be created that have all the laser channels on a single chip. These channels are aligned in parallel in close physical proximity. A small amount of optical energy is exchanged between adjacent channels. This process causes all the channels to radiate in phase at the same wavelength. Coherent radiated power levels from this structure can be much higher than from a single-channel device. Of course, this added performance is not free, as these additional channels make all of the temperature-related problems more severe.

13-5 LASER CONTROL

Bad things happen to good laser diodes when they get hot. Radiated power, wavelength, required current drive levels, and useful life are all changed, usually adversely, by temperature change. Because of this problem, many driver modules

FEATURING:

- HIGH EFFICIENCY AT LOW DRIVE CURRENTS.
- UP TO 1200 WATTS PEAK POWER OUTPUT.
- 904 NANOMETER PEAK EMISSION WAVELENGTH AT 25°C.
- CASE NEGATIVE, REVERSED POLARITY AVAILABLE.
- CUSTOM ARRAYS AVAILABLE.
- HERMETIC LDL-5 PACKAGE.

DESCRIPTION:

The LD-200, 300, 400 series are single heterostructure Gallium Arsenide injection laser diode arrays designed for pulsed operation. They offer peak output power from 50 to over 1100 watts. Emission wavelength is 904nm. Units can be selected to operate to 75°C and may be driven to 0.1% duty factor with the use of thermoelectric cooling of the package. The LD-200 series and the LD-430 come in hermetic TO-5 packages.

	Symbol	Min.	Typ.	Max.	Units
Wavelength of Peak Intensity	λ		904		nm
Spectral Width at 50% points	$\Delta\lambda$		3.5		nm
Rise Time of Radiant Flux — 10% to 90% pts.	Tr		< 0.5		ns
Pulse Width — 50% points at I_{fm}	Tp			200	ns
Storage Temperature	Ts	−196		+150	°C
Operating Temperature	Tc	−196		+75*	°C

*Selected units

(a)

Figure 13-15 Laser diode array data sheets: (a) general features; (b) device specifics. (Courtesy Laser Diode, Inc., New Brunswick, NJ.)

for laser diodes contain two control loops. One loop, the electrical control loop, is connected to the laser's electrical terminals. This circuitry is designed to protect the laser from destructive current and voltage pulses, to modulate laser current, and to adjust threshold current. A second loop is the thermal control loop, which is thermally connected to the case of the laser. In its most sophisticated form, this thermal control loop includes a semiconductor heat pumping device called a *thermoelectric cooler* or *Peltier device.* These thermoelectric devices are becoming a common component in laser diode head assemblies.

To better understand how these heat pumps work, consider the operating characteristics of the home refrigerator. On a hot summer day the kitchen (the ambient) is 95°F (35°C). The inside of the refrigerator will be 40°F (4.4°C). If you touch the refrigerator radiating coils, you will find that they are noticeably warmer

Type	Total Peak Radiant flux at max. rated I_{fm} (watts)		Typical emitting area (mils)	Total number of diodes	Typ. threshold current (amps)	Max. peak forward curr. (amp)	Typ. peak forward voltage (volts)		Duty factor %	Package
							@50ma	@I_{fm}		
	Min.	Typ.	Max.	#	I_{th}	I_{fm}	V_f	V_{fm}	Max.	
LD-210	50	60	100 x .08	10	7	25	12	45	.02	LDL-5A
LD-211	75	90	150 x .08	15	7	25	18	68	.02	LDL-5A
LD-212	150	180	160 x 25	30	7	25	36	135	.02	LDL-5C
LD-213	300	360	160 x 65	60	7	25	72	270	.02	LDL-5B
LD-214	60	85	156 x .08	12	10	40	14	65	.02	LDL-5A
LD-214S	60	85	45 x 45	12	10	40	14	65	.02	LDL-5E
LD-215	100	120	156 x .08	12	10	40	14	65	.02	LDL-5A
LD-215S	100	120	45 x 45	12	10	40	14	65	.02	LDL-5E
LD-220	200	240	156 x 25	24	10	40	29	130	.02	LDL-5C
LD-224-8S	150	170	45 x 45	8	18	75	9.6	55	.02	LDL-5E
LD-235	350	400	156 x 65	48	10	40	58	280	.02	LDL-5B
LD-330	300	380	115 x 65	36	10	40	44	235	**.02	Copper Block One Port
LD-430	300	375	70 x 70	40	10	40	48	240	.02	LDL-5B
LD-360	600	700	156 x 105	66	10	40	*40	*215	**.02	Copper Block Two Ports
LD-410	1000	1150	170 x 190	120	10	40	*28	*150	**.02	Copper Block Five Ports

* per port

** up to 0.1% with thermal electric cooling to maintain T_c at 25°C.

(b)

Figure 13-15 *(continued)*

than the room. The compressor-driven heat pumping system of the refrigerator pumps heat from the inside of the refrigerator to the radiating coils, which becomes hotter than the ambient and radiate the heat into the ambient.

A typical laser module containing a thermoelectric heat pump behaves similarly. The thermoelectric cooler pumps heat from the laser diode to the outside case of the laser diode head. This case becomes hotter than the ambient and radiates heat into the ambient. In this system temperatures can be measured at five locations:

1. The case of the laser
2. The heat-absorbing surface of the heat pump
3. The heat transfer surface of the heat pump
4. The case of the laser diode head
5. The ambient

The heat pump works to maintain the laser diode's case, location 2, at a fixed temperature (e.g., 20°C). The heat pump's heat transfer surface, location 3, will have a temperature that is the highest of the five temperatures. This maximum temperature will depend on the amount of heat being pumped, the ambient temperature, and the thermal resistance between this location and the ambient. Figure 13-16 illustrates these temperature relationships.

The thermoelectric cooler uses electrons to carry heat from the heat-absorbing "cold" surface to the heat-transfer "hot" surface. The thermoelectric cooler is fabricated with many P and N bismuth telluride semiconductor elements arranged as illustrated in Figure 13-17. The voltage source E drives electrons around the loop in the direction illustrated. As the electrons move from the P material to the N material, they must move from a low-energy state to a higher-energy state. The electrons absorb the energy required to make the change of state at the "cold" surface. The absorbed heat energy is expelled from the "hot" surface. The power required to move this heat energy is derived from the voltage source. Practical thermoelectric heat pumps use many P and N semiconductor elements connected in electrical series parallel combinations.

The amount of the electric power required to drive a semiconductor heat pump depends on the amount of heat being pumped, the temperature difference between the hot and cold surfaces, and the temperature of the hot plate. The graphs of Figure 13-18 illustrate typical characteristics of a low-power thermoelectric cooler. The drive current versus voltage curve [Figure 13-18(a)] illustrates that the current–voltage relationship is essentially linear for fixed-temperature conditions. The drive power versus pumped heat [Figure 13-18(b)] illustrates that the required drive power depends on the amount of heat being pumped and the temperature difference across the heat pump.

Figure 13-19 is a block diagram of a laser head assembly. The laser diode is driven by a current control network that responds to the modulation input signal and a feedback signal from a detector diode. The protection circuits shut the laser diode current off if modulation is removed and provide protection against transient electrical pulses. The temperature of the laser diode is controlled by a heat pump that responds to a temperature-sensing element mounted near the laser diode case. The control circuit actually maintains the sensing element at a fixed temperature.

A representative set of operating parameters for a laser head with this type of control system is illustrated in Figure 13-20. This particular control system and laser head were designed around an operating temperature of 20°C. The tempera-

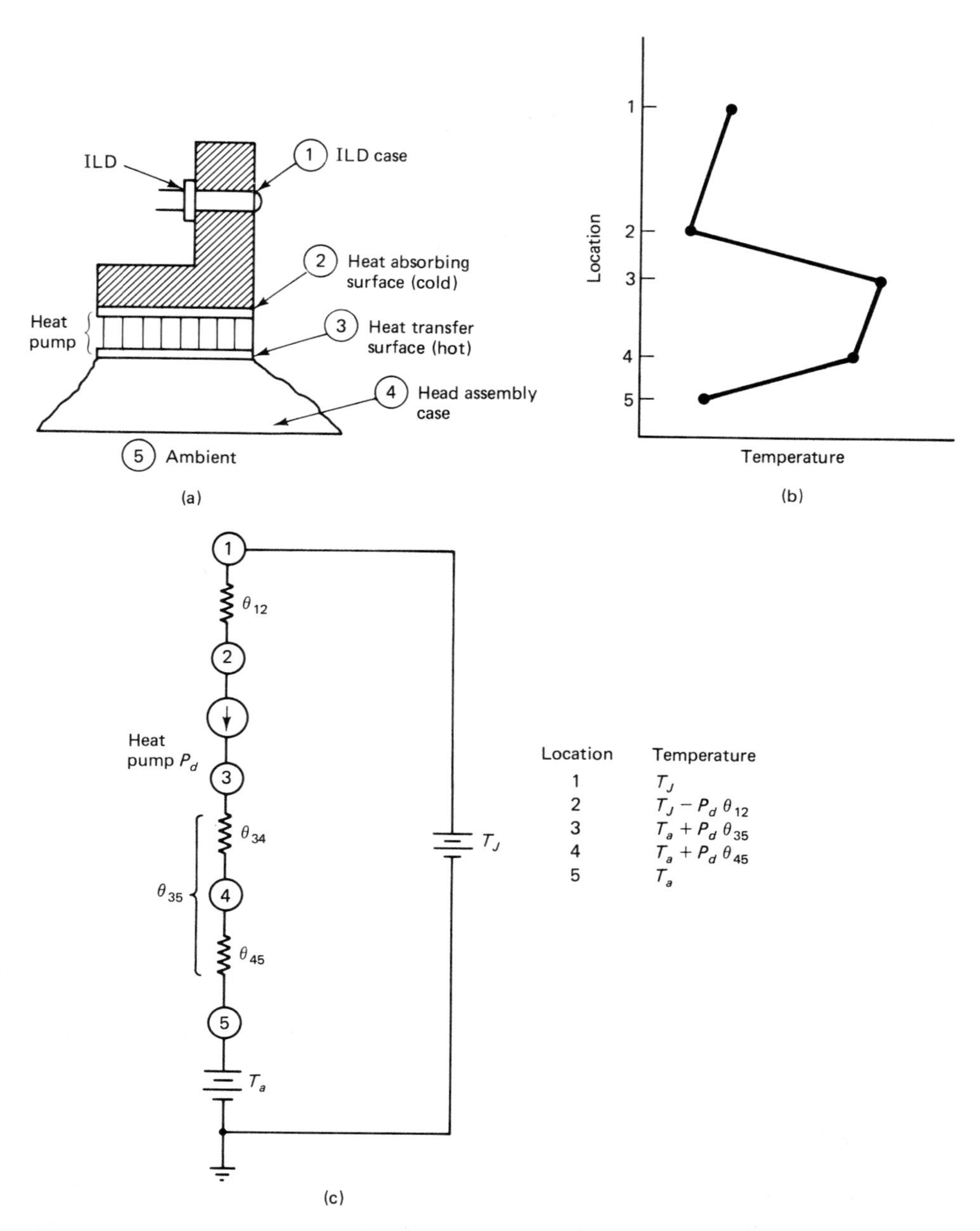

Figure 13-16 Diode assembly with heat pump: (a) mechanical assembly; (b) temperature distribution graph; (c) thermal schematic diagram.

ture used as an independent variable on the graph is the temperature of the outside of the laser head assembly, not of the laser diode. This temperature is directly proportional to the ambient temperature and the pumped heat. At high ambient temperatures the thermoelectric cooler is using a lot of current to pump the diode heat

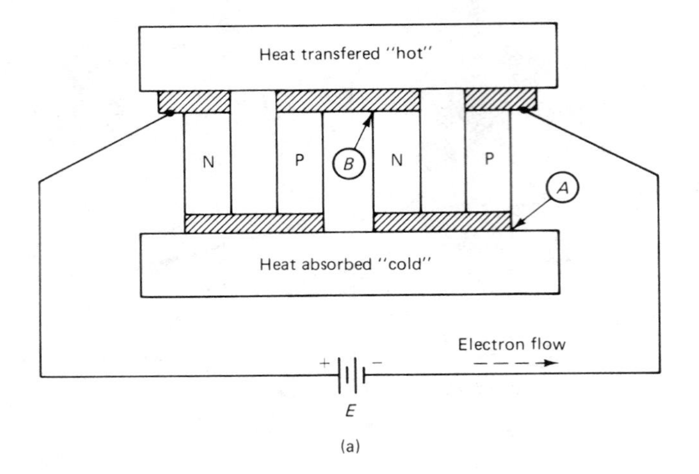

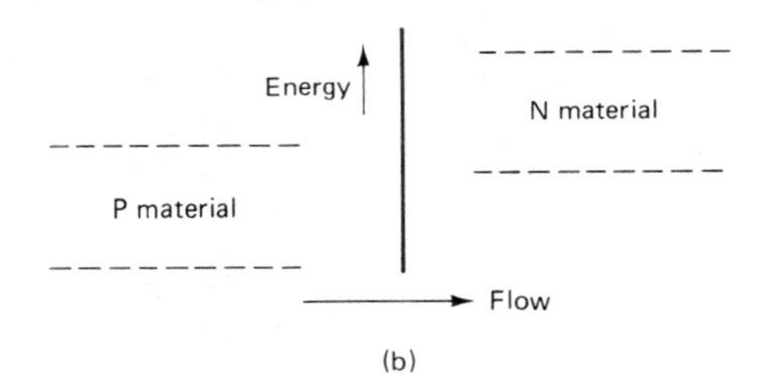

Figure 13-17 Thermoelectric cooler module: (a) diagram of module; (b) energy diagram for absorption, cold side; (c) energy diagram for radiation, hot side.

out because the temperature differential across the cooler is large. This keeps the temperature of the sensing element, a thermistor, constant, which in turn keeps the diode temperature almost constant. Apparently, the diode actually runs at a slightly lower temperature than the thermistor, so as the diode is cooled, radiated power tends to increase and the current feedback circuit responds by decreasing bias current. At low ambient temperatures the head and diode temperatures decrease as the ambient decreases, which results in a decreasing bias current. The resultant radiated output power is not perfectly constant over the temperature range shown, but it does charge slowly and in a linear manner. Laboratory-grade laser heads are

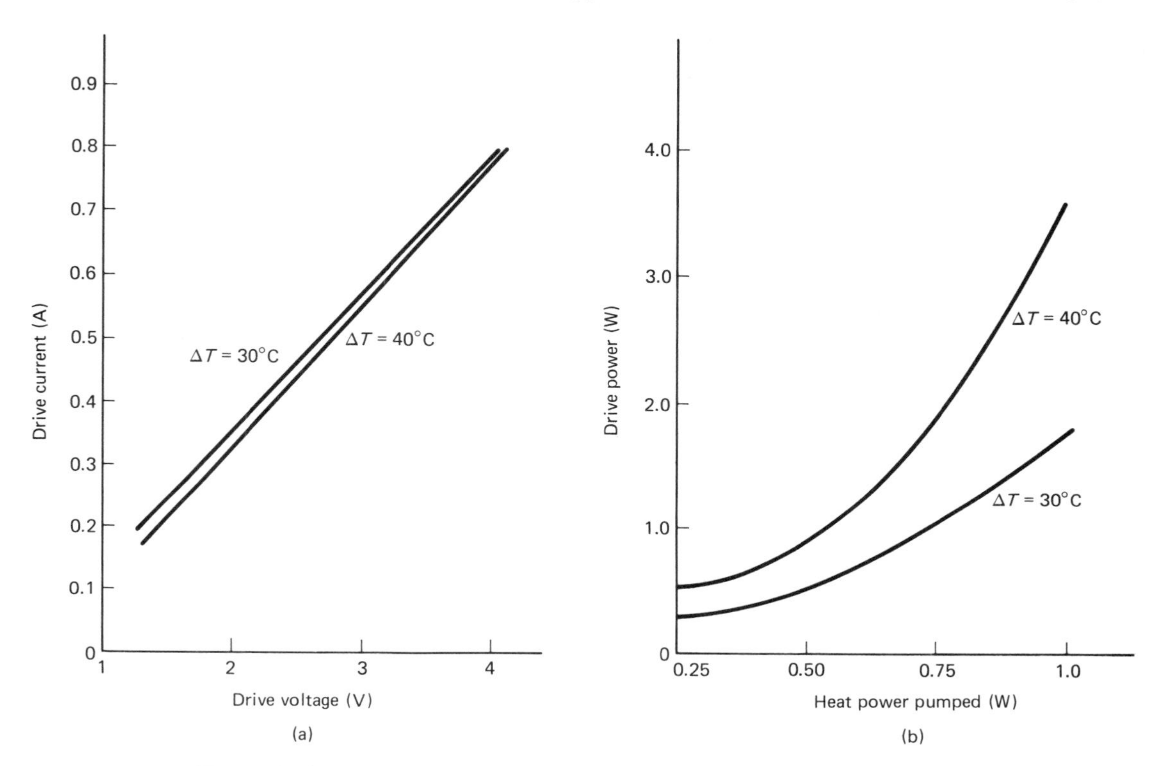

Figure 13-18 Thermoelectric cooler module operating characteristics: (a) current versus voltage; (b) drive power versus pumped heat.

available that control the laser diode temperature to within a few hundredths of a percent for ambient temperature swings of 10 degrees or more. The bias current graphs and radiated power graphs for that kind of a system would appear to be flat straight lines. The laser head assembly, current, and temperature control circuits are the methods used to keep bad things from happening to good laser diodes.

13-6 SEMICONDUCTOR LASER APPLICATIONS

Semiconductor lasers are used in fiber optic systems, optical disk systems, scanning systems, and a variety of instrumentation and research applications. Fiber optic applications are discussed in Chapter 14. In this section, an optical disk reader and a scanning system are discussed.

Figure 13-21 illustrates the optical components of an optical disk read head. This system reads digital data from the optical disk. The laser diode is operating in a continuous radiation mode at a constant average power level. This is achieved by either biasing the diode with a direct current of the appropriate level or by using a very high frequency signal to turn the diode on and off continuously. If the high-

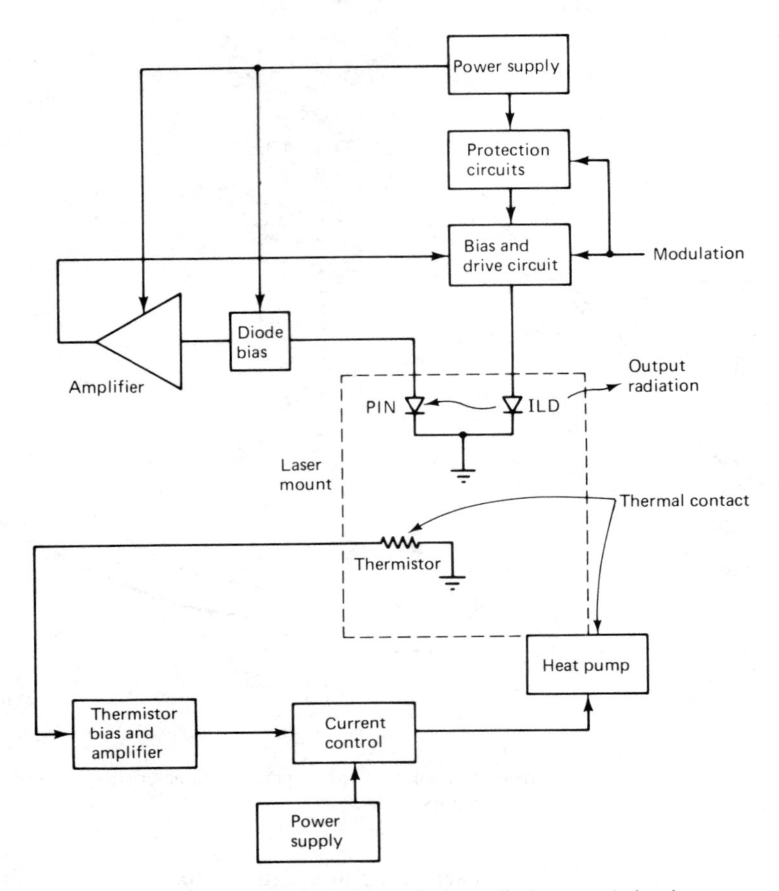

Figure 13-19 Block diagram of a laser diode control circuit.

frequency method is used, the switching frequency will be much larger than the frequency of the digital data. Some systems use a combination of direct-current bias and high-frequency switching to achieve the desired average radiated power output.

In the disk reading system, the laser beam is shaped and collimated, traverses a beam splitter, and is reflected down onto the digital disk through a polarization plate and a focusing lens. On the disk are regions of differing reflectivity representing digital ones and zeros. When a reflective surface is irradiated, the reflected radiation travels back up through the system and is directed by the polarization beam splitter onto the detector. The detector converts the reflected radiation into an electric signal. In addition to converting reflected pulses into electrical pulses, the diode provides information about beam position and focus. This is accomplished by using a detector containing four symmetrically arranged detectors. When these detectors are illuminated equally, they put out equal currents and the beam is correctly centered. If the illumination becomes asymmetrical, the detected currents become unequal and a control signal is generated that repositions the focusing lens.

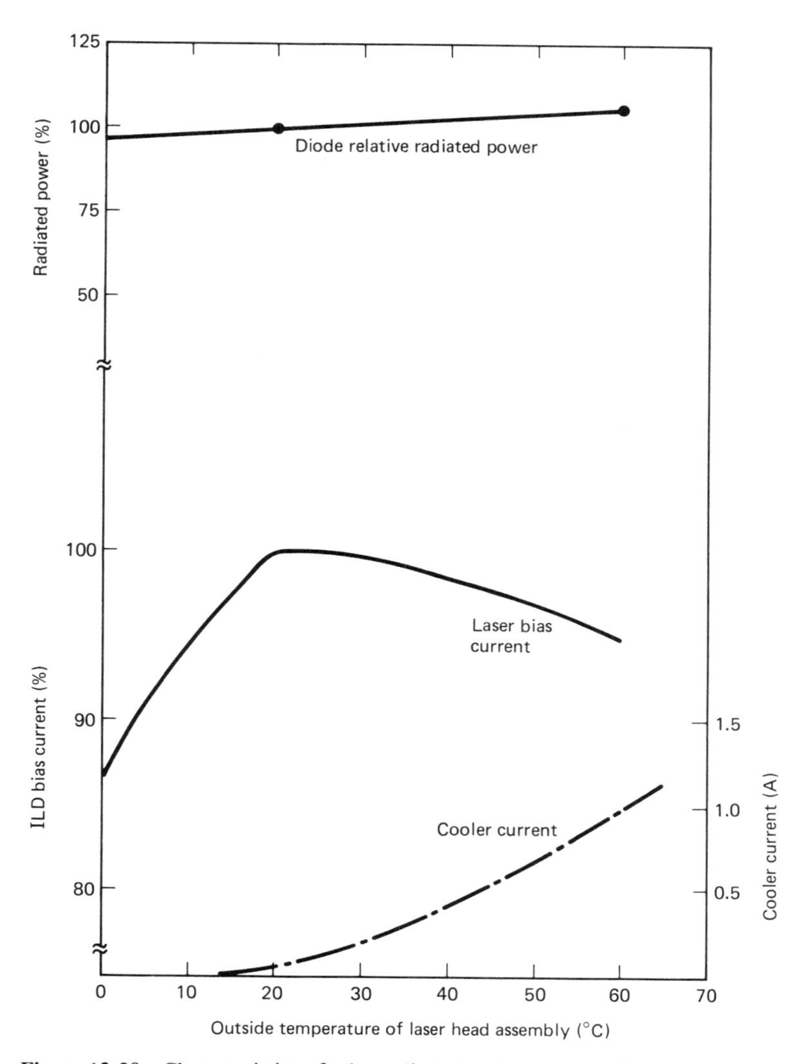

Figure 13-20 Characteristics of a laser diode head assembly with automatic current and temperature control.

Scanning systems sweep or scan a laser beam through space. Two type of systems that use this technique are laser printers and noncontact metrology (measurement) devices. Figure 13-22 illustrates a typical scanning system. The laser diode beam traverses a collimating and beam-shaping lens system that collimates the beam and changes its cross section from elliptical to circular. The beam is reflected off a rotating mirror and traverses another lens, which aims the sweeping beam coming off the mirror. In Figure 13-22 the scanning lens causes the sweeping beam pattern to cover an arc. This is the kind of setup that would be used in a laser

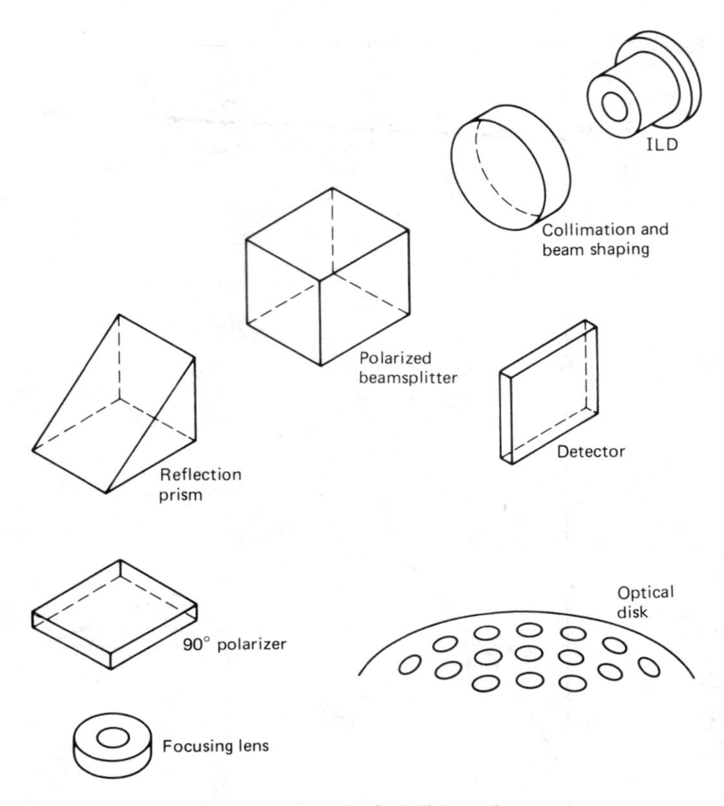

Figure 13-21 Optical disk reader.

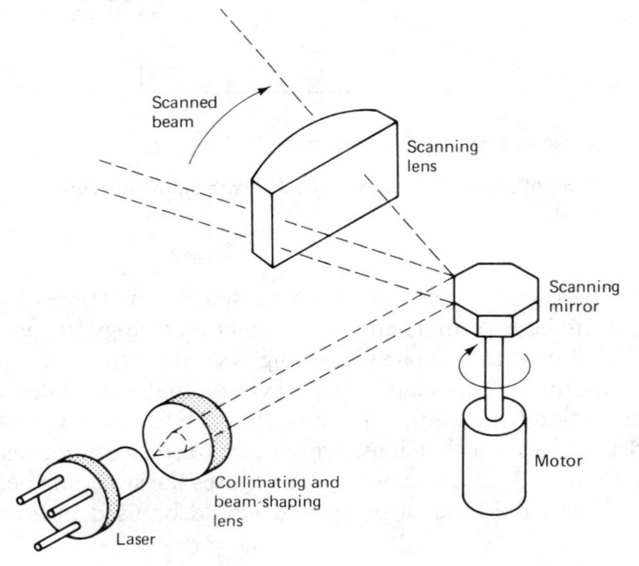

Figure 13-22 Laser scanner.

printer. The beam in a printer is amplitude modulated and light and dark spots are caused on the print drum.

In a measurement (meteorological) application a continuous-wave laser is used. Scanning lenses shape the swept beam into a straight parallel pattern. The part to be measured is inserted into the swept beam, where it blocks a portion of the beam. A detector array senses which portions of the beam are blocked and this information is used to calculate the part's dimensions.

The optical systems used with a laser diode can get quite sophisticated. Because of the small dimensions involved and the precision required of the systems, the positions of key optical components are often controlled by a feedback system. A common feature of laser diode optical systems is a set of components that convert the beam shape from elliptical to circular and which collimate the beam.

13-7 SUMMARY

Laser diodes are devices that provide the monochromatic, coherent radiation characteristic of all laser devices. Currently, individual laser diodes can provide only small amounts of average power. Higher power levels can be obtained by using arrays of individual diodes, but these devices do not have coherent outputs. Coherent arrays created by fabricating multiple channels on a single chip are another possibility. The wavelength, power level, electrical characteristics, and useful life of a laser diode are all affected by device temperature. A common approach to the temperature problem is to employ a thermoelectric cooler to maintain device temperature at a desired level.

Lasing action occurs only when device current exceeds a critical value called the *threshold current*. The required value of threshold current varies from device to device and increases with temperature. Drive circuitry for laser diodes maintains bias currents through the diode at values equal to or greater than the threshold value. The required bias current is commonly adjusted by a feedback system which senses the radiated output of the diode. The magnitude of the bias current affects radiated power levels, spectral purity, wavelength, and response times. Larger currents result in faster response, longer wavelengths, greater spectral purity, and higher radiated power. Large currents also cause greater self-heating of the laser, which must be controlled to obtain reliable performance.

Commercial laser diodes are fabricated from gallium- and indium-based semiconductor compounds. Gallium-related devices operate at wavelengths from 740 to 910 nm. Indium-related devices operate at wavelengths from 1180 to 1580 nm. Scanning, instrumentation, and digital disk applications use the short-wavelength devices to obtain high resolution. Visible-light semiconductor lasers are now coming out of the laboratory and into commercial channels for these applications. The fiber optics industry prefers devices operating at 1300 nm and 1550 nm for long-distance applications because these wavelengths experience low losses in long cables.

Materials are being researched continuously, and undoubtedly the number of material selections and wavelengths will increase as the technology matures.

QUESTIONS

1. How are photons created in a semiconductor laser?
2. Describe two types of feedback used in semiconductor lasers.
3. How are semiconductor energy levels different from the energy levels in a low-pressure gas?
4. What are the Fermi energies?
5. What is threshold current?
6. What is a double heterojunction structure?
7. What is a buried heterojunction structure?
8. How is lasing energy confined to the lasing channel?
9. Describe the beam shape radiated from a laser diode.
10. What is a typical range of continuous forward voltages for GaAs diodes?
11. What effects does raising current drive have on radiated power, wavelength, and spectral purity?
12. How can bias current be used to decrease delay times in a diode laser?
13. What effect does bias current have on modulation bandwidth?
14. Describe modal noise.
15. Describe reflection noise.
16. Can you infer from a spectrum analyzer display that the display components always have the relative magnitudes displayed?
17. What steps should be taken to prevent damaging laser diodes during handling?
18. What effect does temperature have on the useful life of a laser diode?
19. What effect does a high radiation level have on the laser diode structure?
20. Define accumulated failure rate.
21. Describe how a thermoelectric cooler works.
22. If you could maintain device's case temperature fixed and independent of the ambient:
    - **(a)** What would a plot of drive current and radiated power versus ambient temperature look like if the total time involved were 1 hour?
    - **(b)** What would a plot of drive current and radiated power versus time look like if the time period were 50,000 hours?
23. Describe how a digital disk reader works.

PROBLEMS

1. A laser diode has a threshold current of 50 mA at 25°C and a threshold current of 58 mA at 50°C. Calculate the empirical constant K. Calculate the threshold currents be at 0°C and 75°C?
2. Based on the experimental results described in Section 13-3, how much would you expect bandwidth to increase if bias current were increased from twice the threshold current to four times the threshold current?
3. Determine the size and shape of the beam image that would fall on a flat surface mounted 1 cm in front of a LT015MD laser diode.

4. An LT015MD laser diode is radiating with a wavelength of 830 nm at 25°C and current is controlled to maintain radiated power at 30 mW. What will the radiated wavelength be at 45°C?
5. Estimate the threshold current of a LT015MD at 45°C. Refer to diagram 90-3 in Figure 13-12(b).
6. Estimate the electrical power dissipation of a LT015MD when the case temperature is 25°C and the radiated power is 20 mW. What is the conversion efficiency?
7. What operating current would be required to maintain a LT015MD radiating at 30 mW with a case temperature of 50°C? What would the device dissipation and efficiency be?

14

Optical Waveguide

14-1 INTRODUCTION

Two major commercial systems have pushed developments of electro-optics in the last decade: digital disk and fiber optic communications. These two technological areas represent the major share of sales dollars for both components and systems. A great deal of the optoelectronics research and development effort for systems, components, and materials is devoted to improving the performance of these two systems. Many other areas of the technology have benefited from spin-offs of this research and development effort. In this chapter the general characteristics of optical waveguides and the special case of the optical fiber waveguide are considered.

An *optical waveguide* is a structure designed to confine the propagation of radiation within the boundaries of the structure. The radiation can be envisioned as splitting into a number of rays that follow different paths through the structure. The rays into which the radiation is split are called *modes.* A typical waveguide has a uniformly shaped central region, the *core,* which is fabricated from a homogeneous transparent material. The region surrounding the core must have a lower index of refraction than the core. The core and the surrounding material acting together form the wave-guiding structure. A major portion of the propagating energy moves through the core material with a small but finite amount (the "tail") extending into the surrounding material.

An optical fiber is a particular type of optical waveguide. An optical fiber is a thin cylindrical filament of flexible dielectric with a uniform cross section. The core material will have an elliptical or circular cross section. The material surrounding the core is called the *cladding.* The fibers are made of flexible glass, plastic, or glass and plastic combinations. A fiber optic cable has one or more fibers inside a single protective covering, like a multiple-wire cable. Fiber optic cables are used in a variety of medical, military, communications, and industrial applications. Bundles of optical fibers are also used as light guides and image guides.

Optical waveguides can also be formed in or on glass and semiconductor substrates. This type of waveguide is used to fabricate a whole variety of optical devices, the most common of which are couplers, switches, and phase shifters. A number of physical transducers have been developed using both types of optical waveguides.

14-2 PROPAGATION IN OPTICAL FIBERS

Electromagnetic energy is confined within the fiber core by reflection and refraction mechanisms. The electromagnetic energy from a source enters one end of the fiber and propagates through the fiber to the other end (Figure 14-1). When the energy can propagate by many different paths in the fiber, the fiber is said to be a *multimode fiber,* and the different paths possible represent the various permitted modes of propagation. If there is only one possible path for energy, that along the fiber's central axis, the fiber is said to be a *single-mode fiber* (Figure 14-2).

The most common fiber types are manufactured with cores having circular cross sections. The index of the core material n_1 must be greater than the index of the cladding n_2. Multimode propagation can be modeled using the ray optics concept of total internal reflection. Total internal reflection will occur when a ray hits the interface between the core and the cladding at an angle equal to or greater than the critical angle θ_c. Rays that hit the interface at angles smaller than the critical angle will transmit a portion of their energy across the interface at each successive reflection. Eventually, all of this energy will be lost. The following equation defines the critical angle in terms of the two indices of the fiber. For $n_1 > n_2$,

$$n_1 \sin \theta_c = n_2 \sin 90^\circ$$

$$\sin \theta_c = \frac{n_2}{n_1} \tag{14-1}$$

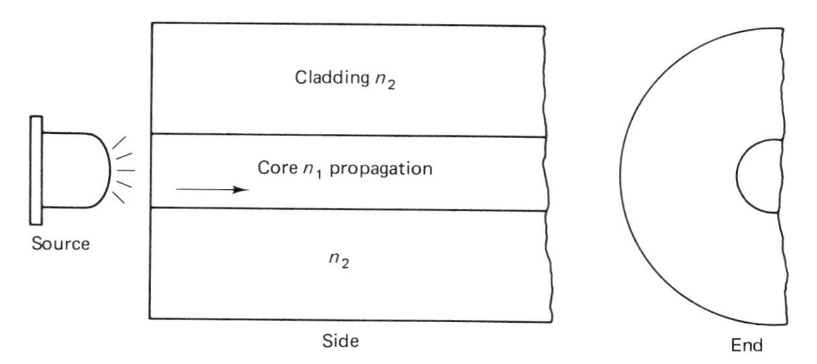

Figure 14-1 Optical fiber cross section.

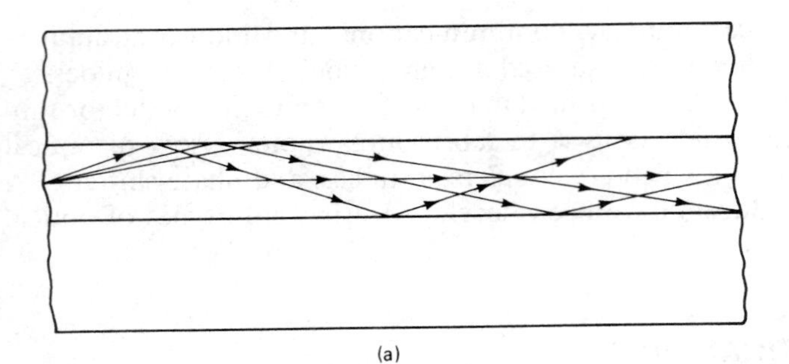

(a)

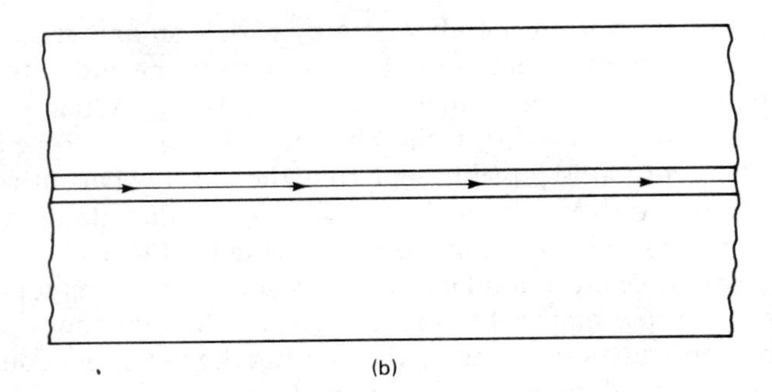

(b)

Figure 14-2 Multimode and single-mode waveguide: (a) multimode, multipath; (b) single mode, single path.

where n_1 is the core index
n_2 is the cladding index
θ_c is the critical angle

The source energy enters the fiber from an external medium with an index n_0. The angle at which a ray enters the fiber determines the angle at which it will hit the core–cladding interface. The entrance angle that just causes a ray to assume the critical angle is called the *acceptance angle.* Equation (14-2) relates the acceptance angle θ_a and the indices of the system to the critical angle θ_c.

$$n_o \sin \theta_a = n_1 \sin (90° - \theta_c) \tag{14-2}$$

or

$$n_0 \sin \theta_a = n_1 \cos \theta_c$$

Recall that $\cos x^2 = 1 - \sin x^2$; therefore,

$$n_0 \sin \theta_a = n_1 (1 - \sin \theta_c^2)^{1/2}$$

$$n_0 \sin \theta_a = n_1 \left[1 - \left(\frac{n_2}{n_1}\right)^2\right]^{1/2}$$

$$\sin \theta_a = \frac{1}{n_0}(n_1^2 - n_2^2)^{1/2} \tag{14-3}$$

and for the special case of air when $n_0 = 1$,

$$\sin \theta_a = (n_1^2 - n_2^2)^{1/2} \tag{14-4}$$

For small angles, $\theta_a < 20°$,

$$\theta_a = (n_1^2 - n_2^2)^{1/2} \qquad \text{rad} \tag{14-5}$$

$$\theta_c = \frac{\pi}{2} - \frac{\theta_a}{n_1} \qquad \text{rad} \tag{14-6}$$

Within the fiber the largest angle with respect to the central axis that a ray can assume and still cause total internal reflection is called the *maximum angle of propagation,* θ_p. This angle is the complement of the critical angle. All rays that successfully propagate through a long length of fiber by way of reflections that pass through the central axis of the fiber will do so at propagation angles between zero degrees and the maximum angle of propagation.

$$\theta_p = 90° - \theta_c$$

Therefore,

$$\cos \theta_p = \frac{n_2}{n_1} \tag{14-7}$$

and

$$\sin \theta_p = \left(\frac{1}{n_1}\right)(n_1^2 - n_2^2)^{1/2}$$

For small angles,

$$\theta_p = \frac{1}{n_1}(n_1^2 - n_2^2)^{1/2} \qquad \text{rad} \tag{14-8}$$

The sine of the acceptance angle is called the fiber's *numerical aperture* (NA). The numerical aperture specification is a way of describing the maximum external angle of incidence that will cause energy to be propagated within the fiber. For communication-grade fibers with small acceptance angles, the NA will be numerically equal to the acceptance angle in radians. Figure 14-3 illustrates the geometric relationships that exist among the acceptance angle, the critical angle, and the maximum angle of propagation.

There are three basic fiber types: step-index multimode, step-index single mode, and graded index (Figure 14-4). *Step-index fibers* exhibit a sharp change in index at the interface between the core and the cladding. This sharp change in index is called a *step change,* hence the name *step index. Graded-index fibers* have

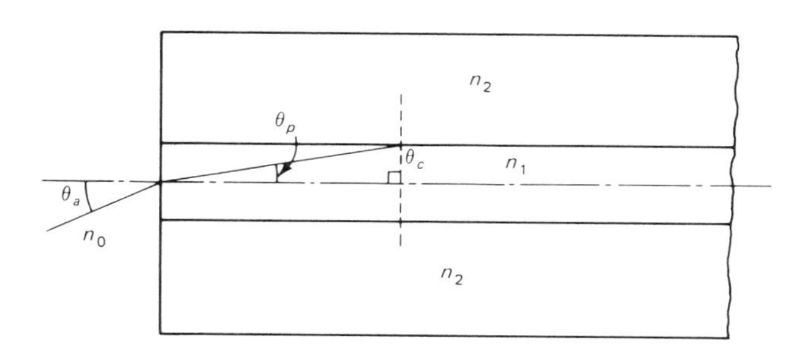

Figure 14-3 Ray angles. θ_a, acceptance angle; θ_p, propagation angle; θ_c, critical angle.

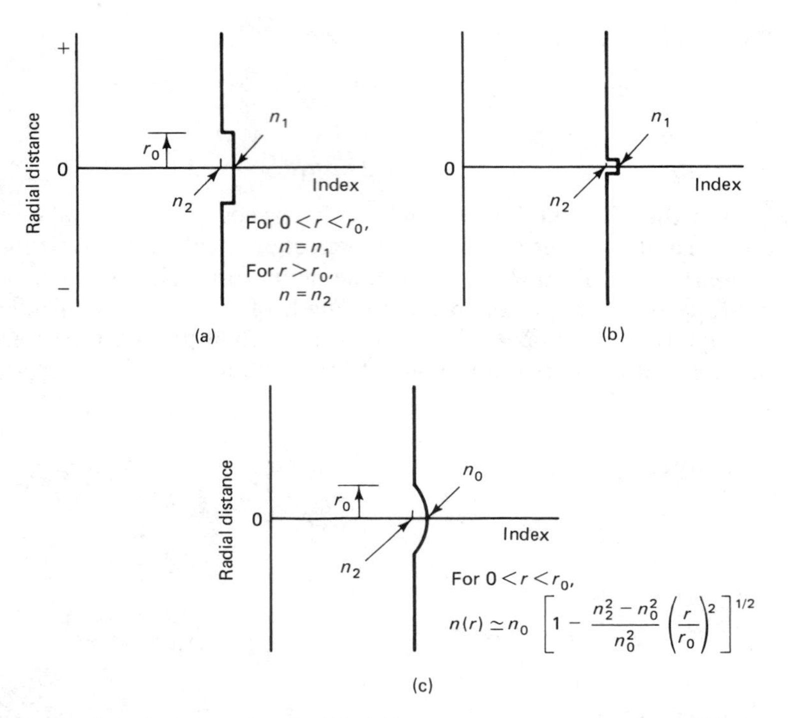

Figure 14-4 Index profiles of optical fibers: (a) step-index multimode; (b) step-index single mode; (c) graded-index multimode.

a core index that varies from a peak value at the center of the core to a value equal to the cladding index at the core–cladding interface. *Step-index multimode fibers* have core diameters of 50 to 1000 μm and numerical aperture values of 0.2 to 0.5. Fiber outside diameters are available from 125 to 1100 μm. Graded-index multimode fibers will have core diameters of 50 to 100 μm and numerical apertures of 0.2 to 0.3. Fiber outside diameters are available from 125 to 150 μm. Graded-index fibers are used primarily in medium and long-distance communication systems.

Single-mode fibers are step-index fibers with very small core diameters of 4 to 10 μm. The numerical aperture of these fibers is typically 0.1 to 0.15. Fiber outside diameters are large compared to core diameters, typically 75 to 125 μm. Single-mode fibers are used in long-distance communication applications. The number of paths or modes of propagation affects the integrity of the information transmitted. If the source radiates a single monochromatic pulse of optical power, each fiber mode will carry a portion of this power.

Because the path length of each fiber mode is different, the power pulses carried along the different paths will arrive at the opposite end of the fiber at different times. The axial mode pulse will arrive first and the pulse following the path defined by the maximum angle of propagation will arrive last. The detector at the receiving end of the fiber will sum all the modal pulses and generate a detected current pulse. Because of the modal delays, the detected current pulse will be wider than the original radiated pulse. This pulse broadening process is called *modal dis-*

tortion. Single-mode fibers do not exhibit modal distortion because there is only one modal path. Graded-index fibers exhibit less modal distortion than step-index fibers of the same core diameter because the lower index values near the outside of the fiber core cause less propagation delay for the higher-angle modes.

Fibers also affect the amplitude of the transmitted pulse. As the pulse propagates through the fiber, its amplitude is attenuated. Three physical processes account for this phenomenon: absorption, scattering, and radiation. *Absorption loss* mechanisms convert the radiant energy to heat within the fiber. The base material from which the fiber is made will have a particular absorption spectra. In addition, impurities in the fiber materials will cause absorption.

The *scattering loss* mechanism is one in which the rays of energy are diverted from the desired path. Scattering losses can be caused by reflections from minor fiber defects, but the major cause is *Rayleigh scattering* by the fiber material. Rayleigh scattering is a manifestation of the electromagnetic properties of radiation. As it propagates through the fiber, the wave interacts with the electrons of the fiber material, and these electrons absorb and reradiate the wave energy. Most of the reradiated wave energy is just delayed in time—in other words, phase shifted from the original—so it continues to contribute to the propagating wave. Some of the reradiated energy, however, is launched away from the direction of propagation; this is the scattered radiation. The scattering loss is proportional to the reciprocal of the fourth power of the wavelength.

Radiation loss is caused by energy leaving the fiber; this loss will occur if the critical angle relationship is exceeded by having too sharp a bend or kink in the fiber. Changes in the fiber core diameter or index characteristics also cause radiation losses. In addition, radiation loss takes place at the input end of the fiber when the incident energy from the source has rays at angles larger than the acceptance angle. These high-angle rays enter the fiber core but are only partially reflected from the core–cladding interface. The unreflected energy transmits into the cladding and is lost by absorption in the cladding or by radiation from the cladding. After a very short distance of travel in the fiber, all the rays that entered the core at angles larger than the acceptance angle have been lost through radiation out of the core. The acceptance angle and the maximum angle of propagation therefore define those rays that can be successfully propagated through a long length of fiber.

Figure 14-5 illustrates a typical attenuation versus wavelength graph for a glass-on-glass multimode graded-index fiber. The dominant attenuation effect at the shorter wavelengths is Rayleigh scattering. The major attenuation peak occurring at 1400 nm is due to oxygen and hydrogen impurities in the glass. The glass materials of the fiber become significant absorbers above 1650 nm. At short wavelengths the fiber exhibits a small amount of excess loss above the Rayleigh scattering curve, due to impurities and imperfections in the glass material and the geometric variations of the fiber core–cladding structure. An additional minor peak in attenuation due to oxygen and hydrogen impurities also occurs at 950 nm.

Fibers fabricated from materials other than glass have attenuation characteristics that differ significantly from those shown in Figure 14-5. Acrylic fibers, for example, have a characteristic such as that illustrated in Figure 14-6. This acrylic fiber is intended for use over very short distances using a red LED as a source. Note the extremely high levels of attenuation per kilometer.

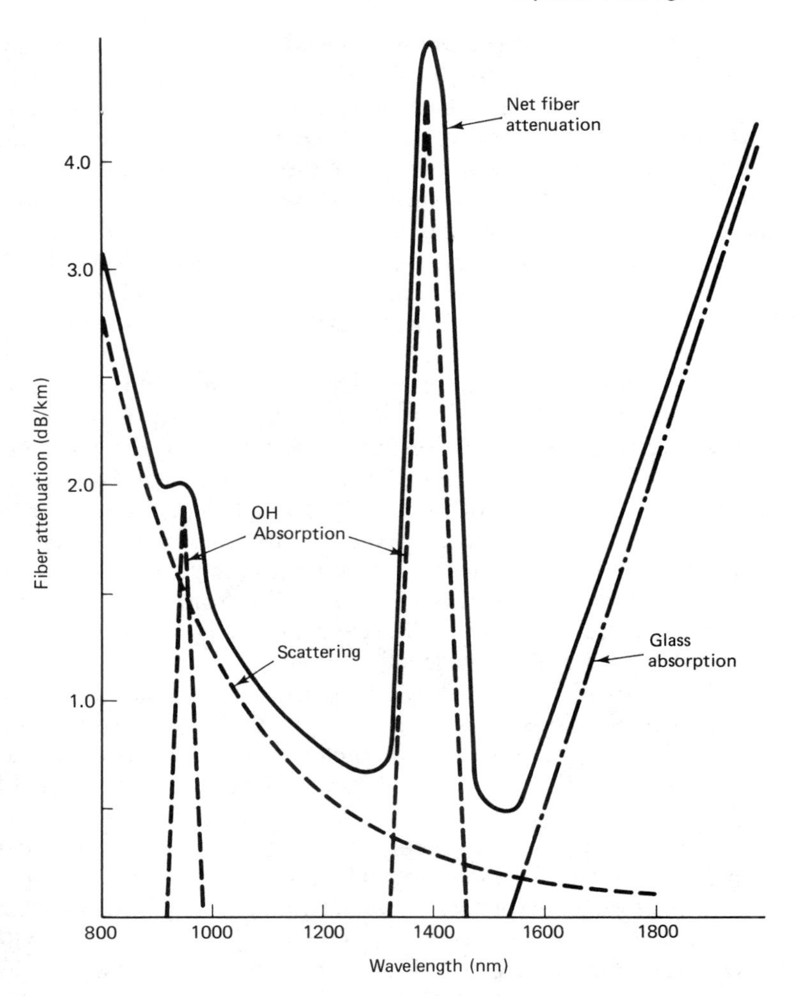

Figure 14-5 Attenuation versus wavelength for a glass multimode graded index fiber.

In this section the three basic types of optical fiber—step-index multimode, step-index single mode, and graded index—have been introduced. Modal distortion and attenuation characteristics were discussed. Finally, the differences in attenuation characteristics that exist between glass and acrylic fibers were illustrated.

14-3 FIBER PHYSICAL PARAMETERS AND FIBER MODES

The maximum angle of propagation defines the limit of the range of angular values that propagated rays may assume. Within this range there are only certain discrete possible angular values. The angular separation between these permitted angular

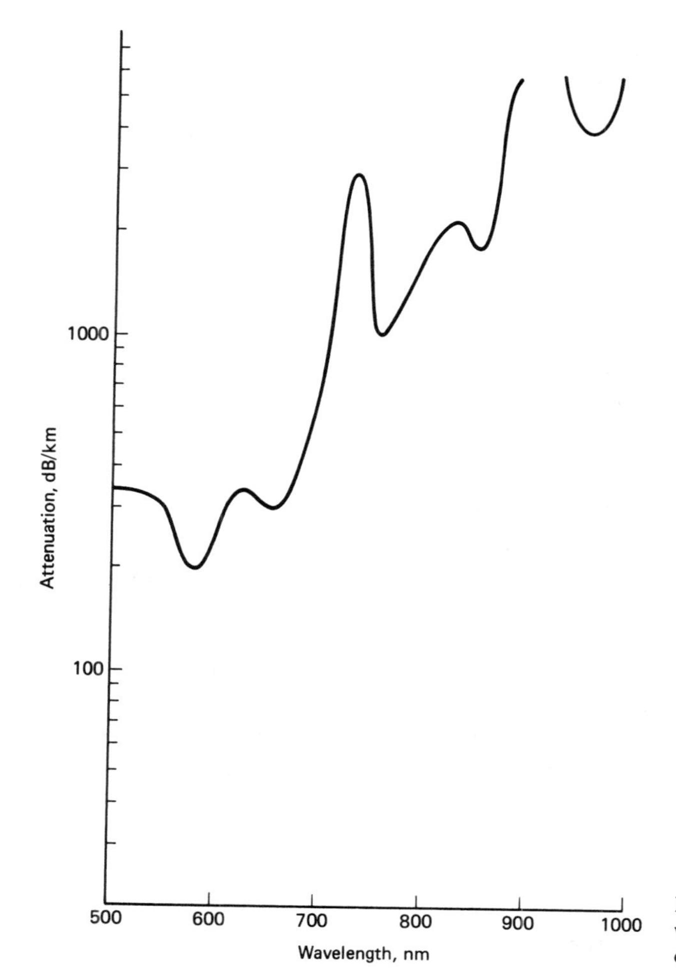

Figure 14-6 Attenuation versus wavelength for an acrylic multimode optical fiber.

values has a constant value, which is established by the wavelength of the radiant energy in the fiber and the diameter of the fiber.

$$\Delta\theta = \frac{\lambda}{d}$$

or

$$\Delta\theta = \frac{\lambda_0}{n_1 d}$$

where $\Delta\theta$ is the permitted angular separation, in radians
λ is the wavelength in the fiber
d is the fiber diameter
λ_0 is the free-space wavelength
n_1 is the index of the fiber core

The number of modes that can exist in the fiber is related to the permitted angular separation and the angle of propagation. A circular fiber can support reflections around its entire circumference, giving rise to many possible modes of propagation. An estimate of the number of permitted modes can be made using the following procedure:

$$T = \frac{\theta_p}{\Delta\theta}$$
$$n = \frac{(\pi T)^2}{2} \tag{14-10}$$

where θ_p is the maximum angle of propagation
$\Delta\theta$ is the permitted separation angle
T is the ratio of the angles
n is the number of modes when πT is greater than 2.405

Equation (14-10) provides a good estimate of the number of permitted modes for multimode fibers. As the core diameter gets small, however, this equation becomes less accurate. Optical fiber will support only a single mode when the product πT is less than or equal to 2.405. This would correspond to a calculated value of n of 2.89 using equation (14-10).

In advanced textbooks a term called the *V parameter* or *normalized frequency* will be encountered. This term is numerically equal to the πT product. When $V < 2.405$, the fiber will operate as a single-mode waveguide.

$$V = \pi T = \frac{\pi 2r[(n_1^2 - n_2^2)^{1/2}]}{\lambda_0} \tag{14-11}$$

where V is the normalized frequency
T is the ratio of the permitted angles
r is the radius of the fiber core
n_1 is the core index
n_2 is the cladding index
λ_0 is the wavelength of free space

At the beginning of this discussion, we envisioned the ray's paths as being defined by the propagation angle, which is measured with respect to the central axial line. Rays following these paths bounce from side to side of the fiber, passing through the center after each reflection and giving rise to *transverse* electromagnetic modes. It is possible for rays to propagate along paths that follow chords of the circular core but do not pass through the fiber center; these rays are called *skewed rays.* The transverse rays and the skewed rays, taken together, give rise to a set of modes called *linearly polarized modes.* Transverse modes and linearly polarized modes are identified by subscripts that describe the energy distribution that would exist across the circular cross section of the fiber if the mode existed in isolation. The transverse modes were discussed in Chapter 6.

Linearly polarized modes, identified as LP modes, utilize the subscripts C and N, where C represents one-half of the number of energy peaks counted around a circumferential path and N represents the number of radial positions at which these circumferential paths occur. Figure 14-7 illustrates the intensity patterns for three low-order linearly polarized modes in a circular waveguide. Example 14-1 is presented as a way of summarizing this discussion of modes.

EXAMPLE 14-1

Estimation of the Number of Modes, the Acceptance Angle, and the Numerical Aperture of a Silica Fiber

Given:

FIBER SPECIFICATIONS

Specifications	Fiber 1	Fiber 2	Fiber 3
Core diameter (μm)	7.4	50	100
Core index	1.500	1.500	1.500
Cladding index	1.495	1.486	1.472
Free-space wavelength	1300	1300	1300

Find: **(a)** The number of modes.
(b) The acceptance angle.
(c) The numerical aperture.

Use the data for fiber 2.

Solution: **(a)**

$$V = \frac{\pi d}{\lambda_0}(n_1^2 - n_2^2)^{1/2}$$

$$= \frac{\pi(50\ \mu\text{m})}{1300\ \text{nm}}(1.5^2 - 1.486^2)^{1/2}$$

$$= 24.70$$

$$n = \frac{V^2}{2} = 305 \text{ modes}$$

(b) $\theta_a = 11.83$

(c)

$$\text{NA} = (N_1^2 - N_2^2)^{1/2}$$

$$= 0.205$$

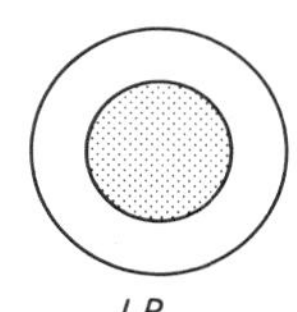

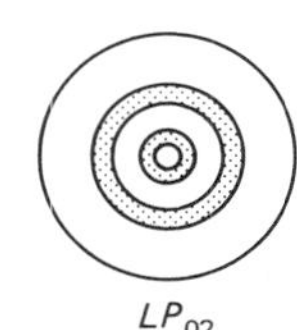

LP02

Figure 14-7 Intensity patterns for linear polarized modes in a circular waveguide.

SUMMARY OF RESULTS

Fiber	V	n	NA	θ_a (deg)
1	2.188	1*	0.1224	7.03
2	24.70	305	0.205	11.83
3	69.60	2422	0.288	16.74

* $V < 2.405$.

14-4 MODAL DISTORTION AND DISPERSION

The angle of propagation can also be used to estimate the difference in delay time that exists between a pulse traveling down the axis of the fiber and a pulse traveling along the path defined by the maximum angle of propagation. This time delay difference is derived below (see Figure 14-8). Let

t_0 be the axial delay for distance L
t_m be the delay along the path defined by θ_p
Δt be the time-delay difference

$$t_0 = \frac{n_1}{c} L$$

$$t_m = \frac{n_1}{c} \frac{L}{\cos\theta_p}$$

$$\frac{t_m}{t_0} = \frac{1}{\cos\theta_p}$$

$$\Delta t = t_m - t_0$$

$$= t_0 \frac{1}{\cos\theta_p} - 1$$

But

$$\cos\theta_p = \frac{n_2}{n_1}$$

Therefore,

$$\Delta t = \frac{Ln_1}{c} \frac{n_1 - n_2}{n_2} \tag{14-12}$$

Once a fiber is fabricated, all of the terms on the right side of the equation (14-12) are constants except for the distance L. The difference in time delay caused by the fiber modes is equal to the distance traveled multiplied by a constant delay parameter of the fiber, which has dimensions of time per unit length. This difference in

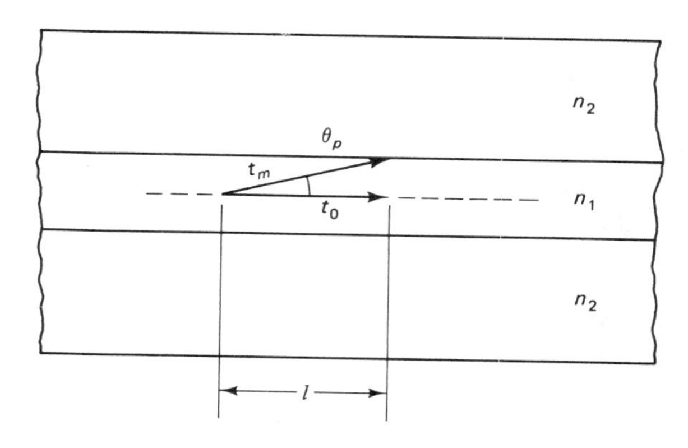

Figure 14-8 Propagation times for a mode at the propagation angle θ_p and a mode on axis.

time delay is called *modal delay spreading* or *modal distortion.* You will also see it referred to as *modal dispersion,* but this is poor practice because dispersion refers to differences in propagation times related to wavelength dispersion, which cannot occur in a system operating at a single wavelength.

Dispersion phenomena occur when the source radiates multiple wavelengths, that is, when the radiation occurs over a range of wavelengths, $\Delta\lambda$. Dispersion occurs because the index of the fiber is not constant as a function of wavelength. At short wavelengths the index of the fiber will be larger than it is at longer wavelengths, with the result that the shorter wavelengths will travel more slowly and at different propagation angles than the longer wavelengths. This process can cause pulse spreading even in single-mode fibers. The following equation illustrates how pulse spreading is related to fiber length and the spectral width of the source:

$$t = K(\lambda)\,\Delta\lambda\,L \tag{14-13}$$

where t is the pulse spreading time
$K(\lambda)$ is the dispersion factor, a material parameter with a value that varies with wavelength
$\Delta\lambda$ is the spectral width of the source centered at the wavelength used to determine $K(\lambda)$
L is the fiber length

As equation (14-3) and Table 14-1 illustrate, there are two ways to minimize the effects of dispersion: (1) use devices with small spectral widths, and (2) work at

TABLE 14-1 REPRESENTATIVE VALUES OF THE DISPERSION FACTOR $K(\lambda)$ FOR SILICON-BASED FIBERS

λ (nm)	$K(\lambda)$ (ps/nm · km)
800	110
1000	40
1270	0

wavelengths near 1270 nm. The modal distortion and material dispersion widen the transmitted pulse width. To maintain the integrity of the data, this pulse broadening must be insignificant in terms of the original pulse width. Fiber delay times are often specified in terms of frequency and distance. This specification, called the *fiber bandwidth,* lumps together all the delays and is given for a particular source wavelength and source spectral width. The units of the bandwidth specification are megahertz-kilometers. The bandwidth of a given length of fiber is obtained by dividing the specification value by the length of the fiber in kilometers.

The multimode graded-index fiber was developed to reduce modal distortion and increase bandwidth. In a graded-index fiber the higher-order modes, those with high propagation angles, encounter lower indexes as they move away from the fiber axis. In the lower-index material these modes begin to travel faster, so they experience less delay than they would in a step-index multimode fiber. The multimode graded-index fiber can have bandwidths that are two to five times as large as those of comparable step-index fibers. Single-mode fibers which do not exhibit modal distortion can have bandwidths that are 20 times as large as those of multimode step-index fibers. Table 14-2 lists typical fiber specification ranges.

TABLE 14-2 COMPARISON OF FIBER SPECIFICATIONS[a]

Specification	Multimode step index	Fiber graded	Single-mode fiber	Units
NA	0.2–0.5	0.2–0.31	0.1–0.15	—
θ_a	11.5–30	11.5–18	5.7–8.6	degrees
	0.2–0.52	0.2–0.32	0.1–0.15	rad
Core diameter	50–1000	50–100	4–10	μm
Fiber diameter	125–1100	125–180	75–125	μm
Materials	G/G P/G P/P	G/G	G/G	
Bandwidth	3.5–100	100–1000	2000	MHz · km

[a] G, glass; P, plastic (cladding core).

14-5 DETECTION AND TRANSMISSION OF WAVEGUIDE POWER

The purpose of a waveguide is to transport optical power from its input to its output. The power will be introduced from an optical source such as an IRED or laser diode. The power will be detected by an optical detector such as a PIN diode or an APD. The radiated power will exit the waveguide in a cone pattern like the light exiting a pinhole. This cone of radiation will cause a circular energy pattern on a surface located opposite the fiber end. If a single circular detector with an area equal to or larger than the pattern is mounted on this surface, all of the energy can be detected. For the moment we ignore the losses due to reflection at the fiber and detector surfaces. If the detector is smaller than the projected spot, some of the energy will be lost. As the distance between the detector and the fiber end decreases, the projected spot size approaches the fiber core diameter. If the detector's area is smaller than the spot area, the ratio of the radiant flux intercepted by the detector to the flux leaving the fiber is equal to the area of the detector divided by the area of the projected spot. If the detector area is larger than the radiated spot area, all the flux is intercepted and the ratio of the detected flux to the radiated flux is equal to 1.

Actual waveguide detector packages are more complicated than a simple planar detector. These packages normally have the detector mounted at a recessed location in the detector package with a lens system between the detector and the package aperture. Such a package will have a numerical aperture specification or a "port" specification. An estimate of the percentage of the radiant flux intercepted can be made if these detector specifications and the fiber core diameter and NA are known.

$$\frac{\theta_e}{\theta_o} = \left(\frac{Dd}{Dc}\right)^2 \left(\frac{\text{NA}_{\text{det}}}{\text{NA}_{\text{fiber}}}\right)^2 \tag{14-14}$$

where θ_e is the intercepted radiant flux
θ_o is the flux from the fiber
D_d is the detector aperture diameter
D_c is the fiber core diameter
NA_{det} is the detector's numerical aperture
NA_{fiber} is the fiber's numerical aperture

(*Note:* If either ratio is greater than 1, set it equal to 1.) Equation (14-14) can be modified to estimate the power that a source can introduce into a multimode fiber by substituting the fiber specifications for the detector specifications and the transmitter specifications for the fiber specifications. Finally, this equation can be used to estimate coupling between butted fiber ends.

In all of these estimates, reflections were ignored. If we assumed glass-to-air-to-glass transitions for all three cases, the reflection losses would be on the order of 8%. Also, these estimates assume perfect central axial alignment. As a practical matter, these calculations provide estimates of performance that can identify possible useful combinations of components and eliminate incompatible combinations. After you have selected fiber, detector, and source combinations, you will still have to conduct bench testing to evaluate average loss magnitudes and the frequency of occurrence of various alignment losses.

14-6 BANDWIDTH MEASUREMENT

The actual measurement of the pulse broadening and the bandwidth caused by a length of fiber is difficult and the methods used can be quite sophisticated. One approach measures the 50% pulse width of the pulse coming out of a known length of fiber and compares it to the 50% pulse width of the pulse introduced into the fiber. These pulse-width values are used to calculate the bandwidth specification by the following method:

$$T = \frac{(t_2^2 - t_1^2)^{1/2}}{2} \tag{14-15}$$

$$\mathrm{BW} = \frac{0.35}{T} \tag{14-16}$$

where T is the pulse broadening factor
t_2 is the output pulse width
t_1 is the input pulse width
BW is the bandwidth

A more esoteric approach employs the characteristics of the Fourier transforms of the input and output pulses to determine fiber bandwidth. On an oscilloscope a voltage pulse is represented as a graph of voltage versus time. The same pulse can be represented mathematically as an array of spectral components, which start at zero hertz and extend to very high frequencies. Each unique pulse has its own unique spectral array. The mathematical process of going from the time representation of a pulse to a frequency representation of a pulse is called the *Fourier transform.*

In a bandwidth measurement setup employing the Fourier transform the amplitudes of the transmitted pulse and the detected pulse being applied to the measurement instruments are adjusted to be the same. This amplitude adjustment nulls out the effects of fiber attenuation. Any remaining differences in pulse shape and therefore in pulse spectrum are due to the fiber time-delay processes. The amplitude values of the spectral components of the input pulse are divided by the amplitude values of the corresponding spectral components of the output pulse. If the output pulse and the input pulse were identical, the quotient of each of these divisions would be equal to 1 and the resulting spectral display would be perfectly flat, indicating a system with infinite bandwidth. Because the output pulse is wider than the input pulse, the individual quotients will not be equal to 1 and the calculated spectral display will not be flat. Instead, the spectral display will have a shape that looks like the frequency response plot of a low-pass filter. The low-frequency spectral components will have amplitudes of 1 and the higher-frequency spectral components will exhibit decreasing amplitudes. The spectral component that is 3 dB less than the zero-frequency spectral component is defined as the cutoff spectral component. The bandwidth of the length of fiber under test is equal to the frequency of the cutoff spectral component.

Whichever test method is used, the resultant bandwidth specification is dependent on both the modal distortion and the chromatic dispersion characteristics of the fiber. The effects of these characteristics depend in turn on the wavelength of the source and the spectral width of the source. Specified fiber bandwidth applies only to systems where the system's source is essentially the same as the source used to determine the bandwidth specification.

14-7 COMMUNICATIONS LINKS

A communication link consists of a source, a detector, and the interconnecting optical fiber. The source may be an LED, and IRED, or an injection laser diode. Although these sources can be modulated with analog signals, they are usually driven with digital pulses. The detectors used are PIN or avalanche photo diodes. Communication links can be thought of as being short-, medium-, or long-distance links. A short link will run at most a few meters. Short links are used to connect:

1. Process control equipment, to industrial machines and processes.
2. Medical sensors mounted on a patient, to recording equipment.
3. Microcomputers, to peripherals such as printers.
4. In a recent advertisement, to interconnect high-fidelity components.

Medium-length systems are longer than a few meters and shorter than a kilometer. Medium-length systems are used primarily to interconnect computers, computer terminals, and peripherals in systems called *local area networks.* Long-distance systems with lengths of 1 km to tens of kilometers are currently being used for long-distance phone traffic and to interconnect large computer systems.

Design of a communications link revolves around the issues of attenuation, bandwidth, and cost. *Cost* refers not only to the price of the system parts but also to the labor costs required to install and maintain the system, including the useful life of the components selected. In an industrial control setting the data rates are typically very low. It literally takes hours to change the temperatures of large mixing vats a few degrees. In addition, the distances involved are very short. These two physical factors make it feasible to use large-diameter multimode plastic fiber and LED sources. A side benefit to this design approach is that this type of fiber is easy to work with and labor and connector costs are low.

A medium-distance local area network will use multimode glass on glass fiber or multimode plastic on glass fiber. An IRED operating around 850 nm would be a typical source in this type of system. The glass-on-glass fiber tends to have a wider bandwidth per unit length and a lower loss per unit length than the plastic on glass fibers. The numerical aperture of fibers for these applications can vary from 0.2 to 0.5, and core diameters are typically 50 to 100 μm. These core diameters and numerical apertures are well suited to coupling radiation from an IRED into the fiber. The relatively large core diameters also make alignment errors less critical. Essentially, labor and connector costs go down as core diameter increases, but unfortunately, bandwidth also decreases. There are many combinations of fibers, sources, and detectors that will meet the system specifications of this type of link.

Long-distance systems can actually be considered easier to design than medium-distance systems because the choice of components is much more restricted. Long-distance fibers carry wide-bandwidth data. Graded-index fiber will work in some systems. But in very long systems only single-mode fiber will provide acceptable levels of bandwidth and attenuation. Most communications-grade laser diodes are designed to couple energy effectively into a single-mode fiber. A few edge-emitting IRED devices can also effectively couple energy into these fibers.

The very small core diameters of the single-mode fiber makes it difficult to connect or splice fiber ends and to align fiber ends with sources and detectors. The skill levels of the installers must be higher, the equipment used in the installation is more expensive, and the time required to get the link up and working will be longer for single-mode fiber than for the other fiber types. The good news is that the wide bandwidth of this type of link allows many simultaneous messages and the link can therefore produce high revenues that will offset these higher installation costs.

All of these links require connections at the ends of the fibers to sources and detectors, and often intermediate connections are required between fiber lengths or between fibers and signal splitting devices. Fibers can be connected using splicing techniques or with mechanical connectors. *Splicing* techniques align the ends of the fibers and then permanently attach the fibers together, hopefully fixing the alignment; they can be made using adhesives or by butt welding the fiber ends. The welding operation is called *fusion splicing.*

Mechanical connectors attempt to bring the fiber ends into close physical contact, with good core alignment. The mechanical connector attempts to maintain the fiber end contact and alignment via some mechanical clamping system. Mechanical connectors for optical fibers, like the mechanical connectors used with electrical wires, employ a variety of threaded, spring-loaded, and pressure-clamping mechanisms to hold the fibers in place. Splices can be expected to have lower initial losses and greater long-term stability than mechanical connectors. Mechanical connectors should be used only when it is likely that the connection will have to be opened on a regular basis or when the equipment you are using permits connection only through a connector.

14-8 OTHER WAVEGUIDE STRUCTURES AND DEVICES

Integrated optics is the term used to describe the fabrication of optical waveguides and other optical components in a single package on a small substrate of material. Batch processing techniques similar to those used in the fabrication of semiconductor devices are employed. The ultimate goal is to be able to fabricate a number of electro-optical functions on a single substrate.

Commercially available integrated optical devices include signal splitters (couplers), phase shifters, modulators, and switches. All of these devices use waveguiding structures that are formed by creating paths in the substrate which have a higher index than the substrate material. These waveguides work like optical fibers. Most commercially available integrated optics devices are designed for applications in fiber optic systems. These device packages generally contain the integrated optics device and connecting optical fibers. The device user integrates the device into the larger system by making connections to these fiber pigtails.

A common and relatively easy to understand device is the *signal splitter* or *coupler,* which can be fabricated in two different ways, as illustrated by Figure 14-9. Figure 14-9(a) illustrates a technique that uses a direct optical path between the input and output ports. The geometry of the branches is adjusted to achieve the desired level of power division. The waveguide core penetrates down into the substrate, and the fibers are connected to the waveguide at the waveguide ends on the edge of the substrate. The Corning Inc. manufactures a coupler of this type in a glass substrate.

The operation of the coupler illustrated in figure 14-9(b) is less straightforward. In this structure the driven waveguide connected to port 1 is not obviously connected to port 3. This coupler exploits a characteristic of waveguide propagation, which requires that there always be a small amount of energy traveling in the cladding. As the driven waveguide cross section is decreased near the center of the coupler, more energy moves out of the driven waveguide core, and it is coupled into the adjacent waveguide core. By controlling the waveguide cross section, material index, core spacing, and the length of the coupling region, the amount of coupled energy is established. This type of coupler can also be fabricated by drawing and welding optical fibers together. There is a reasonably well-established method for specifying the operation of a coupler at a fixed wavelength. Table 14-3 defines these specifications.

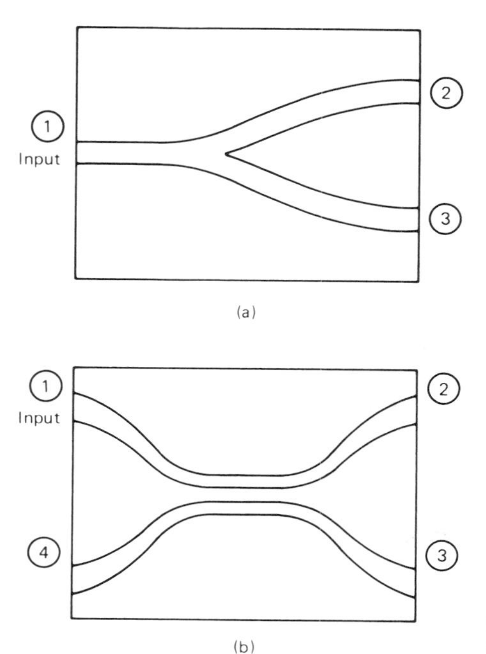

Figure 14-9 Coupler on substrate: (a) three-port branched coupler; (b) four-port coupler.

TABLE 14-3 DEFINITION OF OPTICAL COUPLER PARAMETERS

Parameter	Four-port coupler [Figure 14-9(b)]	N-part coupler[a]
Coupling ratio	$P_2/(P_2 + P_3)$	$P_N P_o$
Excess loss	$P_2 + P_3/P_1$	$P_o P_i$
Insertion loss	P_2/P_1	P_N/P_i
Uniformity	P_2/P_3	P_h/P_s
Directivity[b]	P_4/P_1	P_x/P_i

[a] N is any integer, P_N is the power out of any port N, P_i is total input power, P_o is total output power, P_h is largest output, P_s is smallest output, P_x is the power out of the uncoupled port.

[b] For the three-port structure of Figure 14-9(a), the directivity would be a measure of the power reflected back out port 1. Typical values of directivity are −40 to −60 dB. It is common practice to express these parameters in decibels.

EXAMPLE 14-2

Determination of Coupler Specifications for Measured Power Data

Given: A four-port coupler with the following data:

(1) Input power = 100.0 μW
(2) Output, port 1 = 20.8 μW
(3) Output, port 2 = 21.3 μW
(4) Output, port 3 = 20.3 μW
(5) Output, port 4 = 20.9 μW

Find: The coupler specifications.

Solution:

1. Determine the coupling ratio for port 1.

10 log 20.8 μW/(20.8 μW + 21.3 μW + 20.3 μW + 20.9 μW)
−6.03 dB

2. Calculate the excess loss.

10 log(20.8 μW + 21.3 μW + 20.3 μW + 20.9 μW)/100 μW
−0.79 dB

3. Calculate the insertion loss for port 1.

10 log 20.89 μW/100 μW
−6.82 dB

Note: The insertion loss minus the coupling ratio is equal to the excess loss.

4. Determine the uniformity value.

10 log 21.3 μW/20.4 μW
0.21 dB

The coupling process used by the four-port coupler described in Example 14-2 can be combined with a phenomenon called the *electro-optic effect* to create an optical switch. Materials that exhibit the electro-optic effect have an index of refraction which will change when an electric field is present; this can be achieved by applying a voltage. The degree of coupling between the adjacent waveguides can be controlled by altering the index of the waveguide and the surrounding material.

Figure 14-10 schematically illustrates an electro-optic switch. The combined action of the fixed bias voltage and the switching voltage determines which of the two outputs will transmit radiation. For example, with the switching voltage at zero, radiation will exit port 3. With the switching voltage on, radiation will exit port 2. While a number of crystalline materials, including gallium arsenide, exhibit the electro-optic effect, most commercial devices are fabricated on substrates of lithium niobate, LiNbO, which exhibits a very strong electro-optic effect. Switching voltages for waveguide devices because of the very small physical dimensions of the waveguide are modest, on the order of 5 to 10 V. A small amount of radiation is always present at the "not"-selected output, and relative power ratios of 100:1 to 300:1 between the selected and nonselected outputs are common.

The index of the waveguide material also affects the velocity of propagation through the material. By altering the index of the waveguide, the relative phase of the transmitted wave is altered. Phase shifters and phase modulators are produced by placing a single waveguide in an electro-optic crystal between two electrodes. The amount of phase shift achieved depends on the magnitude of the voltage and the length of the waveguide. Typical commercial products will yield 180° of phase shift, in response to a 10-V input signal. Phase-shifting devices can be used to introduce a fixed phase shift in response to a dc voltage or to provide a continuously

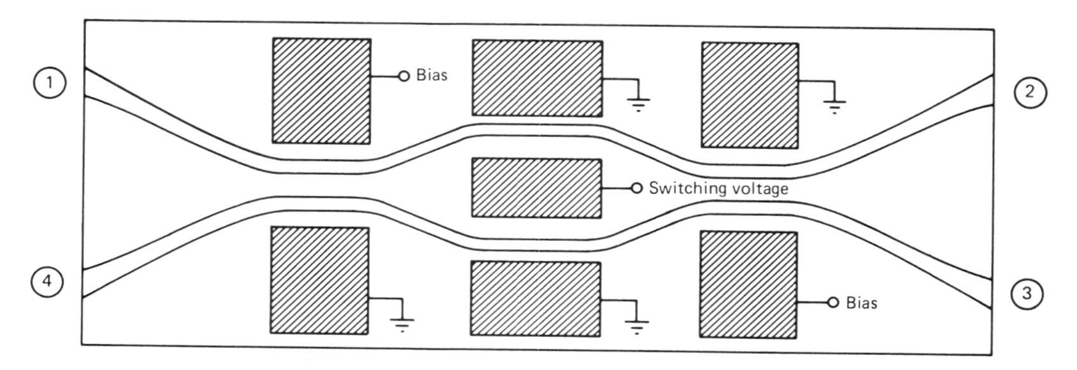

Figure 14-10 Switching coupler.

varying phase output in response to analog or digital signals. Devices driven by sine waves or digital signals are called *phase modulators.*

Electro-optic devices can also be used to alter the polarization of a beam. This type of structure has been used in conjunction with fixed polarizers to obtain amplitude modulation. As the polarization is altered, the transmission though the fixed polarizers varies. This process is rather like that in the liquid-crystal systems discussed in Chapter 11.

In addition to index changes caused by electric fields, lithium niobate substrates will exhibit a change in refractive index when subjected to mechanical strain. The strain can be introduced into the substrate with a vibrating electromechanical transducer. The transducer vibration causes a pressure wave to move through the material. The peaks and nulls of pressure cause high and low regions of strain. This type of wave phenomenon is called an *acoustic wave.* Normally, we think of the term *acoustics* as referring to frequencies in the audible range 20 Hz to 20 kHz. The acoustic waves in the crystal, however, are at RF and microwave frequencies that can range from tens of megahertz to several gigahertz.

As the acoustic wave propagates through the crystal, the alternating high and low regions of strain cause alternating regions of high and low index creating an effective optical grating. As you will recall from Chapter 2, a grating causes the incident radiation to be diffracted at a number of different fixed angles, which are related to the grating spacing and the wavelength. The acoustic wave causes a grating spacing that is proportional to the frequency of the acoustic wave. If the acoustic frequency is held constant, incident radiation of different wavelengths will be diffracted at different angles and therefore separated in space. If the acoustic frequency is shifted slowly and the wavelength of the incident radiation is held constant, the beam of radiation can be moved to different angular positions; thus the acoustic grating can be used as a wavelength separator or a beam steering mechanisms. If the amplitude of the acoustic wave is changed, the amplitude of the index variations will also change. The acoustic amplitude modulator uses this process to vary the amplitude of the optical beam.

If the optical wavelength is held constant and the acoustic wave contains several different frequency components, the transmitted optical power will be diffracted at several different angles. These angles will be proportional to the different acoustic frequencies, so that this technique can be used to determine the frequency components of the microwave signal that causes the acoustic wave. If the acoustic wave frequency is shifted rapidly over a wide range of frequencies (e.g., 50 MHz), the radiation passing through the device will exhibit a shift in wavelength equivalent to a shift in wavelength equivalent to a shift in frequency of the same magnitude. The acoustic grating can therefore be used to frequency modulate the radiation. This means that optical beams can be modulated in frequency just like radio waves.

Acoustic waves can be launched in most crystalline and glass materials, including gallium arsenide and flint glass. In most materials the acoustic losses are high, however, and lithium niobate and lead molybdate appear to be the most efficient at the acoustic frequencies of interest. With waveguide structures of very small dimension located near the surface of the substrate, the total energy required, even with high losses, can be reduced significantly, so it may be possible to build commercial acoustic diffraction devices in a variety of materials.

At the present level of the technology, commercial devices are available that will perform the various optical functions described in this section. An obvious goal is to integrate optical functions such as these in a single substrate that also contains semiconductor sources and detectors. At present, "hybrid" packages are available that contain separate substrates containing optical and semiconductor devices.

14-9 LOW-VOLUME APPLICATIONS

Although the devices and phenomena that have been discussed so far are primarily aimed at communication applications, these are not the only uses for fiber optics and optical waveguides. There are a variety of applications in which optical fibers are used to carry light for illumination and signaling; these cables are called *light guides.* Bundles of fibers called *image guides* can also be used to carry images in cables. In perhaps the biggest area for growth, fibers in conjunction with other electro-optic components can be used to create a variety of transducers which are used to make measurements of physical and biological phenomena.

Light-guide applications use fibers that are fabricated out of glass, polymers, and glass–polymer combinations. These fibers often have very large core diameters and numerical apertures. Fiber losses as a function of distance will also be much higher than those specified for communications-grade fibers. This is due in part to the higher scattering loss incurred at the visible wavelengths but also to the higher absorption characteristics of the materials used.

A light guide is a cable with a large number of fibers tightly grouped together. Depending on its diameter, the light guide may have several hundred to tens of thousands of fibers packed together in the cable. Some applications for light guides include:

1. Distributing light from a single bright source to a number of locations on an instrument panel.
2. Monitoring a flame or pilot light in a furnace.
3. Providing illumination to a microscope or other instrument where the heat or size of a lamp would be objectionable.
4. Providing illumination in environments where an electrical device might be a hazard.

An *image guide* is a light guide in which the relative position of the fibers is maintained from one end of the cable to the other. The fiber count in image guides is of necessity large, typically between 3000 and 50,000, to provide resolution of the image. Flexible image guides allow the user to inspect areas that might not otherwise be accessible. Examples of image guide applications would be the inspection of the insides of pipes and storage tanks. Inspection instruments using image guides often incorporate light guides to provide the illumination necessary for the desired inspection. Image guides are available in lengths from 1 to 2 m and light guides in lengths from 0.20 to 6 m.

If this were a text on fiber optics and waveguides, we could spend several chapters discussing all the many and varied transducers that can be made using optical waveguides. We will consider only one transducer here, as an illustration. The reflection transducer is actually a small system; it consists of a source, a detector, and lengths of light guides. Schematically, the system looks as shown in Figure 14-11. The arrangement of the sensor head containing the optical fibers would actually be more complex than this simple representation. An actual sensor head would be a bundle of light-guiding fibers, some of which are transmitting fibers and some of which are receiving fibers. Figure 14-12 illustrates some possible fiber arrangements.

The output of the transducer system is the detector current. The detected physical parameter is the distance between the end of the fiber bundle and the reflecting surface. Different fiber bundle arrangements will yield different transducer sensitivities. Figure 14-13 illustrates typical relative responses for the bundle types illustrated in Figure 14-12. Each of the responses illustrated has a relatively linear detection region extending from the origin to a peak detected current value. In this region of operation, the transducer could be calibrated to measure the physical separation between the fiber end and the surface being monitored. The primary problem with this approach is that it assumes a predictable and uniform reflection characteristic for the surface being monitored. If the assumption is justified, distances on the order of tens of micrometers can be measured.

A less stringent application would be to use the transducer as a proximity sensor. In this application the reflecting surface is initially far away from the detector,

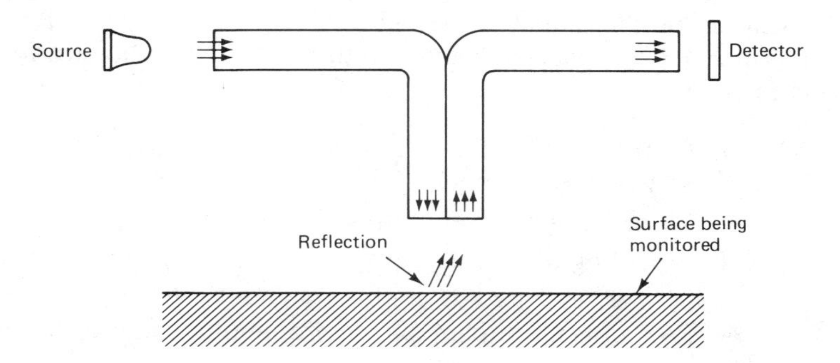

Figure 14-11 Light-guide optical sensor.

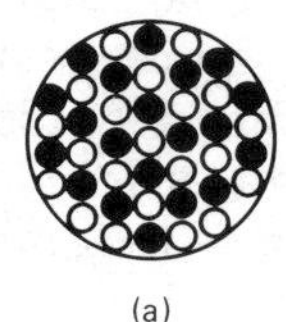

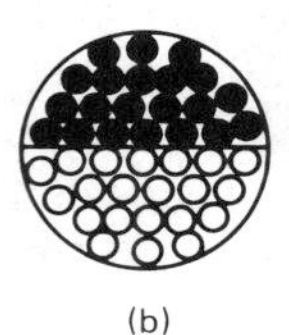

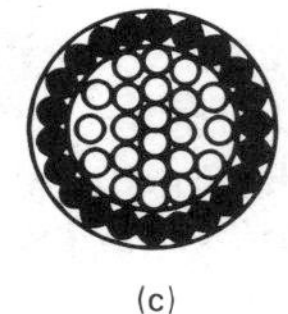

Figure 14-12 Fiber arrangements at detection end of a light-guide optical sensor: (a) interleaved fibers; (b) half-circle; (c) coaxial fibers. Dark fibers receive, light fibers transmit.

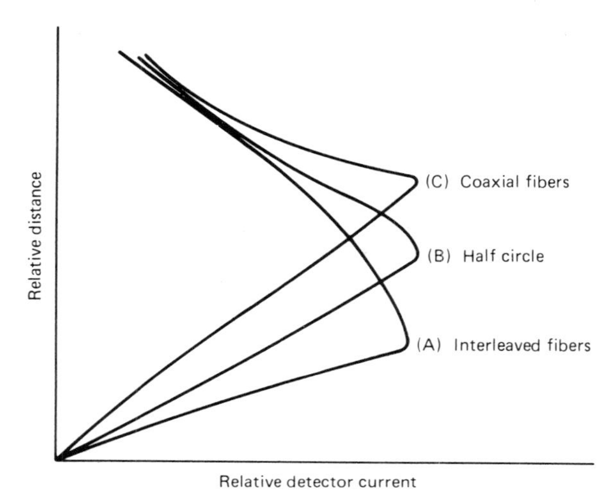

Figure 14-13 Response of light-guide optical sensor with different fiber arrangements.

and as it gets progressively closer, detector current increases. At some point a preestablished current level is exceeded and an electronic proximity indication is given. Sensors like this are used as level detectors in tanks containing liquids. This optical sensor head illustrates characteristics common to most optical transducers:

1. The measurement is a noncontact measurement. There is no mechanical contact between the surface being monitored and the detector.
2. The transducer package uses a source, a detector, and optical waveguides as components.

The ability to make measurements without being in physical contact with the test object is one of the most attractive features of optical transducers. Practical systems require some source of radiation, which may come from the object under test, but usually is an integral part of the transducer. Detectors have to be selected which are compatible with the wavelengths and magnitudes of the radiation available. The optical paths in the transducer system may be simple paths in free space directed by lenses, prisms, and mirrors. Many newer systems, however, use optical waveguides and optical fiber to create all or part of these optical paths.

14-10 SUMMARY

Radiation both light and infrared can be guided by structures called waveguides. These waveguides have a central region called the core which has a higher refractive index than the surrounding material. The possible paths that the radiation can take through the waveguide are called propagation modes, and the number of modes that can exist in the waveguide is determined by the wavelength of the radiation in the waveguide and the diameter of the waveguide core. Both single-mode and multimode waveguides are used. Over the cross section of the waveguide, each mode causes a specific energy distribution pattern, which is called a linearly polarized modal pattern.

The number of modes propagated affects the integrity of transmitted information, which is typically in the form of pulses of radiation. During transmission the width of the pulse increases. Two processes, modal distortion and dispersion, cause this widening of the pulse. Modal distortion takes place in multimode fiber and is the result of different modes traveling along different path lengths. Dispersion occurs in multimode and single-mode fiber and is the result of the source transmitting multiple wavelengths through a waveguide which presents a different index to each wavelength. The net effect of these two processes is lumped into a single specification called the bandwidth.

Waveguides are fabricated in glass and semiconductor substrates and as long flexible fibers. The fibers come in three basic types: step-index multimode, step-index single mode, and graded index. Step-index multimode fibers are available in the greatest variety of sizes and material combinations. They are used both in communication and light-guiding applications. Step-index single-mode fibers are made of glass, have extremely small core diameters, and are used primarily in communication systems. Single-mode fibers have higher bandwidths and lower losses than other fiber types.

Graded-index fibers are also used for communications. They are multimode fibers with core diameters of the same size as step-index multimode communications fibers. Graded-index fibers are made of glass and have higher bandwidths than step-index glass fibers of the same core diameter.

Waveguides manufactured in substrates can be affected by electro-optic and acoustic effects which can change the phase, frequency, amplitude, polarization, and direction of the propagating radiation. All waveguides cause some attenuation. The major factors causing attenuation are scattering, absorption, and radiation losses.

PROBLEMS

1. The core refractive index of an acrylic optical fiber is 1.495 and the cladding index is 1.402. What is the critical angle in the fiber?
2. What is the acceptance angle in degrees and radians for the fiber described in Problem 1?
3. What is the angle of propagation for the fiber described in Problem 1? Express the answer in degrees and radians.
4. What is the numerical aperture of the fiber described in Problem 1?
5. A light guide has a core index of 1.62 and a cladding index of 1.52. What is the numerical aperture of this light-guide fiber?
6. A step-index communications grade fiber has a core index of 1.52 and a specified numerical aperture of 0.21. What is the index of the cladding? How much reflection loss occurs as radiation exits the fiber?
7. The fiber being used in a system has a rated loss of 5 dB/km. Over 50-m runs the received power is typically 15 μW. If a new fiber that has the same core diameter and

numerical aperture and a rated loss of 2 dB/km is substituted, what will the received power be for 50-m lengths?

8. One manufacturer suggests that fiber bandwidth specification can be estimated by dividing the constant 0.35 by the fiber dispersion. If the measured dispersion is 200 ps/km, estimate the fiber bandwidth specification. Based on the values of dispersion and bandwidth specification, what kind of fiber do you think this is?
9. The dispersion factor for a fiber at 960 nm is 66 ps/mn · km. The core index is 1.520 and the cladding index is 1.505. Estimate the delay spreading and dispersion time per kilometer. Assume that λ_o is 960 nm and that the spectral width is 5 nm.
10. A fiber has a specified bandwidth of 50 MHz · km. What will the bandwidth of 1.5 km of fiber be?
11. A fiber has a specified bandwidth of 25 MHz · km. What will the bandwidth of 50 m of fiber be?
12. The flux introduced into a waveguide by an LED is proportional to the numerical aperture of the waveguide squared and the waveguide core area. Waveguide 1 has a core diameter of 50 μm and a NA of 0.3, and waveguide 2 has a core diameter of 100 μm and a NA of 0.3. What is the ratio of the largest value of introduced flux to the smallest value of flux?
13. Two waveguides are being tested for use with an IRED source. All of the waveguide specifications are the same except for NA. Waveguide 1 has a NA of 0.25, and waveguide 2 has a NA of 0.3. Into which waveguide will the IRED introduce the most flux? What is the ratio of the largest to the smallest introduced flux?
14. The specifications for two optical fibers are listed below. The typical power received at 50 m with fiber 1 is 15 μW. What would the typical power received be when fiber 2 is used?

	Specifications		
Spec.	Fiber 1	Fiber 2	Units
Loss	15	12	dB/km
NA	0.3	0.28	—
Core diameter	125	100	μm
Cost	1.75	3.25	$/m

15. Even when a fiber connector is perfectly aligned, it will exhibit reflection loss at each fiber surface. What will the typical irreducible connector loss be if the core index is 1.5 and the gap between the fiber ends contains air?
16. If a typical connector loss is found to be 1.5 dB, estimate how much it would be when the space between the fiber ends is filled with an index matching fluid that eliminates all reflection losses.
17. A coupler is designed to split the optical signal into two equal paths. Experience indicates that 20% of the power will be lost using the coupler. Estimate the coupling ratio, the excess loss, and the insertion loss for this coupler.
18. The power out of each port of a four-port coupler is measured and found to have an average value of 8 μW. The input power is measured as 40 μW. Estimate the coupling ratio, the excess loss, and the insertion loss for this coupler.

19. A six-port coupler has a rated insertion loss of 8.8 dB and a rated uniformity of 0.5 dB. If the minimum acceptable output power is 4 μW, how much input power is required?

20. A fiber installation has one transmitter connected to six terminals through a coupler. The fiber used has a loss factor of 0.01 dB/m. The longest distance from the transmitter to a terminal is 200 m. Connectors are used at the transmitter output, the terminal input, and on both sides of the coupler. Typical connector loss per connector is 1.5 dB. The coupler has an insertion loss of 8.8 dB. The minimum acceptable power level at a terminal is 4 μW. How much power must the transmitter launch into the fiber to ensure that all terminals receive at least the minimum acceptable level of power?

Answers to Odd-Numbered Questions and Problems

Chapter 1 Basic Optical Devices

1. $v = 185 \times 10^6$ m/s
3.

θ	$\varnothing$
10°	6.52°
20°	12.92°
30°	19.08°
40°	24.84°

5.

n	r
1.53	0.0439
1.60	0.0533
1.80	0.0816
2.00	0.1111

As the index increases, r increases.

7. $F = 250$ cm
9.

s (cm)	M
6	5
7	2.5
10	1.0

11. $F = 472$ cm

13.

n	F (cm)	m
1.53	472 cm	5.36
1.62	403 cm	2.57
1.76	329 cm	1.42

15. $T_R/T_o = 0.4655$, 46.55%

Chapter 2 Interference and Diffraction Devices

1. 1.818×10^{-15} s
3. **(a)** 261 nm **(b)** 196×10^6 m/s
   (c) 0.75×10^{12} Hz **(d)** Same as part (c)
5. Material B, minimum index change
7. 438.9 nm
9. 4.00 μm
11. **(a)** $m = 2$; four off-axis lines **(b)** $\pm 22.3°$ and $\pm 49.4°$
13. Schott glass B: the square root of the index is closest to 1.33.
15. Achromat
17. Half-wavelength
19. 0.155

Chapter 3 Radiation and Radiometry

1. 2.845 eV, visible; 4.889 eV, not visible; 3.385 eV, not visible; 3.397 eV, not visible; 3.064 eV, visible
3. **(a)** 31.83×10^3 mW/cm^2 **(b)** 0.007958 mW/cm^2
5. **(a)** 1×10^{-4} sr **(b)** 0.000796%
7. **(a)** 1 μw/cm^2 **(b)** 10 nW/cm^2
9. 3.848×10^6 W/m^2
11. 279.3×10^3 W/m^2
13. **(a)** 3.562×10^6 W/m^2 **(b)** *IR* 92.56% **(c)** 7.26%, Visible **(d)** UV 0.18%
15. **(a)** Tungsten **(b)** 17.5%
17. 197 W
19. 799 W

Chapter 4 Photometry and Incandescent Lamps

1. 11.61 lm
3. 622 mlm
5. 588 mW/cm^2
7. 225 mW/sr

9. Calculated: 10.2 lm/W, 7.5 lm/W, 11.4 lm/W, 6.0 lm/W; from Figure 4-3: 18 lm/W, 18 lm/W, 22 lm/W, 11 lm/W.
11. 64.7 cm
13. $T_N \simeq T_R \left(\frac{V_N}{V_R}\right)^{0.388}$
15. 0.643 A, 22.34 lm, 3040 K, 2.48 W
17. At $\frac{1}{2}$ m, 80 μW/cm², 964 μlm/cm², 12 lm/W at all locations; at 80 cm, 31 μlm/cm², 377 μlm/cm².
19. 36,000 lm · s; specified, 35,500 lm · s; percent difference, 1.4%

Chapter 5 Gas Lamps

1. 10
3. 284×10^{-21} to 497×10^{-21} J
5. **(a)**

Model number	V_{dc} (V)	I_{dc} (mA)	F_v (lm)	Efficiency (lm/W)
6269	23	43.5	30×10^3	29.99
6271	23	43.5	30×10^3	29.99
6277	23	100	81×10^3	35.22
6279	23	100	81×10^3	35.22
6293	38	33	40×10^3	31.90
6295	38	33	40×10^3	31.90
6278	55	55	120×10^3	39.67
6297	55	55	120×10^3	39.67

(b) 6278 and 6297 are the same and the most efficient.
7. To get the mercury and sodium into gaseous form.
9. A flashtube with a current through it and a voltage across it both of which are below breakdown levels.
11. Probably electrode spacing, possibly final gas pressure.
13. Calibration wavelength standards, monochromatic light sources for test plating, and as sources of ultraviolet radiation.

Chapter 6 Gas, Solid-State, and Liquid Lasers

1. Light amplification by stimulated emission of radiation.
3. The emitting medium and the tuned cavity.
5. The emission of energy over a narrow range of wavelengths (frequencies).
7. Atom mass and cavity temperature.
9. It will increase, get wider.
11. Frequency spacing decreases.
13. The cavity resonant frequencies are the longitudinal modes.
15. The lifetime of the emitting state. Short lifetimes cannot be used.

17. No. It is possible to have more than one cavity resonance fall inside the permitted line width. Radiation will occur at each of these resonances.

19. By adjusting the optical gain of the cavity and by changing the dye used.

21. 32 μm

23. Refer to Section 6-4 and Figure 6-8.

Chapter 7 Vacuum Photodetectors

1. Graph
 (a) 10-MΩ plot is most linear
 (b) 50-MΩ plot is most sensitive

3. 125 V

5. **(a)** $F_v = 16.7$ μlm.
 (b) $F_v = 6.67 \times 10^{-12}$ lm using typical values. If worst-case dark current is used, $F_v = 50 \times 10^{-12}$ lm.

Chapter 8 Thermal Detectors

1. $R = 40$ mV/mW

3. $t_r \simeq 15.4$ s

5. $D = 16.78$ cm

7. $R_c = 11.28$ cm

9. $R_c = 4.70$ cm a far-field problem,
 $L_e = 3667$ μW/cm² · sr

11. $\theta_v = 29.97°$ average of three values. I would call it 30°.

13. Detector A: $D^* = 3.46$ mm · Hz$^{1/2}$/μW; Detector B: $D^* = 2.93$ mm · Hz$^{1/2}$/μW. Best D^* is biggest; therefore, Detector A is best.

15. Device A: $R_o = 13.15$ kΩ, $\alpha = -0.022$/C°; Device B: $R_o = 6.04$ kΩ, $\alpha = -0.022$/C°. Comment: Because α is the same for both devices, they are probably fabricated from the same semiconductor material.

Chapter 9 Semiconductor Concepts and Photoresistors

1.

	R (kΩ)	λ (nm)	S_r
(a)	25.0	475	0.4
(b)	11.1	550	0.9
(c)	13.9	660	0.7
(d)	33.3	700	0.3

3.

Cell	R_o (Ω)	N
A	27.9	−1.242
B	6.1	−1.324 most sensitive

5. At 25°C, $R = 200$ kΩ; at −20°C, SPEC. = −7%, $R = 186$ kΩ.
7. $t_r = 2.2T = 17.6$ ms
9. 300 mW
11. (a) Resistance minimum at 25°C based on tolerance of 331/3% is 1.733 kΩ, temperature SPEC. = +2%, $R = 1.768$ kΩ at 50°C
    (b) Zero volts because the maximum safe rated operating temperature is +50°C and the ambient and therefore the device are already at that temperature.
13. (a) Cell A $R = 1.6$ kΩ at 50 mlm/cm² yields $I \simeq 462$ μA.
    (b) A meter current of 2 μA corresponds to a cell resistance of 599 kΩ, which corresponds to an illumination of about 250 μlm/cm².
15. Relative response 58%, $R = 1.72$ MΩ [substituting the standard resistance 1 MΩ for R_p in equation (9-6)]

Chapter 10 PN Junction Detectors

1. 443 mA/W
3. 483.6 mA/W
5. 125 Ω calculated
7. 100.8 μW
9. 3.04 μA
11. 9.0 pA
13. $I_s = 1.64$ nA, $I_t = 60.7$ pA, $I_n = 1.64$ nA
15. C: 12 pF, "ratio of areas used to make estimate"
    t_r: 4.9 ns, "ratio of areas used to make estimate"
    R_s:130 Ω, small area $R_s \simeq$ constant
17. 1.128 A/W at 2000 nm, 2.82 A/W at 5000 nm

Chapter 11 Radiating Diodes and Display Devices

Questions

1. Increase proportionately.
3. 900 nm and 1300 nm nominal.
5. It increases.
7. It crosses the element with no change in polarization.
9. To allow photons from the depletion region to escape.

Problems

1. 5 mA
3. 2.9 mcd
5. 2.56 mlm/sr; 38.84°
7. 1.1 mlm
9. 4.76 nA/dot; 572 μA total

Chapter 12 Transistors, Phototransistors, and Opto-Isolators

1. $I_{sat} \simeq$ 9.3 to 10 mA
3. $V_2 \simeq$ 1.43 V, $V_E \simeq$ 0.73 V, $I_E \simeq I_C \simeq$ 3 mA, $V_{CE} \simeq$ 12 V
5. $P_H = P_T =$ 396.25 mW
7. If you considered only the specified diode-to-transistor coupling ratio, the peak collector current would be 32 mA. But 32 mA is in excess of the circuit's saturation current of 9.3 mA, assuming that $V_{sat} = 0.7$ V. Therefore, the collector current that is flowing is 9.3 mA, not 32 mA, so $P_T = I_{sat} V_{CE}(1/2) = 3.26$ mW
9. 100 nA

Chapter 13 Semiconductor Lasers

1. $K =$ 168.44°C, 43 mA at 0°C, 67.3 mA at 75°C
3. Ellipse: vertical dimension 0.69 cm, horizontal dimension 0.25 cm
5. About 70 mA
7. 110 mA [refer to diagram 90-2 in Figure 13-12(b)]
   1.8 V [refer to diagram 90-1 in Figure 13-12(b)]
   $p = IE =$ 198 mW
   efficiency = 15.2%

Chapter 14 Optical Waveguide

1. 69.68° or 1.216 rad
3. 20.32°, 0.355 rad
5. 0.560
7. 15.53 μW
9. Spreading 50.5 ns/km, dispersion 330 ps/km
11. 500 MHz
13. 1.44; waveguide 2 greater
15. $r = 0.04$ net for *one* surface; 0.355 dB
17. 3.01 dB, 0.97 dB, 3.98 dB
19. 34 μW

Index

D

E

F

G

H

I

J

L

M

N

O

Q

R

S

T

U

V

W

Y